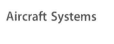

Aircraft Systems

Instruments, Communications, Navigation, and Control

Chris Binns
Egnatia Aviation
Chrysoupolis, Greece

Registered Office
John Wiley & Sons, Inc., 111 River Street, Hoboken, NJ 07030, USA

Editorial Office
111 River Street, Hoboken, NJ 07030, USA

For details of our global editorial offices, customer services, and more information about Wiley products visit us at www.wiley.com.

Wiley also publishes its books in a variety of electronic formats and by print-on-demand. Some content that appears in standard print versions of this book may not be available in other formats.

Library of Congress Cataloging-in-Publication Data

Names: Binns, Chris, 1954– author.
Title: Aircraft systems : instruments, communications, navigation, and control / Chris Binns, Egnatia Aviation, Chrysoupolis, Greece.
Description: First edition. | Hoboken, NJ : John Wiley & Sons, Inc., 2019. | Includes bibliographical references and index. |
Identifiers: LCCN 2018023898 (print) | LCCN 2018026687 (ebook) | ISBN 9781119259862 (Adobe PDF) | ISBN 9781119262350 (ePub) | ISBN 9781119259541 (hardcover)
Subjects: LCSH: Aeronautical instruments.
Classification: LCC TL589 (ebook) | LCC TL589 .B484 2019 (print) | DDC 629.135–dc23
LC record available at https://lccn.loc.gov/2018023898

Cover design by Wiley
Cover images: GPS Satellite © iStock.com/BlackJack3D; Flying Airbus A380 © iStock.com/rusm; Radar © iStock.com/Petrovich9; Close-up of aircraft infrared navigation display © iStock.com/olyniteowl; Jet Airbus A380 © iStock.com/rusm; Airplane Instruments Detail © iStock.com/erlucho

Set in 10/12pt Warnock by SPi Global, Pondicherry, India

MIX
Paper from
responsible sources
FSC
www.fsc.org   FSC® C013604

Contents

Acknowledgments

I do not think it is possible for a person, at least, a person with a family, to write a book without a good deal of support as a lot of time is taken up being away from people. I would like to thank everyone for putting up with the necessary absence, especially my wife, Angela, who has been a constant source of inspiration.

I thank my extended family accrued from two marriages, that is, Callum, Rory, Connor, Edward, Tamsyn, and Sophie for bringing inspiration and joy to my life.

Finally, I would like to thank my current employer, Egnatia Aviation, for maintaining a passion for excellence in aviation.

About the Companion Website

This book is accompanied by a companion website:

www.wiley.com/go/binns/aircraft_systems_instru_communi_Navi_control

The website includes site for **Instructor** and **Student**.

For Instructor

Contains the answers to all the problems at the end of the chapters as a resource for instructors using the book.

1

Historical Development

1.1 Introduction

If you board a commercial flight in 2016, you will step onto an aircraft that has a significant redundancy of electrical power and safety systems with a high level of automation. The instruments in the cockpit show the pilots via highly ergonomic displays the attitude, height, climb rate, speed, and Mach number of the aircraft as well as the state of the engines and other factors such as the outside air temperature and the wind speed and direction. On-board weather radar informs the pilots of storms in the path with detailed information about the precipitation, turbulence, and the lateral and vertical extent of the storms. The navigation system takes inputs from GPS satellites, an inertial reference system (IRS), and VHF radio beacons; filters the information; and provides a precise indication of the position of the aircraft in three dimensions to within a few meters. These same instruments and navigation systems provide information to the autopilot, which can control the aircraft in height and position to follow a specific flight plan and land the plane at the destination airport if the latter has the necessary ground systems installed. The navigation computer contains a detailed database of all man-made and natural potential obstacles and provides warnings of approaching terrain or structures. The flight is conducted via a comprehensive air traffic control (ATC) system that tracks the aircraft and maintains communication links with the pilots throughout the flight to ensure safe separation with other aircraft. In addition, the aircraft will communicate automatically with others in the local area and build a three-dimensional map of all nearby flights to provide a traffic avoidance system that is independent of ground air traffic controllers. The system will not only warn the pilots of nearby traffic but also in extreme cases will inform them what evasive action to take. These and other systems have led to an unprecedented level of safety in commercial air travel which, if represented as fatalities per km traveled, is safer than any other type of transport on water or land [1]. This parameter does not

Aircraft Systems: Instruments, Communications, Navigation, and Control,
First Edition. Chris Binns.
© 2019 John Wiley & Sons, Inc. Published 2019 by John Wiley & Sons, Inc.
Companion website: www.wiley.com/go/binns/aircraft_systems_instru_communi_Navi_control

necessarily provide the fairest comparison between different modes of travel since air travel will naturally do well with any safety assessment that uses distance traveled as the criterion. For example, when using fatalities per hour traveled, aviation drops to third on the list below rail and bus transport. It remains true, however, that air travel safety has made vast improvements by any measure in the last few decades and this is largely due to the incorporation of the systems listed above. All of these will be described in detail in subsequent chapters but before delving into the technical complexity it is worth exploring, in this chapter, a condensed history of the development of some of those instruments and systems.

1.2   The Advent of Instrument Flight

In the earliest days of aviation, the pilot's senses were the main aircraft instruments, with vision being used to estimate speed, height, and flight attitude while hearing and smell were used to monitor the state of health of the engine. The Wright flyer did have instruments installed including an anemometer, a stopwatch, and a revolution counter but these were used exclusively to analyze the performance of the flyer and the engine post landing. The flight itself was conducted entirely utilizing the senses of the pilot. Flying by senses alone dominated aviation throughout the First World War and led to the myth of instinctive balance when flying an aeroplane. This remained while flights were rarely conducted in bad weather and small rolled attitudes that developed inadvertently while flying through an individual cloud went unheeded. After the war, airmail and the first passenger services started to be developed but initially these flew under the weather sometimes at very low altitude resulting in many fatalities.

It was realized by the end of the First World War that flying in cloud with pilot vision completely removed from the available information could quickly lead to spatial disorientation and the aircraft spiraling out of the cloud with complete loss of control. The problem is that inner ear senses, which measure linear and angular acceleration, are evolved for life on the ground and provide misleading sensations in aircraft. Examples include the Somatogyral illusion in which an established banked turn is undetected as there is no angular acceleration but rolling out of the turn produces the illusion of a bank in the opposite direction, and the Somatogravic illusion where accelerations and decelerations are interpreted as pitches up and down, respectively. Gyroscopic turn coordinators were available by 1918 but without instrument training pilots still tended to favor their senses over the indications of the instrument.

Early pioneers in the development of instrument flight were two US army pilots, William Ocker and Carl Crane. By 1918, the first gyroscope-based attitude indicators (AIs) (see Section 1.4 and Section 3.1.9), invented by Elmer Sperry, were available and Ocker was one of the first to attempt an extended flight in cloud using the instrument. The flight still ended up with the aircraft in a spiral dive but Ocker realized that the main reason was his failure to put complete faith in the instrument and to pay too much attention to his erroneous balance senses. Ocker was one of the first to correctly identify the misinformation coming from balance organs and became somewhat of an evangelist for using instruments in flight. Crane was nearly killed in 1925 when he dropped into a spiral dive out of cloud while flying a congressman's son to Washington and was acutely aware of the problems of maintaining control while blind. Ocker and Crane teamed up in 1929 and conducted a comprehensive study of flying in clouds, which led, in 1932, to the publication of their book, *Blind Flight in Theory and Practice*, which is the first systematic exploration of instrument flight. By the late 1920s, a full range of pressure and gyro instruments were available as well as some radio navigation devices (see below) and in 1929 Jimmy Doolittle demonstrated a "blind" takeoff, aerodrome circuit, and landing in an aircraft whose dome was covered [2]. The cockpit in Doolittle's NY-2 Husky biplane is shown in Figure 1.1a and contains the six main glass instruments that are to be found in a current general aviation (GA) light aircraft. That is, an altimeter (i), an AI (ii), an airspeed indicator (iii), a turn indicator or turn coordinator (iv), a direction indicator (DI) (v), and a vertical speed indicator (VSI) (vi). By the 1950s, the layout of these six instruments was standardized into what was deemed to be the most ergonomic arrangement (the so-called "6-pack") and they were mounted as shown in Figure 1.1b, which shows a Piper PA28 cockpit.

(a)   (b)

1. Altimeter   3. Airspeed indicator   5. Direction indicator
2. Attitude indicator   4. Turn indicator   6. Vertical speed indicator

Figure 1.1 (a) Flight instruments in the cockpit of the NY-2 Husky biplane used in the first "blind" takeoff and landing flight by Doolittle in 1929. *Source:* Reproduced from Ref. [2] with permission of ETHW. (b) The same six instruments in the standard layout in a 1960's light aircraft (Piper PA28).

In older large aircraft with traditional instruments, the standard 6-pack is also evident directly in front of the pilot though it is embedded in an extended array of engine and navigation instruments. The main change to this layout came in the transition to "glass cockpits" in the late 1960s where several instruments are displayed on a single electronic screen. The term is slightly misleading as there is probably less glass in a glass cockpit than a traditional one with a large array of glass-fronted instruments but basically it means information is displayed on electronic screens rather than individual instruments. The change to glass cockpits marked the transition from direct-sensing to remote-sensing instrumentation. In the case of older direct-sensing pressure instruments, the pressure being measured is brought via tubes directly into the back of the instrument, which then converts it into a reading on the instrument face as described in Chapter 2. This leads to a large amount of tubing mixed in with all the wiring behind the instrument panel. In remote sensing, a transducer measures the quantity required remotely and converts it into an analog or digital electrical signal, which is conveyed by wires, either to an individual instrument, or to a computer and display generator. In the most modern systems, a digital data bus is used to convey information from all the sensors, which significantly reduces the complexity of the wiring. Figure 1.2 compares a cockpit with traditional direct-sensing instruments in a twin piston engine aircraft (Figure 1.2a) and a glass cockpit in which the remote-sensing instruments communicate via a computer to the electronic displays, again in a twin piston engine aircraft (Figure 1.2b). Note, however, that even in the glass cockpit there are some direct-sensing instruments provided as backup in case of a total power failure. The glass screen immediately in front of the left pilot seat is referred to as the Primary Flight Display (PFD).

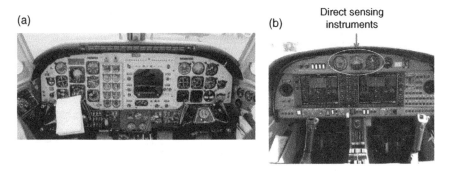

Figure 1.2 (a) Instrument display in the cockpit of a twin piston engine aircraft using entirely analog direct-sensing instruments. (b) Instrument display in the cockpit of a twin piston engine aircraft utilizing remote-sensing instruments and digital displays connected by a digital data bus (glass cockpit). Note, however, that some direct-sensing analog instruments are also provided as backup.

An important aspect of instrument flight is training and although the technology for flight without external references was in place by 1930, pilot training in instrument flying was not standardized internationally until after the International Civil Aviation Organisation (ICAO) was set up in 1947. The modern day "instrument rating" is a separate rating applied to the pilots' license allowing the holder to fly on instruments only and in addition to navigate and land using radio navigation aids (see Chapter 7). This chapter will now describe the historical development of individual instruments and also the evolution of the communication and navigation systems essential to modern aviation.

1.3 Development of Flight Instruments Based on Air Pressure

1.3.1 The Altimeter

Measuring how high an aircraft is off the ground became a necessity as soon as flights over high ground started to become commonplace. Since the beginning of aviation, a number of methods have been tested including sonar, variation in gravity, capacitance, integrating accelerometers (IRSs), cosmic ray detection, and hypsometry (measuring the change in the boiling point of water). The method that became established early on and continues to this day is to measure the air pressure and convert this to an altitude. The French physicist Blaise Pascal first confirmed the decrease of air pressure with increasing altitude in 1648 using a mercury barometer invented by Torricelli four years earlier. He measured the pressure at the bottom and top of a church bell tower in Paris and was able to observe a measurable decrease produced by climbing 50 m to the top. Soon after the first balloon flight by the Montgolfier brothers in 1783, portable mercury barometers were being used to estimate altitude in free balloons, but practical altimeters were not available till Bourden produced an improved design of aneroid barometer in about 1845. From then on altimeters based on the Bourden design were commonly used in free balloons and airships.

The Wright flyer did not have an altimeter and strangely there is no confirmed record of altimeters in use in aircraft before 1913. The first operational aircraft altimeters, available from about 1912, had a single pointer that completed one revolution in 0–10 000 ft (Figure 1.3a). By about 1925, the altitude range had been extended to 0–30 000 ft, still employing a single pointer but in this case, it completed one and a half revolutions with an inner scale used to read altitudes above 20 000 ft. At about the same time, adjustable scales were introduced so that the altimeter could be adjusted for changes in air pressure that occur day to day and also for differences in elevation between departure and arrival aerodromes. The development of radio links (see Section 1.5) enabled altimeters to be set for the measured pressure at a landing field and by the late 1920s,

(a)　　　　　　　　　　(b)

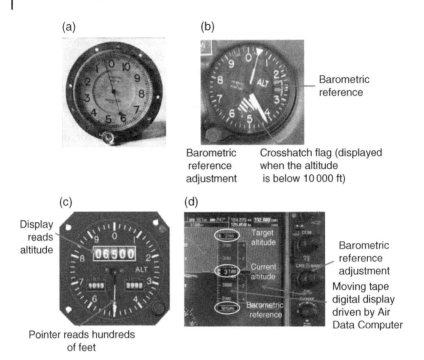

Figure 1.3 Historical evolution of altimeters (a) Circa 1912, single pointer and a range of 10 000 ft. *Source:* Reproduced from Ref. [3]. (b) Three-pointer display seen after 1935 and still in common use today. The large, small, and thin line pointers indicate hundreds, thousands, and tens of thousands of feet, respectively. (c) More modern presentation in which the barrel is a digital display of altitude and the pointer reads hundreds of feet. (d) Moving tape display driven by the Air Data Computer in a glass cockpit PFD.

temperature compensation was built in, which made altimeters accurate enough to be useful for landing. In order to produce a high resolution but still maintain a large altitude range, three-pointer altimeters were introduced in 1935 (Figure 1.3b). In these, the largest pointer makes one revolution in 1000 ft pointing to a number that represents hundreds of feet while a smaller pointer moves one revolution in 10 000 ft pointing to a number that represents thousands of feet. The third hand, which is usually a thin line, indicates tens of thousands of feet. This type of altimeter is still found in most light aircraft flying today. A major improvement was the introduction of the servo-altimeter in which the aneroid capsule used to measure pressure does not directly drive the indication mechanism, but its distortion is measured electrically and this electrical signal drives the display (see Section 2.8.3). More recently, the problem of the easy misreading of the three-pointer dial has resulted in the display shown in Figure 1.3c in which the altitude is displayed directly on a digital barrel and the pointer indicates hundreds of feet with 20 ft divisions. In modern commercial airliners and GA aircraft with a glass cockpit, the air pressure is measured by a transducer and the electrical signal along with other data such as the outside air temperature is passed into an air data computer (ADC). The output of the ADC is then used to drive a digital moving tape-type display shown in Figure 1.3d). Although measurement of air pressure remains the primary method of determining altitude, Radio Altimeters (see Section 5.6) are the primary input into the autopilot during landing. In addition, now that accurate models of the surface of the Earth are available (see Section 6.4), satellite and inertial reference systems can also provide accurate altitude information.

1.3.2 The Vertical Speed Indicator (Variometer)

As in the case of the altimeter, the history of the VSI begins before the advent of heavier than air flight. Balloonists used instruments called *statoscopes*, which were sensitive aneroid barometers to detect climb or descent though they did not indicate the rate. In the twentieth century, gliders drove the development of the VSI or *Variometer*, as it is referred to in gliding, because of its importance for finding optimum lift in updrafts. The human body is insensitive to small gradual changes in barometric pressure that are typical of normal climb rates so to optimize endurance and range a variometer is critical. The earliest variometers (Figure 1.4a) were based on statoscopes but later designs measured directly the rate of change of air pressure to provide a quantitative reading. The gliding pioneers Alexander Lippisch and Robert Kronfeld are credited with the invention of the first truly quantitative variometers in 1929, which greatly improved glider performance. Their design was based on having a sealed

(a)　　　　　　　　　(b)　　　　　　　　　(c)

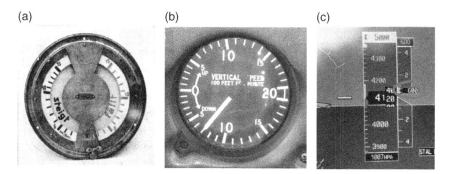

Figure 1.4 Development of the Vertical Speed Indicator (VSI). (a) Atmos variometer from circa. 1922. *Source:* Reproduced from Ref. [3]. (b) Direct reading VSI found in a General Aviation aircraft showing a 700 ft min^{-1} rate of descent. (c) Digital indication and pointer (highlighted by the red oval) next to the altitude moving tape in a glass cockpit PFD showing a 600 ft min^{-1} climb.

container connected to a diaphragm, the other side of which was at ambient pressure. As the air pressure changed, air would flow into or out of the container and the rate of height change was determined by the flow rate.

The current design of direct reading VSIs used in GA (see Section 2.9) is based on an aneroid capsule whose internal part and surround are both connected to ambient air but a controlled leak between the inside and outside of the capsule maintains a pressure differential if the air pressure is changing. The first report of this design was by Wing Commander Roderic Hill in his book *The Baghdad Air Mail* [4] first published in 1929. He described how hard it was to climb in the Vickers Vernon biplanes used for the mail runs in the Middle East when they were taking off fully laden from hot aerodromes well above sea level. The pilots resorted to using updrafts (termed *dunts*) to climb the first few hundred feet and initially would try and find them by observing birds. One of them built a *dunt indicator* by drilling a small hole in a two-gallon petrol can and connecting the can to one side of a pressure gauge the other side of which was exposed to ambient air. If the pressure changed due to a climb or descent, because of the time lag required for the pressure inside the can to equalize, the gauge would indicate a pressure differential. A current design of direct reading VSI is shown in Figure 1.4b.

More recently, the different requirements of powered flight and gliders has led to a divergence in the design of VSIs and variometers. The most important quantity to a glider pilot is the change in the total energy (potential plus kinetic) and modern glider variometers have total energy compensation. Thus, they distinguish whether climb is coming at the expense of kinetic energy, which would happen, for example, by simply applying back pressure to the stick. In powered flight, on the other hand, the pilot needs to know the absolute rate of climb or descent irrespective of speed. In a glass cockpit, the vertical speed indication comes directly from the ADC (see Section 2.14), which analyzes digitally the pressure measured by the static source and determines if there is an upward or downward trend. The vertical speed indication is normally given by a pointer and a digital display next to the altitude moving tape as shown in Figure 1.4c.

1.3.3 The Airspeed Indicator

The *true airspeed* (TAS) of an aircraft is defined by its speed relative to the still air around it that is close enough to be in its immediate environment but sufficiently distant to remain undisturbed by its passage. Thus, it is the speed of the aircraft relative to something invisible and all devices that determine airspeed measure the force generated by the pressure of the passing air on some surface, that is, the *dynamic pressure*. The direct measurement of dynamic pressure leads to a quantity known as the *indicated airspeed* (IAS), which will vary relative to the TAS depending on the air density around the aircraft. For example, if the aircraft climbs at a constant IAS, that is, maintaining a constant dynamic pressure at the measurement device, its TAS will steadily increase as the air gets thinner. The relationship between TAS and IAS is described in detail in Section 2.10, but given that IAS almost never equals TAS it may seem surprising that it is always IAS that is primarily presented by the cockpit instruments to the pilot. In addition, the operating handbooks of aircraft give speeds for safe operation of procedures such as lowering flaps or landing gear in terms of the IAS. The reason is that the IAS is essentially a measure of dynamic pressure and all the important flight parameters, such as lift, stall conditions, critical speeds for lowering landing gear and flaps, etc. depend on the dynamic pressure, irrespective of the TAS. Thus, the IAS can be used by the pilot to maintain safe flying conditions without having to worry about converting to TAS. The latter is only important to determine ground speed (after correcting for wind), which is required for navigation. In transonic and supersonic aircraft, the Mach number is another important indication required for safe flight, which in older aircraft is measured from the static and dynamic pressure by a separate instrument known as a Mach Meter (see Section 2.11). In more modern aircraft, the Mach number is calculated by the ADC and displayed on the PFD.

The earliest airspeed indicators were anemometers, similar to those used to measure wind on the ground, attached to a revolution counter as illustrated in Figure 1.5a or a metal plate perpendicular to the airflow on a spring (Figure 1.5b).

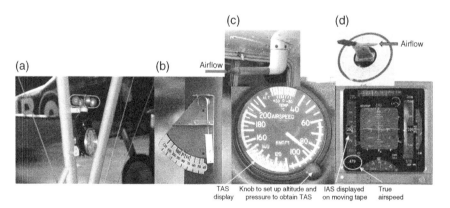

Figure 1.5 Evolution of airspeed indicators. (a) Anemometer-type airspeed indicator on the wing strut of a WW1 Albatross D.V. biplane. *Source:* Reproduced with permission of 20thcenturybattles.com. (b) Plate on a spring used to indicate airspeed on the wing strut of a De Havilland 60 Moth (1930s). *Source:* Reproduced with permission of James Knightly. (c) Pitot tube and direct reading Airspeed Indicator (ASI) on a General Aviation aircraft. (d) Pitot tube mounted on a rotating table on the fuselage. *Source:* Reprinted under Creative Commons license 3.0 [5] and PFD on a current commercial aircraft (Boeing 737NG). The moving tape indicates the IAS and the highlighted digital display shows the TAS or Mach number. *Source:* Reproduced with permission from Chris Brady from http://www.b737.org.uk.

By the end of the First World War, the majority of aircraft used a pitot tube, that is, a closed tube in which air is brought to rest (Figure 1.5c). By measuring the difference between the *stagnation pressure* in the tube and the static air pressure (i.e. the *dynamic pressure*), it is possible to determine the airspeed. The pitot tube is a very simple measuring device, originally invented in 1732 by Henri de Pitot to measure water flow and improved on by Darcy in 1856. It is still in widespread use and has remained fundamentally unchanged though improvements have been made by adding internal heaters to prevent icing and installing them on rotating platforms with vanes so that they are always presented in the same direction to the airflow (Figure 1.5d).

In direct-reading airspeed indicators (Figure 1.5c), the pressure difference generated by the pitot tube is sensed by an aneroid capsule whose distortion drives the indication as described in Section 2.10. In aircraft with glass cockpits, the stagnation pressure in the pitot tube and the static pressure are measured by transducers and the signals are processed by the ADC (see Section 2.14), which outputs the IAS to a moving tape on the PFD as illustrated in Figure 1.5d. The ADC has all the other necessary parameters (outside air temperature, altitude, etc.) to calculate the TAS or Mach number and this is also displayed on the PFD as highlighted in Figure 1.5d.

1.4 Development of Flight Instruments Based on Gyroscopes

The properties of gyroscopes and their use in aircraft instruments are described in detail in Section 3.1, but a brief introduction is given here. One of the most important characteristics for their use in instrumentation is the tendency of a spinning rotor to maintain the same spin axis (its *rigidity*). The observation of the stability of a spinning top must precede recorded history and the earliest examples discovered are clay tops from the Middle East dating back to 3500 BCE. In order to make use of this stability for orientation and build a practical gyroscope, however, the spinning top (or *rotor*) needs to be mounted in gimbals as shown in Figure 1.6. In the case of the rotor mounted in a simple cage (Figure 1.6a), the cage can rotate freely about the spin axis but torque applied about any other axis will be resisted and generate a torque about a perpendicular axis (see Section 3.1). If the cage is mounted in a gimbal that pivots about a perpendicular axis (Figure 1.6b), the support structure can now rotate freely about two axes without disturbing the rotor. Adding a second gimbal (Figure 1.6c) allows the support structure to rotate about all three axes so if the support is clamped to the frame of an aircraft the rotor will maintain its spin axes irrespective of changes in the pitch roll or yaw. If the gyroscope rotor is initially aligned with the horizon, for example, its rigidity will always indicate

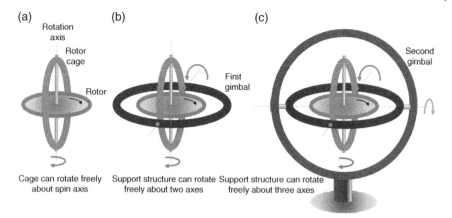

(a)
Rotation axis
Rotor cage
Rotor
Cage can rotate freely about spin axis

(b)
First gimbal
Support structure can rotate freely about two axes

(c)
Second gimbal
Support structure can rotate freely about three axes

Figure 1.6 (a) In a simple caged rotor the cage can rotate freely about the spin axis without applying torque to the rotor. (b) After adding a single gimbal the support structure can rotate about the two axes shown without applying torque to the rotor. (c) After adding a second gimbal the support structure can rotate about all three axes without applying torque to the rotor. The outer gimbal can be clamped to the frame of an aircraft and the spinning rotor will maintain its axis of spin as the aircraft maneuvers.

the plane of the horizon as the aircraft maneuvers. How this characteristic is used to build a practical instrument to show the orientation of the horizon, that is, the *artificial horizon* or AI is described in Section 3.1.9.

The first reported use of a gyroscope in navigation was in 1743 by the English sea captain John Serson who invented his *whirling speculum* to locate the horizon on misty mornings – an early form of artificial horizon. The name gyroscope (from the Greek *gyros* for circle and *skopeein* to observe) was first introduced by Léon Foucault who used one in 1852 to demonstrate the rotation of the Earth. The German inventor Hermann Anschütz-Kaempfe patented the first practical gyrocompass in 1904 and an improved design was presented by the American entrepreneur Elmer Sperry later that year. Initially, gyroscopes were used in ships but the Sperry Gyroscope company, founded in 1910, pioneered the use of gyroscopes in aircraft instrumentation. Lawrence Sperry, Elmer's son, used gyroscopes to build the first functional autopilot prior to World War 1, as described in Section 1.9. The first artificial horizon was installed in an aircraft in 1916 and by the time Doolittle made the first blind instrument flight in 1929 (see Section 1.2) the full set of gyro instruments found in a modern 6-pack, that is the DI, the turn indicator, and the AI were all available. These remain in widespread use today, largely unchanged, in most GA aircraft and some older commercial airliners.

The introduction of glass cockpits in the 1960s changed the display of information given by gyroscopes from individual direct reading instruments to the

(a)   (b)

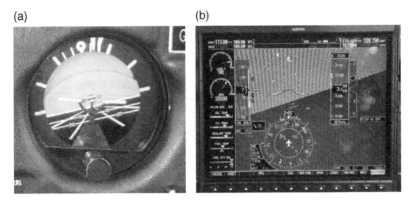

Figure 1.7 (a) Roll and pitch indications in direct reading instruments in a general aviation aircraft performing a climbing turn (10° pitch up, 15° bank). (b) Roll and pitch indications on the PFD in an aircraft with a glass cockpit performing a climbing turn (10° pitch up, 15° bank).

multi-instrument PFD but the spinning mechanical gyroscopes were still used to determine the attitude information. This all changed with the introduction of Micro-Electrical–Mechanical Systems (MEMS) gyroscopes in the early twenty-first century. These are based on the rigidity of the plane of vibration of a tuning fork (see Section 3.2.5) and their main advantage is that they can be micromachined into silicon wafers so that a three-axis gyroscope can now be mass-produced cheaply on a piece of silicon a few millimeters across. On the same chip can be mounted a number of other devices as described in Section 3.4 so that an entire attitude and heading reference system complete with gyroscopes, magnetometers, and accelerometers can be mounted on a single chip. Apart from savings in cost and weight, MEMS devices are more reliable than mechanical rotating gyroscopes. Figure 1.7 shows the AI display found in the traditional "6 pack" of direct-reading instruments in GA aircraft performing a climbing turn and the same information displayed on the PFD in a glass cockpit.

Gyroscopes are also used in IRSs, which were initially developed in rockets starting from the 1930s and then were widely incorporated into aircraft navigation systems post World War 2. Initially, these were also based on mechanical spinning gyroscopes but from the 1970s the mechanical gyros were replaced with laser-based systems (see Section 3.2).

1.5 Development of Aircraft Voice Communications

In the earlier days of aviation, communication between pilots and people on the ground was either by hand signals, aircraft maneuvers (rocking wings, etc.), or by large symbols on the ground. Interestingly, all three forms of communication have survived in aviation into the twenty-first century, having been standardized internationally and are still in the training syllabus for commercial pilots. Hand signals survive in the form of a set of hand signals for communication between pilots and ground marshallers. Large symbols prepared on the ground following a plane crash have been devised to communicate with search and rescue (SAR) aircraft. Many GA airfields still have a "signals square," which is a set of large symbols visible from the air that communicate information about the airfield and the runway in use. These are for use either by vintage aircraft with no radio or by aircraft that have suffered a radio failure. In the case of loss of radio communication at night, signals using aircraft lights can be used. Finally, a series of aircraft maneuvers for communication between commercial airliners and intercepting military aircraft has been standardized.

The world's first radio transmission from an aircraft was a Morse code message sent in August 1910 by the Canadian aviator, James McCurdy, as he flew over a receiver in Brooklyn, New Jersey, trailing a long aerial wire from his aircraft. Within a few weeks a Morse message from air to ground was transmitted in excess of one mile in England. Leading into the First World War, the Marconi company in collaboration with the Royal Flying Corps began a research program on air to ground and ground to air communications, which was driven by the need to communicate observations from the air to artillery on the ground. Prior to radio communication this had been achieved by dropping hand-written notes from the aircraft. Initial attempts at wireless transmission used Morse code and significant problems were soon encountered such as the weight of the early transmitters – 35 kg or more plus 250 ft or so of aerial wire that had to be unwound from a spool. This created significant drag and sometimes the wire would get caught around the control surfaces creating a serious hazard. A plane was a death trap if it was attacked while set up for transmission. In addition, it was very hard for pilots to tap out a message on a Morse key while flying the plane. Things were not much better for ground to air transmissions as it was found that hearing a Morse message above the roar of the engine and the wind and sometimes gunfire was challenging to say the least.

The goal soon became to send voice transmissions over a significant range and the first step in this quest was achieved in April 1915 when Captain J.M. Furnival allegedly heard the following message in his headphones from Major Prince while flying over Brooklands Park in England:

Hello Furnie. If you can hear me now it will be the first time speech has ever been communicated to an aeroplane in flight.

Within a few months, voice communication in both directions was achieved and Marconi started manufacturing the world's first production aviation radio weighing just 9 kg. Across the Atlantic, the technology was developing at the

Figure 1.8 AT&T employees and military personnel watch an early aircraft-radio test. *Source:* Courtesy of the AT&T Archives and History Center.

same rapid pace and by July 1917, AT&T engineers demonstrated two-way voice communication at Langley Field in Virginia (see Figure 1.8). By August of that year they had also produced two-way communication between aircraft in flight.

The radio communications developed during the First World War were specific to the requirements of the aircraft used to observe the position of landing artillery shells and the messages were transmitted over quite short ranges. The radios of the time worked in the Low Frequency (LF) band (30–300 kHz) in which the radio waves are refracted by the ground and tend to follow the curvature of the Earth (see Section 4.2.2.1). The maximum range in this band is given by the transmitter power and for practical transmitters that could be carried on aircraft the range was limited to about 20 miles. After the war, the requirements changed and the ability to communicate with aircraft over a long range became important as the first air mail routes were established.

Initially, messages were bounced from airfield to airfield leading to anecdotes of aircraft arriving at a destination airfield before the radio message that it was on its way had been received. In the longer term the range of radios had to be

Figure 1.9 Imperial Airways Short S23 *Caledonia* flying boat at Felixstowe in 1936. In July 1937, *Caledonia* flew from the Foynes seaplane port in Ireland to Botwood, Newfoundland maintaining radio contact with Foynes for the entire 1900-mile journey.

increased and this was achieved by developing transmitters working in the Medium Frequency (MF: 300 kHz to 3 MHz) and High Frequency (HF: 3–30 MHz) bands. At these frequencies, it is possible to utilize the ionosphere to reflect radio waves around the curvature of the Earth (so-called *skywaves* – see Section 4.2.3) producing very long ranges. By 1932, Marconi had developed the AD37/38 radio that used both the MF and HF bands and could achieve two-way voice transmission with the ground at a range of 1000 miles and communication between aircraft at a range of 200 miles. In addition, the radio could be used as an Automatic Direction Finder (ADF: see Section 1.7.1 and Section 7.1). During the 1930s radios transmitting in the MF/HF band continued to improve and became the norm for aircraft communications in the rapidly developing commercial aviation sector throughout the 1930s. In July 1937, an Imperial Airways Short S23 flying boat (*Caledonia*, Figure 1.9) took off from the Foynes seaplane port in Ireland and flew to Botwood in Newfoundland maintaining radio communication with Foynes for the entire 1900-mile flight.

Thus, at the start of World War 2, MF/HF long distance radios were highly developed for commercial air travel and were also used in military aircraft. It was decided that improvements were required in the performance, frequency range, and ease of use and so in 1939, new specifications were drawn up and presented to Marconi who developed the T1154/R1155 transmitter/receiver combination. The transmitter had a continuous wave output power of up to 70 W and the entire combination weighed about 36 kg. In addition to communication, the installation could be used in an ADF mode providing a radio navigation capability. Bomber command had the radio sets installed in their aircraft by June 1940 (see Figure 1.10), and by the end of the war over 80 000 sets had been manufactured. After the war, many sets found their way into civilian planes

Figure 1.10 T1154/R1155 installation at the radio operators station in an Avro Lancaster, circa 1943.

and they continued to be used in military aircraft (Vickers Varsity and Handley Page Hastings) till these were withdrawn in the 1970s.

The T1154/R1155 installation was fine for large planes but was not suitable for fighters because, apart from its size, the MF/HF frequency bands used presented a more fundamental problem. Generally, the lower the frequency (i.e. longer the wavelength) of radio waves used, the longer the antenna needs to be (see Section 4.4). The Lancaster had two antennas – a fixed one for HF transmission running from above the cockpit to the tail and a trailing 290-ft cable that could be deployed as an MF antenna. A trailing antenna was not an option for a fighter and the relatively short fixed antennas that could be used were inefficient in the HF band severely restricting the range. In addition, the HF band is prone to a lot of interference resulting in a high level of static noise in the crew's headphones. It was decided at the beginning of the war to equip Spitfires and Hurricanes with VHF (Very High Frequency: 30–300 MHz) radio installations [6] so that an efficient short antenna could be used and transmissions would suffer less static noise. The result was the Marconi TR1133 and subsequently TR1143 transmitter/receiver in a single box weighing 21 kg and with a transmitter power of 10 W. Radio waves in the VHF range are not reflected by the ionosphere, nor do they follow the curvature of the Earth due to refraction like LF waves so the range is limited by the distance to the horizon and thus the range depends on the height the aircraft is flying. As shown in Figure 1.11, the distance, d, from an aircraft to the horizon is one side of a right-angle triangle

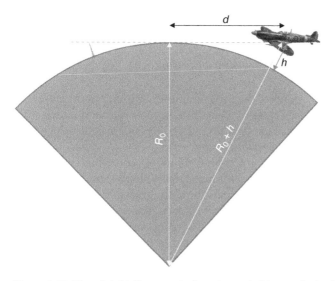

Figure 1.11 Line of sight distance, *d*, of an elevated object to the horizon forms one side of a right-angle triangle whose other sides are R_0 and $R_0 + h$, where R_0 is the radius of the Earth and *h* is the height of the object. To get line of sight to a second elevated object the two line of sight distances are added.

whose other sides are of length R_0 and $R_0 + h$, where R_0 is the radius of the earth and *h* is the height the aircraft is flying. Thus, we can write:

$$d^2 + R_0^2 = (R_0 + h)^2$$

and since h^2 is insignificant compared to R_0^2 this becomes:

$$d = \sqrt{2R_0 h}$$

Normally we would want to enter the value of *h* in feet and calculate *d* in nautical miles. So, putting in the value for $R_0 = 3440$ nautical miles and converting *d* from feet to nautical miles, that is dividing it by 6076.12 gives:

$$d(\text{nm}) = 1.064\sqrt{h(\text{ft})}$$

This is always increased by about 15% to take into account the fact that VHF waves can be picked up slightly beyond the horizon giving:

$$d(\text{nm}) = 1.23\sqrt{h(\text{ft})}$$

(some sources specify the pre-factor as 1.25). This also assumes the ground station antenna is at ground level but if the antenna has a significant elevation or the aircraft is talking to another aircraft then it is simply a matter of adding the second station's line of sight distance to the horizon, so for two elevated objects, the range can be written:

$$d(\text{nm}) = 1.23 \left(\sqrt{h_1(\text{ft})} + \sqrt{h_2(\text{ft})} \right) \tag{1.1}$$

where h_1 and h_2 are the heights of the two communicating entities. As an example, even at 2000 ft, the range is about 55 nautical miles to a ground station with no elevation and double that for two aircraft at the same height, which was adequate for normal fighter operations.

Thus, emerging from World War 2, aircraft communications had evolved into dual modalities with VHF being used for high fidelity short-range communications and HF band skywave transmissions for long-range messages. Remarkably, these two types of transmission are still in place 70 years later and the main evolution in the equipment has been due to advances in electronics. From the late 1950s, transistors started to replace valves and from the mid-1960s integrated circuits replaced individual transistors and other circuit components. Integrated circuits have continued to evolve with increasing levels of integration from about 10 transistors on a chip in the mid-1960s to over seven billion in 2015. Note, however, that it is the control aspects that can be handled by integrated circuits while the main transmitter power still has to be derived from individual power transistors. Nevertheless, a typical transmitter box on a GA aircraft with a similar power to the TR1143 unit in a spitfire weighs less than 2 kg and would fit in a coat pocket. In addition, it is multichannel with both communication and VHF radio navigation capabilities (see Section 1.7.3 and Section 7.2).

VHF radio transmissions (at similar frequencies to those used in World War 2) remain the main communication mode for air to ground and air to air for all terminal operations and also for traffic over land, with aircraft switching to different stations as they travel to stay in range. For trans-Oceanic flights, VHF does not provide coverage for most of the flight and from World War 2 till the mid-1970s this gap continued to be filled by HF skywave communications but the disadvantages of this started to become increasingly apparent. The optimum frequency to use to achieve long ranges depends on the state of the ionosphere, which varies considerably in the diurnal cycle, so using HF equipment properly requires significant training. The reliability of transmission is also at the mercy of natural phenomena such as sunspot cycles and solar flares (so-called *space weather*) leading to the setting up of space weather monitoring services to try and predict periods when HF communication is likely to be interrupted. From an operational point of view, there is a high background noise level on HF radio frequencies and listening for transmissions for long periods of time was tiring for aircrews. This led to the setting up of the SELCAL system where each aircraft has a four-letter code, which, for a trans-oceanic flight, is included in its flight plan. If a ground station operator wants to communicate with a specific aircraft, they will transmit the corresponding code, which all aircraft on frequency will receive but the one that has been selected will activate a chime alerting the crew that a message is coming in. Thus, the crew can turn down the volume of their headsets until an alert is received.

1.6 Development of Aircraft Digital Communications

1.6.1 Communication Via Satellite (SATCOM)

The SELCAL system used with HF communication is an early example of digital communication as opposed to voice, but as aviation moved into the digital age the main problem with HF communication was its inability to transmit data at a high rate. For transmission of digital data, the carrier wave needs to be at a higher frequency than the modulation frequency and a high data rate demands a very high carrier wave frequency, which restricts the range to line of sight. From the early 1970s, this range limitation of VHF transmissions was removed by using satellites. A satellite orbiting at a height of 22 236 miles (35 786 km) above the Earth's equator is *geostationary*, that is, it has the same angular velocity as the spinning Earth below and so it stays fixed at one point in the sky as seen from the surface. It has line of sight with everything below and a constellation of a few geostationary satellites can cover the globe. A schematic of the system used to communicate between aircraft and ground via satellites is shown in Figure 1.12. An operator who can be an air traffic controller or from airline operations staff will send a message to the network run by the satellite operator (for example, Inmarsat), which will provide not only voice communication but digital services such as the Aircraft Communications Addressing and Reporting System (ACARS: see Section 1.6.3) or the internet. The message (along with all the other services) is sent to a ground station satellite dish, which is pointing permanently at the geostationary satellite. Data and voice are transmitted to the satellite via a Super High Frequency (SHF: 3–30 GHz) band signal and then relayed by the satellite to the aircraft in flight via an Ultra-High Frequency (UHF: 300 MHz to 3 GHz) link. This mode of communication is now the norm and its capabilities are constantly expanding to include, for example, monitoring of aircraft in flight via ACARS. The need for ever higher data transmission rates is pushing the radio frequency higher and in the latest system under development this will reach 40 GHz, corresponding to 1 cm waves normally used in radar.

Hence, is HF skywave communication now redundant? Far from it. Satellites have their own problems, for example, their inaccessibility for repairs if they fail. Most airlines maintain the capability for HF transmission as a failsafe backup to their normal satellite services for long-range communications and the system infrastructure has been continued. This has been proven to be a wise policy,

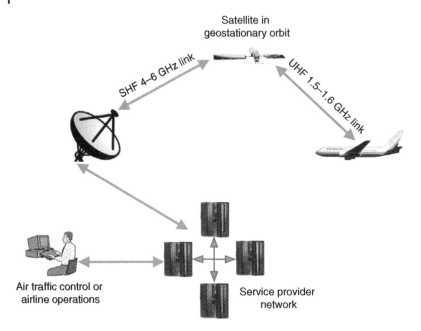

Figure 1.12 Typical system for satellite communication (SATCOM). The operator (ATC or airline staff) sends a message to the service provider network that also transmits digital data (e.g. ACARS or internet). The message is sent to a ground station whose dish is aligned with the satellite in geostationary orbit. The ground station transmits the voice message and data via an SHF link and the satellite relays everything to the aircraft in flight via a UHF link.

especially in the case of emergencies and natural disasters. There are numerous instances including the 2004 Indian Ocean tsunami, in which damaged or overloaded satellite and telephone systems have meant that humanitarian and emergency agencies have had to make extensive use of HF radio to coordinate their missions. In addition, the polar regions still have unreliable satellite coverage making it necessary to use HF radio communication. The HF communications equipment in aircraft and the ground infrastructure required to support it will stay for the foreseeable future.

1.6.2 Secondary Surveillance Radar (SSR) and Traffic Alert and Collision Avoidance System (TCAS)

Primary radar was first developed in the mid-1930s and by 1936 a chain of radar stations along the South and east coast of England called the Chain Home system (Figure 1.13) was installed in time for the outbreak of World War 2. Primary radar works by emitting a pulse of radio energy along a specific direction and measuring the time taken for the reflected pulse from an aircraft to

Figure 1.13 Three of the transmitter masts of the Chain Home system photographed in 1945. The antenna itself is the wire array just visible to the right of the masts.

return giving the distance along that direction and hence the aircraft position (see Section 5.1). Problems that were recognized immediately from the birth of the technology included the weakness of the return pulse and the lack of other information about the aircraft including whether it was friend or foe. The original radar patent granted in April 1935 [7] contained an idea for overcoming these problems by having a transmitter installed on the aircraft that would detect the primary pulse from the radar and be triggered into transmitting its own signal. This would greatly increase the strength of the return signal, which could also contain information to identify friend from foe (IFF). The device on the aircraft was called a transponder (combination of transmitter and responder) and the system was known as Secondary Surveillance Radar (SSR: see Section 5.4). SSR was also developed independently in Germany and the United States, and by 1939 all three countries had their own IFF systems for aircraft and ships. Originally, transponders responded with a radio signal that had the same frequency as the primary pulse but as radar frequencies changed, it became necessary to transmit the signal in a separate band and this was set at 1000 MHz, which is very similar to the frequency used by transponders today.

After the war, the system was adapted for civilian use and in the first manifestation of civil SSR (Mode A), an aircraft transponder would transmit a four-digit octal *squawk* code set by the pilot to the radar station to help identify it. The term squawk originates from the codename "parrot," which was used when discussing the secret IFF system and is now a permanent part of aviation nomenclature. Another problem with primary radar is that it can only measure the slant distance and apart from nearby traffic is unable to determine the altitude. From the mid-1960s a second mode (Mode C) was introduced into SSR that would transmit a second four-digit octal code in addition to the squawk code that would report the aircraft altitude. It was soon realized that it would be possible for transponders to transmit much larger quantities of digital data and still maintain compatibility with both Mode A and Mode C and from 1990 a new mode called Mode S (for "Select": See Section 5.4.5) started to be installed.

Mode S eliminates some of the problems encountered with Modes A and C caused by overlapping signals in regions of heavy traffic and can transmit a code that is unique to a specific aircraft (assigned to it at manufacture). In addition, it not only communicates with the ground station but also with other aircraft and can interrogate the transponders of nearby traffic. Research into autonomous collision avoidance systems has been conducted since the 1950s and was given particular impetus in 1956 when a Lockheed Super Constellation and a Douglas DC7 collided over the Grand Canyon killing 128 people. Incorporating a directional SSR antenna so that the position of other traffic could be determined meant that Mode S SSR with its ability to automatically communicate with other aircraft provided a good platform to realize this, which led to the birth of TCAS. There have been various implementations of TCAS since its inception and the current system is TCAS II, version 7.1 but upgrades are planned in the future (see Section 5.5). The method used for the surveillance of the local mode S capable traffic (mode S is compulsory in terminal areas and airways) is that each aircraft broadcasts an unsolicited message every second at 1030 MHz called a *squitter*. This message includes the unique address of the sender and if it is received by another aircraft, the address is stored in the receiver's database of local traffic. Thus, every aircraft in the vicinity builds a list of all other traffic and can address each plane individually. The range of the other traffic is determined by the time lapse between interrogation and response and the bearing is derived from a directional antenna. Also, the responses from other mode S traffic will contain additional information including the altitude, heading, etc. The algorithm used to determine whether any of the local traffic poses a collision risk is described in Section 5.5 but having determined that a hazard is present the system will present to the pilot a "Traffic Advisory" (TA) or "Resolution Advisory" (RA). A TA alerts the pilot to be vigilant and prepare for evasive action if necessary while an RA requires immediate evasive action and the TCAS display will indicate what evasive action to take. Currently, this is limited to demanding a climb or descent and an indication of what rate is required but future implementations of TCAS will incorporate left and right turns as well. The system operates independently of ATC and pilots are now trained to prioritize the information from the TCAS system following a collision over Überlingen in 2002 between two TCAS-equipped aircraft caused by one pilot following the RA and the other following advice from ATC.

1.6.3 Aircraft Communications Addressing and Reporting System (ACARS)

ACARS was first implemented in 1976 essentially as an automated reporting system for times spent in different phases of flight. The system accepts inputs from sensors on the wheels, doors, etc. to determine whether the flight is Out of the gate, Off the ground, On the ground, or In the gate (OOOI) and reports the data back to the airline operations. The communication is digital and uses whatever radio link is in use, that is VHF, HF, or satellite, as well as removing the burden for flight crews to record all this information the system provides for smooth operation, for example, by alerting replacement crews the optimum time to report for flight. The digital capabilities were soon expanded and the system can now be used to obtain additional information from airline operations such as weather reports at destination aerodromes, flight plans, amendments to flight plans, etc. It can also communicate with other ground facilities including the airline maintenance department and the aircraft or engine manufacturer. For example, ACARS can transmit a continuous report on the state of health of the engines to the manufacturer to organize maintenance schedules, troubleshoot problems, or provide data to implement design improvements. ACARS also communicates with ATC and is used to implement a new mode of communication between pilots and ATC known as Controller–Pilot Data Link Communications (CPDLC).

A major problem with voice communications, which were largely developed in an environment of one-on-one between pilot and controller, is that in a busy airspace a large number of pilots are on the same frequency. Any aircraft transmitting will block the frequency for all others and as the traffic increases, the probability that a pilot will accidentally cut off the transmission of another will also increase. In addition, each exchange between the controller and a pilot takes a specific amount of time and a traffic volume can be reached where it is no longer possible to pass the necessary messages to all aircraft within the time required. A solution is to increase the number of controllers and have each one on a separate frequency but that also introduces its own problems including the increased time required to handover flights from one controller to another. The CPDLC system alleviates these problems by using a datalink with ATC rather than voice communication to pass on clearances and other instructions to pilots and for pilots to request level changes, etc. This also relieves some of the stress on flight crews as they do not have to concentrate on picking out their call sign from almost continuous ATC talk and the text information that appears on the screen cannot be misheard. ACARS controllers are also interfaced to a printer so that a hard copy of important information can be obtained.

As an addendum to this section, it is worth discussing the role that satellite communications and ACARS played in the attempts to find the missing Malaysian Airlines flight MH370 that disappeared on 7 March 2014. The system is a commercial service provided by Inmarsat that has to be paid for and Malaysian Airlines had opted only for the engine monitoring service that reports the state of engine health to Rolls Royce. The transponder was switched off and the ACARS system was disabled by switching off the SATCOM and VHF channels, which stopped all ACARS transmissions to the ground station. In this circumstance, the satellite continues to send a simple handshaking signal to the aircraft known as a *ping* every hour to check whether the aircraft is still online and the aircraft continues to respond. From the time taken between the ping and the reception of the aircraft response it is possible to determine the distance from the satellite. Six pings were sent after the loss of contact and from these it was possible to map out an arc traveled away from the geostationary satellite in two directions, North and South, but distance measurements alone could not distinguish which track was flown. Some innovative work by Inmarsat engineers analyzed the shift in frequency of the return signal due to the Doppler effect to determine the aircraft speed relative to the geostationary satellite. By comparing this with other Malaysian airline flights on the same route they were able to say with a reasonable degree of certainty that it was the Southern track that was taken, which was a significant help to the SAR operation. Tragically, the mystery of what happened to flight MH370 has never been solved but the incident has reignited the debate over whether full ACARS implementation should be compulsory to act as a kind of continuous "black box" in flight.

1.7   Development of Radio Navigation

1.7.1   Radio Direction Finding

In 1865, James Clerk Maxwell predicted the existence of electromagnetic waves, which travel at the speed of light and that light itself was such a wave. Between 1886 and 1889, Heinrich Hertz was the first to demonstrate conclusively that the waves generated by his spark gap transmitter were the same electromagnetic waves predicted by Maxwell, which was an important milestone in Physics. Almost immediately after the first detection of radio waves, the basic phenomenon that enables direction finding was also discovered by Hertz in 1888 using a simple loop of wire. He found that the maximum signal was detected when the plane of the loop was aligned along the direction of the transmitter and the signal went to zero when the loop was turned face on. The theory describing this

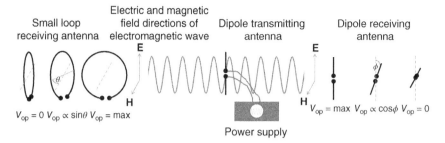

Figure 1.14 Schematic of a simple dipole antenna transmitting a linearly polarized wave. A small loop receiving antenna will give maximum signal when the loop axis is parallel to the magnetic field of the electromagnetic wave while a dipole receiving antenna will give maximum signal when it is aligned with the electric field.

observation is developed in Section 4.4.6, but the phenomenon is illustrated schematically in Figure 1.14.

Assume the radio wave is transmitted by a simple dipole aerial (not the case in Hertz's experiment but the characteristics of the radio wave are similar), which produces a polarized wave, that is, the electric field of the electromagnetic wave is vertical everywhere. This means that the magnetic field is aligned along the horizontal everywhere so that a loop set up as shown in Figure 1.14 is threaded by a time-varying magnetic field. If the loop diameter is small compared with the wavelength, the magnetic field can be assumed to be spatially constant. As shown in Section 4.4.6, in this scenario the time-varying magnetic field generates the maximum signal when it is aligned along the loop axis, that is, when the plane of the loop points to the transmitter. On the right-hand side of Figure 1.14 is shown the response of a dipole aerial and in this case the maximum signal is obtained when the transmitter and receiver dipoles are aligned and goes to zero when they are perpendicular. A simple loop does not give an absolute direction as it could be either side of the transmitter, that is, there is an ambiguity between two positions 180° apart. As shown in Section 7.1 this can be resolved by mixing the signals obtained from a loop antenna and a separate dipole antenna, also known as the sense antenna, to give a unique direction.

Small loops, that is, small compared to the wavelength, simplify the theory of coupling of the radio wave to the antenna but the signal is small and the use of small loops would have to wait till the development of electronic amplifiers. The reason that Hertz was able to detect the variation of signal with the loop orientation was that the transmitter was only a few meters away. Large loops that are resonant with the radio wave (that is, the total length of the loop is about half the wavelength) produce a much larger signal. They were already in use by the end of the nineteenth century and direction-finding patents were being filed by 1902. Given that it was the LF band being used with wavelengths of hundreds of feet, these early devices were very unwieldy requiring massive loops. During experiments in 1907, two Italian engineers, Ettore Bellini and Alessandro Tosi, noticed that they could feed the signal from two large fixed perpendicular loops into two small perpendicular coils and recreate the directional properties of the radio wave in a small space. Thus, the movable coil could be placed in a bench-top device, the size of a saucepan, while the large receiving antennas could remain fixed. The patent for the Bellini–Tosi Direction Finder (BTDF) was filed in1909 and within three years the signal-amplifying abilities of the triode vacuum valve were first discovered allowing the BTDF to be combined with electronic amplification. These devices were in widespread use by the 1920s and continued in operation till the end of World War 2. Bellini and Tosi called their device a *radiogoniometer* and the basic principle, that is, recreating a large-scale electromagnetic field within a small desired space has found its way into several instruments in aviation including the ADF in use today (see Section 7.1).

Bellini–Tosi and other LF-based direction finders were successful instruments in widespread use for decades but the antenna arrays were much too large to install on an aircraft and they had to be ground-based systems with a ground operator reporting bearings to the pilot. One possible solution was to go to higher frequencies/shorter wavelength so that small loops could be used but this introduced a new problem. Moving in to the HF band, for example, where a 1 m loop would be resonant meant that radio waves would be reflected from the ionosphere and arrive at the aircraft from more than one direction. Thus, instead of picking up a single strong minimum in the signal, several weaker minima would occur making it impossible to determine the direction of the transmitter.

By the mid-1920s improvements in receiver and amplifier electronics meant that small multiple wire loops used in the MF band, just below frequencies where the ionosphere would become troublesome were sufficiently sensitive to use as direction finders and installed on aircraft. This meant that the ground station just had to provide an omnidirectional radio signal referred to as a Non-Directional Beacon (NDB). The earliest Radio Direction Finder (RDF) loops were turned manually or remotely by a motor with the operator finding the signal null and obtaining the bearing to the transmitting station (Figure 1.15a). By the end of World War 2 the loops would slew automatically to find the null if the signal was strong enough and pass the bearing indication to a cockpit display. The device was thus described as an ADF and the antenna was installed within a weatherproof aerodynamic housing as shown in Figure 1.15b.

The next development was a goniometer system in which the external rotating loop antenna was replaced by a pair of fixed loop antennae at 90° wound on ferrite cores to boost the field. This was then relocated via a goniometer system so a small rotating loop within the receiver could be used to find the direction of the NDB. In this design, there are no external moving parts on the airframe but since the 1980s it has been possible to implement the ADF entirely with solid-state electronics with no moving parts anywhere as described in Section 7.1.

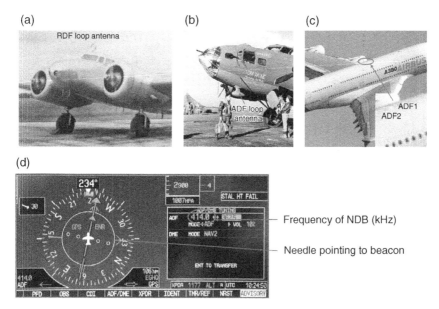

(a) RDF loop antenna

(b) ADF loop antenna

(c) ADF1 ADF2

(d) Frequency of NDB (kHz)

Needle pointing to beacon

Figure 1.15 Evolution of Automatic Direction Finding (a) RDF rotatable loop antenna on Amelia Earhart's Lockheed Electra in 1937. *Source:* USAF. (b) ADF antenna in housing installed on a B17 at RAF Knettishall, England in 1943/44. *Source:* USAF Historical Research Agency. (c) ADF solid-state goniometers on the roof of an Airbus A380 in 2014. *Source:* Adapted from Ref. [8] and reprinted under Creative Commons license 2.0 [8]. (d) The ADF display in the PFD of a G1000 glass cockpit.

The compact ADF "blisters" for two independent ADF cockpit displays on the top of an Airbus A380 are shown in Figure 1.15c and the ADF display of the PFD in a glass cockpit is shown in Figure 1.15d. Generally, ADF used in conjunction with LF NDBs has limited use in radio navigation having been superceded by VHF navigation (see Section 1.7.3 and Section 7.2) and then Global Navigation Satellite Systems (GNSS: see Chapter 8).

Thus, it may be surprising to find a modern version of this radio navigation system that is over a century old implemented in the latest aircraft, but this is a testament to the power of an ADF and the intuitive information it provides. The needle simply points to the transmitter and shows the direction relative to the aircraft heading of the selected NDB. In modern displays, the outer angle scale of the ADF is slaved to the compass, as demonstrated in Figure 1.15d so that the bearing to the station can be read directly from the display. Such a device is referred to as a Radio Magnetic Indicator (RMI: see Section 7.1). Homing to a station is straightforward and many airport instrument approaches specify an approach procedure in terms of an NDB. In addition, NDBs are often used as anchor points for holds, for example, the holds North and South of

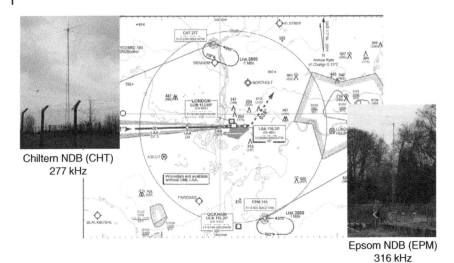

Chiltern NDB (CHT)
277 kHz

Epsom NDB (EPM)
316 kHz

Figure 1.16 2015 approach chart for runway 09L at London Heathrow Airport showing holds North and South of the airport anchored on NDBs operating in the LF band. *Source:* Chart reproduced with permission from National Air Traffic Services and photos of beacons reproduced with permission from Trevor Diamond (td@trevord.com).

London Heathrow Airport shown on a 2015 approach chart in Figure 1.16. These are anchored on the Chiltern NDB (CHT) and the Epsom NDB (EPS) both operating in the LF band. In all current NDBs the signal is modulated with a Morse code identification consisting of the three letters designating the beacon so that the flight crew can confirm that it is operating. In reality, the hold would normally be flown with the autopilot, which would use GPS to follow the racetrack pattern, with the position of the NDB stored in the database. The ADF system is there as a backup, however, and there are plenty of instances where it has been used due to failures of other equipment. Tracking to and from NDBs and flying holds using ADF is still an important part of training for commercial pilots.

1.7.2 Guided Radio Beam Navigation

Another form of radio navigation using the LF band is to set up a directional beam along which the aircraft flies. Systems started to be developed shortly after World War 1 based on a patent originally granted to the German engineer, O. Scheller, in 1904. The system envisaged by Scheller employs four overlapping beams as shown in Figure 1.17a. Each one is amplitude modulated to produce a string of Morse "A"s (• –) or "N"s (– •) as shown in Figure 1.17b and if these are picked up equally, that is, the aircraft is somewhere along the line of intersection of the beams, a continuous tone is heard in the headphones. Off the

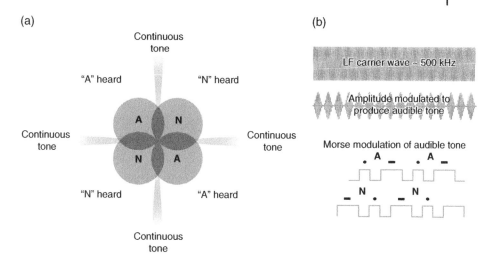

Figure 1.17 (a) Directional 4-beam system originally devised in 1904 by Scheller. A crossed loop or Adcock antenna generates the radiation pattern of overlapping radio beams. (b) Each quadrant is modulated by an audible tone, which itself is modulated by the Morse code for "A" or "N" in neighboring quadrants. If the two signals are picked up equally, that is, the aircraft is along one of the centerlines of the overlapping beams, then a continuous tone is heard. Moving off center produces an audible A or N indicates that the aircraft is off center and also which way to turn to recover the track.

intersection a Morse "A" or "N" is heard so not only does the system indicate whether the aircraft is off the centerline but it also shows which way to turn to recover it. The overlapping beam pattern was originally produced by crossed-loop antennas but subsequently a superior design of antenna invented by the British army officer, Frank Adcock, in 1919 was employed. This consisted of four vertical antennas connected underground with the current in opposing pairs in antiphase and was found to give a "cleaner" signal with reduced interference.

During the development of commercial air transport after World War 1, these beacons became widespread and in the United States an entire network was set up defining airways across the country. This was known as the Low Frequency Radio Range (LFR) and by the mid-1930s it covered the entire country with stations typically 200 miles apart. The transmitted beams defined airways that aircraft would navigate by tracking from one beacon to the next and a chart for the Silver Lake transmitter in California is shown in Figure 1.18. Thus, a pilot would be guided for 100 miles by the beacon behind and then switch frequency to the beacon 100 miles ahead. When flying over the beacon, the signal would disappear altogether momentarily then reappear on the other side indicating to the pilot they had flown directly overhead thus providing an accurate position fix.

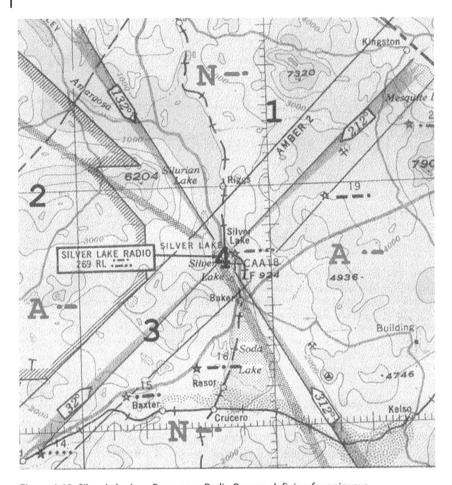

Figure 1.18 Silver Lake Low Frequency Radio Beacon defining four airways.

In addition to the Morse code "A"s and "N"s the stations would generate Morse identifiers every 30 seconds in the "N" quadrant and then in the "A" quadrant to confirm the identity of the station and indicate it was operating. From an operational point of view, listening to the continuous tones could be tiring and by 1930, cockpit instruments were available that would produce a visual "fly right" or "fly left" indication to maintain the centerline [9].

From the late 1920s, LFR networks facilitated the transition to reliable commercial air transport that, apart from extremes, could operate in all weathers and they dominated airways navigation for decades. By the end of World War 2, new VHF-based navigation systems had been developed (see next section) that addressed some of the problems with the LFR system. Being a LF system made it susceptible to interference especially from electrical storms and the ground wave propagation mode of the LF waves could produce reflections and refractions off ground features giving false directional information. Another problem was that the beams from the beacons were restricted to just four directions. In principle, it is possible to produce six or eight beams from more overlapping radio signals but that creates its own problems. On the other hand, the later VHF stations could be used to fly along any chosen direction to or from the beacon.

The LFR network went some way to making navigation in air travel weather-proof but visual conditions were still required to land at the destination airport and from the late 1920s, short-range systems to guide aircraft into land were also being developed. In fact, the biplane that Doolittle performed his blind flight in 1929 was fitted with a prototype system using overlapping beams and a cockpit "fly left," "fly right" indicator to guide him along the runway center line. The instrumentation has many similarities to the Instrument Landing System (ILS) that is in widespread use today (see Section 1.7.3 and Section 7.4).

So, in principle, instrument landings were available from 1929 but Doolittles flight was something of a bold experiment and he also had a safety pilot in the back seat who was visual with the airfield and could take over if necessary. Landings in almost zero visibility (down to 75 m) can be performed by the autopilot in modern aircraft but only at airports that have the top level of ILS system (Category IIIB) installed and there is the requisite equipment on the aircraft including backup autopilots. There has to be some visibility on the ground otherwise the aircraft is unable to taxi off the runway. There is an even higher category of ILS defined (IIIC) with which a true zero visibility landing can be made, but this system also requires that the aircraft can taxi to the gate on the autopilot and at the time of writing this is not available anywhere in the world.

1.7.3 VHF/UHF Radio Navigation Systems

By the late 1930s, it was clear that future airways navigation beacons would need to operate at VHF frequencies to overcome the interference problems of the LF network and would also need to be able to provide flexible vectoring guidance to and from beacons. Since airways are defined by beacons that are 200 miles apart or less, the line of sight radio range needs to be 100 miles so, from Equation (1.1), the horizon is not a limitation for aircraft that are flying above 7000 ft, which is certainly the case for long-distance flights. The network went through a brief evolutionary phase using a VHF visual-aural radio range (see Interesting Diversion 7.1) but the development of an omnidirectional VHF beacon known as a VHF Omnidirectional Range (VOR) started in 1937. The first operational VOR transmitting at 125 MHz was put into service in 1946, and in 1949 the ICAO selected the VOR as the international civil navigation standard.

The operation of VORs is described in detail in Section 7.2, but in summary they determine the direction to or from the station by measuring the phase relationship of two 30 Hz signals modulated on a VHF carrier. The 30 Hz signals are an omnidirectional (reference) signal from a central antenna with the same phase along every azimuth and a directional (variable) signal, whose phase varies continuously around the circle from 0 to 360° relative to the reference signal (Figure 1.19a). The two signals are in phase along magnetic North, so comparing the phase angle between them gives the angle of the bearing to or from the station. A typical VOR transmitter is shown in Figure 1.19b and is installed near Daventry in the United Kingdom. The reference signal from the VOR is modulated to carry a Morse code three-letter identification (for example, Daventry is 'DTY', that is, – • • | – | – • – –) and is sometimes modulated by a voice channel to provide other information.

In the cockpit, there is a tuner to select the VOR frequencies and examples in a traditional and glass cockpit are shown in Figure 1.20a and b. In the latter case the VHF receiver can select two VORs (each driving its own separate display) and two in reserve whose frequency can be switched in by pressing a button. The VOR displays in the two cockpits are also shown in Figure 1.20a and b. Each has a knob known as the omnibearing selector (OBS) that rotates a pointer called the omni-bearing indicator (OBI), which shows the bearing selected to or from the station. This bearing represents the selected radial from the VOR and if the aircraft is on that radial the separate central part of the needle, known as the course deviation indicator (CDI) will be centered. Each division represents either 2° or 5° deviation depending on the type of display but in all cases full-scale deflection represents 10° off the selected radial.

There is also a TO/FROM flag to show whether the selected bearing is to or from the VOR. If the OBS is rotated through 360° the needle will center twice, one null will have the TO flag showing and the other the FROM flag showing. As an example, consider an aircraft on a heading of 045 approaching a VOR along the 225 radial (Figure 1.21a). Rotating the OBS will center the CDI when it is set to 045 with the TO flag showing and when it is set to 225 with the FROM flag showing.

An important aspect of the VOR display is that it is independent of the heading of the aircraft (unlike the ADF), it just gives an indication of where the aircraft is, assuming it is a point object. For example, in Figure 1.21b the three aircraft shown on the 225 radial will all display zero deflection on the CDI but only the aircraft with a heading of 045 will maintain zero deflection on the CDI. For the rest, the CDI will start to deviate as they move away from the 225 radial.

The VOR network achieved the design goals, that is, a VHF system relatively free of interference and flexible vectoring to or from beacons allowing a much greater flexibility of airways. The power of the system was increased further by installing at most VOR transmitters a separate UHF (300–3000 MHz)

(a)

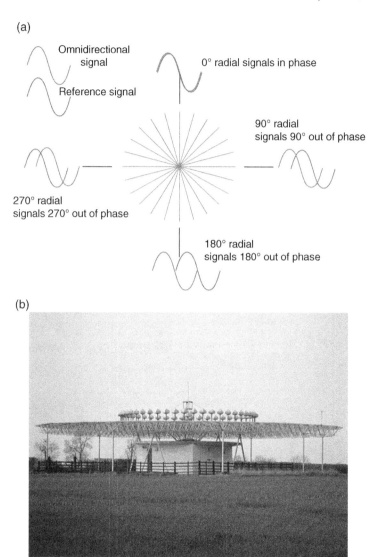

Omnidirectional signal

Reference signal

0° radial signals in phase

90° radial signals 90° out of phase

270° radial signals 270° out of phase

180° radial signals 180° out of phase

(b)

Figure 1.19 (a) The VOR transmits an omnidirectional signal with the same phase in all directions and a directional signal whose phase relative to the omnidirectional signal varies from 0 to 360° around the circle. The two signals are in phase along magnetic North. (b) The DTY VOR in the United Kingdom. *Source:* Photo reproduced with permission from Trevor Diamond (http://www.trevord.com/navaids).

(a)

VOR frequency selected (MHz)

Bearing selected

TO/FROM flag (indicating TO)

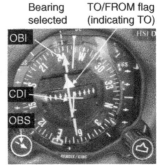

(b)

VOR frequency selected (MHz)

TO/FROM flag (indicating TO)

Figure 1.20 (a) The VHF Navigation tuner and VOR display in a traditional cockpit. (b) The VHF Navigation tuner and VOR display in a glass cockpit.

transponder that would transmit in response to interrogation by an aircraft UHF transmitter. The slant distance to the aircraft can be determined by measuring the time required for radio pulses to be returned. This Distance Measuring Equipment (DME) is described in detail in Section 7.3 and providing the pilot with a radial and a distance from the known position of the beacon enables an absolute position fix. In modern microprocessor controlled receivers the

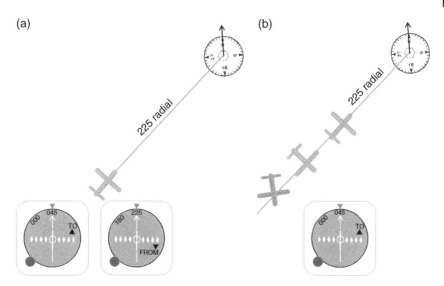

Figure 1.21 (a) An aircraft on the 225 radial will show zero deflection on the CDI for two OBS setting, that is, 045 with the TO flag showing and 225 with the FROM flag showing. (b) The indication of the display is independent of the aircraft heading.

pilot only needs to set the frequency of the VOR and the receiver automatically sets the UHF tuner to the corresponding DME frequency for that station from a database – a process known as frequency pairing.

With increased miniaturization of electronics allowing microprocessors to be incorporated into VHF navigation receivers, it became possible to increase the flexibility of the VOR system even further by allowing VORs to be electronically relocated within the receiver. For example, suppose the aircraft was navigating to a city airport with no VOR beacon but there was a beacon located 50 nautical miles to the West at a bearing 260 (Figure 1.22). The pilot could enter a shift of 50 miles along a radial 080 into the receiver, which would place a virtual beacon over the airport. The internal computer would obtain the radial and DME distance from the real beacon and compute what the phase shift and distance would be from the virtual beacon. The cockpit VOR and DME display would then be driven as if they were receiving signals from the virtual beacon. Thus, the pilot could use the instruments to navigate as normal as if there was a real beacon at the destination.

With this system, it is possible to plan any route between two points and not be restricted to flying from beacon to beacon. This increase in flexibility illustrates the important concept of area navigation (RNAV) to distinguish it from the linear point-to-point navigation that had been the norm since the 1920s. Restricting flights to airways does not normally provide the shortest route

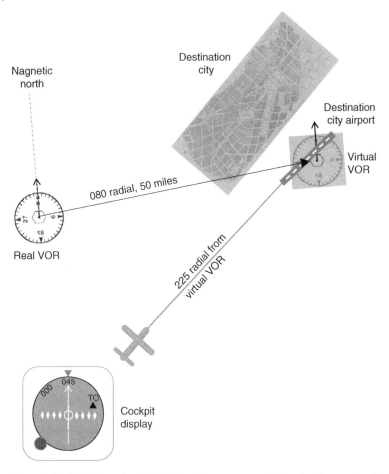

Figure 1.22 RNAV based on VORs. The pilot enters a radial and a distance to electronically shift a real beacon to generate a virtual beacon at the desired point. The cockpit VOR and DME displays are then driven as if they were receiving signals from a real beacon at the virtual beacon position.

between departure and destination points. With increased emphasis on saving fuel, it is rational to utilize the flexibility of the aeroplane to be able to go anywhere and fly routes that minimize the distance traveled, which would normally be great circle routes (see Section 6.1.2). The virtual VOR technology is only one of many RNAV systems, which today include GNSS and inertial navigation (see Section 1.8). Airways are still an important part of the traffic control system and flight plans are submitted using airways but wherever possible controllers will clear flights direct to destinations knowing that the necessary primary and backup RNAV navigation systems are on board.

The worldwide network of VOR beacons grew to over 3000 by the year 2000 with 1033 in the United States alone but this has started to reduce and at the time of writing there are 967 operating beacons in the United States. In 2015, 70 years since its inception, the VOR system has entered its twilight years with GNSS and inertial navigation becoming dominant in commercial aviation. Currently, the Federal Aviation Authority (FAA) in the United States is planning to reduce the network to 500 beacons by 2020 essentially to provide backup in the event of a GNSS outage and for GA aircraft not equipped with a satellite navigation receiver.

The discussion of VORs has focused on long-distance navigation but a VHF/UHF system was also developed for instrument landing. By 1929, the 4-course Low Frequency Range had already been used to provide radio guidance to line up with a runway. That year in the United States tests began on a HF system that could provide guidance for both the runway centerline and the required height at a given distance from the runway threshold, known as the *glideslope*. The system that was eventually developed is known as the ILS (see Section 7.4).

The ILS has the two components illustrated in Figure 1.23, that is, one to determine the deviation from the centerline (the *localizer*: Figure 1.23a) and one to determine the deviation from the required height (the *glideslope*: Figure 1.23b). The localizer antenna situated at the far end of the landing runway transmits two overlapping directional beams in the VHF band, one amplitude modulated at 150 Hz and the other at 90 Hz. The depth of modulation in both beams varies with angle away from the centerline (known as *space modulation*), so by measuring the difference in the depth of modulation (DDM) of the two frequencies the onboard instrumentation can determine the deviation from the centerline. This is passed to the same display as used for VOR tracking and when an ILS localizer frequency is selected the VHF receiver recognizes this and drives the CDI using the DDM measurement. In this mode, the turning of the OBS has no effect as there is only one direction (the runway heading) defined and also the sensitivity of the CDI is increased by a factor of 4 so that full-scale deflection corresponds to an angular deviation of 2.5°.

The glideslope works by a similar space modulation technique but the transmitted overlapping beams are in the UHF band and angled up at the glideslope angle, which is typically 3° relative to the ground. Within the receiver the UHF frequency is paired with the VHF frequency selected for the localizer, so it does not need to be set separately and the display depends on the type of cockpit. In a traditional cockpit, selecting the localizer frequency automatically switches in a second needle on the VOR display that is horizontal and whose vertical deflection shows the angular deviation from the glideslope with full-scale deflection corresponding to 0.7° (Figure 1.23c). In a glass cockpit selecting the localizer frequency automatically brings up an arrow display next to the height tape as

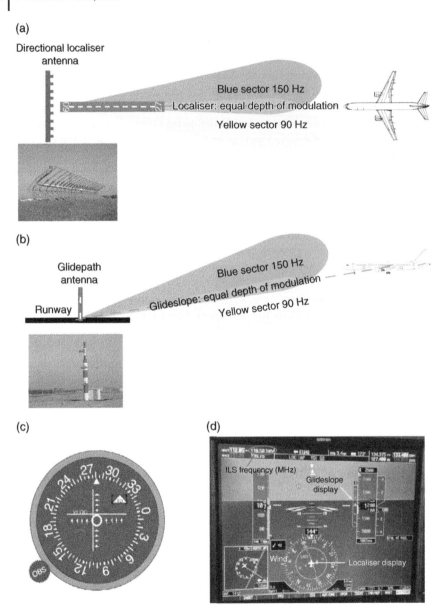

Figure 1.23 (a) Localizer antenna (typical example shown in photo) at the end of the runway transmits two overlapping directional beams amplitude modulated at 150 and 90 Hz with the depth of modulation (DDM) increasing away from the runway centerline. The difference in the DDM is used to determine the angular deviation from the centerline. (b) The glideslope antenna (typical example shown in photo) at the side of the runway produces similarly modulated directional beams and the DDM is used to determine the angular deviation from the glideslope. (c) Localizer and glideslope display in a traditional cockpit during an ILS landing. (d) Localizer and glideslope display in a glass cockpit during an ILS landing.

shown in Figure 1.23d, which also has a sensitivity of 0.7° deviation for full-scale deflection. Normally, the ILS localizer will be colocated with a DME transponder, described above and described in detail in Section 7.3, so that the distance from the runway threshold is known.

The system described is the one finally adopted internationally but in the United States, earlier systems that provided both localizer and glideslope guidance working at different frequencies were available before World War 2. The first landing of a commercial aircraft using ILS was in 1938 when a Pennsylvania Central Airlines Boeing 247D landed in Pittsburgh during a snowstorm. In 1941, the US Civil Aviation Authority authorized 6 ILS systems to be installed and by 1945, 9 systems were in operation with another 10 under installation. During the war, the US army had developed an improved system operating at higher frequency, which was adopted as the international standard by the ICAO in 1949 and is the system we have today. The signals that drive the displays can also act as control inputs to an autopilot and post-war there was a good deal of research effort dedicated to develop the technology to achieve a fully automatic landing without pilot input (see Section 1.9). This also needed the development of other technologies in addition to ILS, for example, the radio altimeter (see Section 5.6), but the first fully automatic landing was demonstrated in 1964 at the Royal Aircraft Establishment airfield at Bedford, England using a Hawker-Siddeley Trident jet airliner. The first fully automatic landing on a commercial flight with passengers was carried out at Heathrow airport in a British European Airways Trident in 1965.

ILS systems have evolved and are now classified into different categories depending on their authorized visibility limits and decision height, that is, the height below which a pilot should not descend if they are still not visual with the runway. The categories are dealt with in detail in Section 7.4.4 but as outlined in Section 1.7.2 above, the top category in operation (IIIB) allows descent to below 50 ft in visibility down to 75 m.

During the 1980s, a new system working in the SHF (3–30 GHz) called the Microwave Landing System (MLS) was developed that offered greater flexibility than ILS. For example, ILS is fixed at a given glideslope angle (usually 3°) along a fixed centerline, whereas MLS enables aircraft to choose a glideslope and approach at an angle to the runway. This is especially useful at military airports where a wide range of aircraft types varying from heavy transport to helicopters are operating. The system has only been implemented at very few civil airports and the number is not likely to increase as GNSS-guided approaches, which offer the same flexibility, are becoming available. ILS is still the dominant radio-guided landing system in use worldwide and its will remain operational for the foreseeable future. It is likely that over the next decade, GNSS-guided approaches will start to become the norm in operational practice but ILS will need to remain in place as a local backup.

1.8 Area and Global Navigation Systems

1.8.1 Hyperbolic Navigation

The concept of Area Navigation (RNAV) was introduced in the previous section and describes navigation systems that can be used to travel to any specific point in two dimensions as opposed to navigating along an airway. The ultimate aim has always been to achieve this globally so that an aircraft can be navigated to any point on the Earth without visual references. The earliest operational method was the British Gee system used by the RAF Bomber Command from 1942 and was based on a concept known as hyperbolic navigation. The basic idea was already well known in the 1930s but to build a practical system required equipment that could measure the timing of radio pulses at the microsecond scale and this came with the development of Radar in the late 1930s.

Hyperbolic navigation relies on the difference in timing between the reception of radio pulses from widely spaced beacons in chains with each chain consisting of a master and at least two secondaries. It is simpler to describe the method with a specific example and Figure 1.24 shows a master and one secondary separated by a distance of 100 nm, which is known as the baseline. The master transmitter emits a series of radio pulses typically at a rate of 10 a second and Figure 1.24a shows the timing of one pulse as perceived by an aircraft, labeled A, at the midpoint on the baseline given that the waves propagate at 6.18 μs per nautical mile. At $t = 0$ the pulse is emitted by the master and at $t = 309$ μs it arrives at the aircraft and is registered on a display. At $t = 618$ μs the pulse arrives at the secondary, which then transmits its own pulse after a precise time delay – 100 μs in this example, that is at $t = 718$ μs. This pulse arrives at the aircraft at $t = 1027$ μs and is registered on the display, which thus shows a time difference between the pulses (TD) of 718 μs. Now consider another aircraft, labeled B, on the same bisecting line. It will take a longer time, say t_1, for the direct pulse from the master station to reach it. The secondary will still emit its pulse at $t = 718$ μs, which will also take time t_1 to reach the aircraft, so when the difference is taken the two t_1's will always cancel. It is clear that anywhere along the bisecting line the time difference is always 718 μs, that is, the time for a pulse to traverse the baseline plus the delay introduced by the secondary.

It is intuitive that any other constant TD value will also be measured along a line, which will be curved as we move away from the bisecting line. It can be shown that the curves are hyperbolae as illustrated in Figure 1.24b and this is the origin of the term hyperbolic navigation. Measuring the TD value will then locate the aircraft on a specific hyperbolic line and to obtain a position fix, another secondary with its own set of TD hyperbolae is used as shown in Figure 1.25a. The pulses from the three stations are encoded in some way (for example, different numbers of closely spaced multiple pulses) so that they

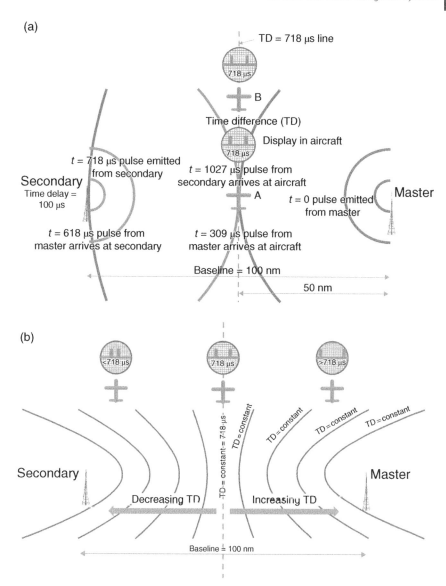

Figure 1.24 (a) Pulse timing for a specific example of a master and secondary station 100 nm apart and a secondary time delay of 100 μs. Anywhere along the bisecting line between the two stations an aircraft will measure a time difference (TD) of 718 μs. (b) Away from the center line the lines of constant TD form hyperbolae.

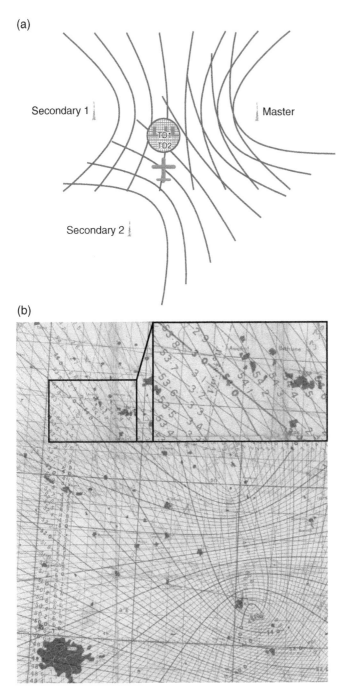

(a)

Secondary 1 | | Master

TD1
TD2

Secondary 2 |

(b)

Figure 1.25 (a) To get a fix at a single point a second overlapping set of constant TD hyperbolae are required from another secondary in the chain. (b) A Gee navigation chart for the Reims chain operating in 1944 with the inset detailing the color-coded TD values. *Source:* Reproduced with permission of Smithsonian Institution office of imaging, printing and photographic services. [10].

can be distinguished. Notice that the entire system is synchronized by just the master station and there does not need to be any physical connection between different transmitters in a chain. During the war, the TDs were measured in the aircraft using an oscilloscope trace and having obtained them the values were plotted on a chart of constant TD lines for a particular chain. An example is the chart for the Reims chain of the Gee system in 1944 shown in Figure 1.25b with the inset detailing the color-coded TDs. The Gee system had an accuracy of a few hundred meters at a range of 350 miles and remarkably, was in operational use by the RAF till 1970.

The British Gee system was operational by 1942 and as such the first genuine RNAV system but it was relatively short range and after the invasion of France in 1944, new chains were set up as the front progressed. It operated at relatively high frequencies around 30 MHz, that is, at the top of the HF band so that sky-wave contamination could be a problem. The US was close behind and by 1943 the US coastguard had a hyperbolic navigation system operating that was known as LORAN (LOng RAnge Navigation). This system, whose designation became LORAN-A to distinguish it from later systems, expanded rapidly and by 1945 included over 70 transmitters that provided navigation over 30% of the Earth's surface. By the late 1940s, experiments had shown that reducing the carrier wave frequencies into the 90–110 kHz (LF) band produced significant improvements in range. This change and other developments resulted in a new system LORAN-C, which by the late 1950s became the global all-weather radio navigation standard for ships and aeroplanes. LORAN-A and LORAN-C were both operated in parallel till the mid-1970s when LORAN-A started to be phased out.

The absolute accuracy (systematic error) of the LORAN-C system was of the order of 200 m while the repeatability (random error) could be as low as 20 m with the onboard receivers automatically measuring TDs and displaying latitude and longitude. In the 1960s, the required instrumentation was expensive and used primarily by the military but following the reduction in cost of solid-state electronics in the 1970s, LORAN-C found its way into widespread civil use and even into the GA market as shown by the installation in Figure 1.26.

After GNSS became widely available in the late 1990s, LORAN-C started to fall into disuse and the decision was taken by the US Government to terminate the signals in February 2010. Shortly afterward the LORAN-C operated by the Canadian government was shut down and a similar hyperbolic navigation system run by the Russian government (CHAYKA) was also terminated. This did not mark the end of hyperbolic navigation systems, however, as increasing fears about the vulnerability of the GNSS system to natural and hostile interference has prompted plans to provide a backup for GNSS and this has led to the emergence of eLORAN (for enhanced LORAN). This is envisaged to reach an accuracy of ±8 m, which is competitive with un-enhanced GNSS. The first chain of

Figure 1.26 LORAN-C installed in a General Aviation aircraft (Grumman AA5) in the 1990s alongside a GNSS display. *Source:* Reproduced with permission of Rob Logan.

eLORAN transmitters became operational in the United Kingdom in October 2014. This technology, however, is emerging at the same time as tests have begun on a cheaper terrestrial navigation system based on tracking transponder signals (see Section 5.4.12).

1.8.2 Global Navigation Satellite Systems (GNSS)

GNSS are the youngest of the global navigation technologies as they were not conceivable before the dawn of the space age starting with the launch of the Sputnik capsule in 1957. Satellite-based navigation systems started to appear shortly after, however, and the first navigation satellite, which formed part of a US Navy system called Transit, was launched in 1959. The complete system consisted of seven satellites in a low polar orbit (see Section 8.2 for a description of different types of orbit) that transmitted stable radio signals, which were used by surface craft to determine their position by measuring the Doppler shift of the radio waves. There was also a series of ground stations to track the satellites and update their orbital parameters for the users. Although Transit worked on a different principle to GNSS, it tested various elements of the infrastructure required for future GNSS. The system was made available for civilian use in 1979, becoming widespread among commercial ships and it continued in operation till the late 1990s. Some drawbacks of the Transit system, including its inability to determine height, the long monitoring times required to obtain a position, and the relatively long off-air periods made it unsuitable for air transport.

A second system developed for the US Navy, called Timation, tested the application of high-stability clocks, the accuracy of time transfer, its ability to determine position, and the initial tests of three-dimensional navigation. The first satellite was launched in 1967 in which the timing was provided by stabilized quartz clocks but later missions flew the first-ever atomic clocks into orbit, which were sufficiently precise and stable to provide accurate position fixes. Thus, the fundamental concept by which GNSS determines the position was developed and tested with the Timation system.

The principle of the position fix is illustrated in Figure 1.27. Each satellite transmits its current orbital position (ephemeris) and since it carries a very accurate atomic clock, it knows precisely when the message is sent. The broadcast is picked up by the vehicle and imagine for a moment that there was also an atomic clock in the receiver synchronized to that of the satellite. Then, knowing the position of a satellite (contained in the message) and the time it was sent, the time for the signal to reach the vehicle and thus the distance to the satellite would be known. A single satellite would define the position of the vehicle as somewhere on the surface of a sphere (Figure 1.27a). Two satellites would fix

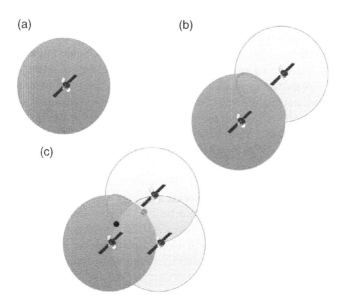

(a) (b) (c)

Figure 1.27 Principle of obtaining a position fix from satellites. If there was an accurate clock on the receiver synchronized to that of the satellite, then the distance to each satellite would be known independently so (a) one satellite would define a point on the surface of a sphere, (b) two satellites would define a point on a circle, and (c) three satellites would define two points, one of which was far out in space and could be discarded. In practice, there is not sufficiently accurate timing available in the receiver so a fourth satellite is required to solve equations for x, y, z, and time.

the position as somewhere on a circle (Figure 1.27b) and three satellites would place the vehicle at one of two points (Figure 1.27c), one of which will be near the surface of the Earth and the other will be far out in space and can be discarded. The problem is that we do not have a sufficiently accurate clock on the receiver so that a fourth satellite is required to solve four simultaneous equations for x, y, z, and time, thus a by-product of getting a position fix is that we also know the time very accurately. If signals can be obtained from more than four satellites, the accuracy of the position fix is improved. The x, y, z coordinates determined by the system define a point in space but to relate this to a geometrical position on the Earth's surface in terms of latitude and longitude, the system needs to know the exact shape of the planet's surface (see Section 6.4).

Simultaneously, the US Air Force was working on its own GNSS and by 1972 had demonstrated the Pseudo-Random Noise (PRN) format of the satellite radio signals that would become the norm for GNSS (see Section 8.3). In 1973, a new program, the Navstar global positioning system (later to become just GPS), emerged by combining the Navy and Air Force systems. Between 1975 and 1985, 11 satellites designated as Block I were launched to begin a program of testing of the system and simultaneously a ground program to test user equipment began. Some of the Block I satellites also carried sensors to detect nuclear detonations partly as a contribution to monitoring of the test ban treaty.

Despite some early wobbles with funding and also, following the Space Shuttle Challenger disaster, a decision to switch to conventional rockets for launch vehicles, 24 satellites designated as Block II and (after further improvements) Block IIA were launched starting in 1989. This constituted the first fully global *constellation* of space vehicles. Only two satellites of this original constellation are still operational having been replaced with new vehicles with additional signaling capabilities designated as Block IIR, Block IIR(M), and Block IIF (Figure 1.28).

From 1976 onward, Russia developed an independent GNSS knows as GLONASS (GLObal NAvigation Satellite System) and began launching space vehicles in 1982. The full constellation was in orbit by 1995 but then went in to a partial decline due to a lack of funding. From 2001, the system became a spending priority and was fully restored by 2011 to provide global coverage. The orbital parameters for GLONASS are slightly different to those of GPS but the signal formats are similar.

The GPS system was made available for civil use at no charge in 1993, though the clock signal was deliberately dithered to limit the accuracy of position fixing to 100 m for civilian users – a process known as selective availability. In 1994, the FAA approved GPS as a stand-alone aircraft navigation system for all phases of flight and non-precision approaches. In 2000, US president Bill Clinton announced the removal of selective availability and overnight the precision of GPS position determination for all users improved to around 10 m.

Figure 1.28 Final GPS IIR(M) satellite being launched by a Delta II rocket in 2009.
Source: Courtesy of gps.gov.

GLONASS and GPS are available to civilian users on a similar basis, but both satellite constellations transmit additional messages that are only available to the militaries of the respective countries. Modern GNSS receivers use both GPS and GLONASS signals to improve the satellite coverage and the precision of position determination. GPS/GLONASS without further improvement is still not accurate enough for precision approaches, but in recent years a number of augmentation systems, effectively add-ons to GNSS (described in Section 8.6), have been implemented to increase the accuracy of position determination and GNSS precision approaches are now available at 1746 airports in the United States alone [11]. Such procedures are known as Localizer Performance with Vertical Guidance (LPV) approaches and are also designated as RNAV procedures in recognition of the fact that Flight Management Systems (FMSs) on modern aircraft are normally using more than one navigation system.

A decision was made by the European Union to develop an independent GNSS amid concerns that the GPS and GLONASS systems could be disabled for non-US/Russian users at any time and an agreed program among participating nations started in 2003. This system is designed from the outset for civilian use unlike the GPS and GLONASS systems that were developed primarily for the military. The system will have its own built-in space-based augmentation system provided by additional geostationary satellites and will also combine position fixing with additional SAR services. Each satellite is equipped with a transponder that will be triggered by a user's emergency locator transmitter (ELT). Signals will be sent to the relevant rescue co-ordination center (RCC) with a precise location of the user in distress. The service will be free to all users at the basic precision level and at a charge for the high-precision augmented service, which has an accuracy of around 1 m. The first satellite launch was in 2011 and the system was declared operational at the end of 2016 but the full constellation will not be complete till mid-2018 so there are still holes in the coverage. Other systems around the world include the fully global Chinese COMPASS/BEIDOU system due to be operational by 2020 and partial coverage systems operated by Japan and India. All of these GNSS systems require a significant ground infrastructure of tracking and communication centers, but these are covered along with the detailed operation of GNSS in Chapter 8.

1.8.3 Inertial Navigation Systems (INS)

Inertial navigation is unique among the systems available in that it determines the position entirely by measuring acceleration within the vehicle itself requiring, in principle, no external input whatsoever. The only restriction is that the starting position needs to be known but then all subsequent positions during travel can be deduced. The basic principle is that if acceleration can be accurately measured along an axis, then integrating the acceleration with respect to time gives the velocity along that axis and integrating again with respect to time determines the distance traveled as illustrated in Figure 1.29. This shows an accelerometer measuring motion in one dimension over a period of 10 minutes and initially there is an acceleration of 0.1 g, which produces an increase in velocity to 111 m s^{-1}, followed by a period of no acceleration (constant velocity). Then there is a deceleration of -0.05 g, which reduces the velocity to 58.5 m s^{-1} followed by another period of no acceleration and constant velocity. Finally, there is a further deceleration that brings the object to rest. During periods of acceleration (deceleration) the distance increases at an increasing (decreasing) rate and during periods of no acceleration (constant velocity) it increases at a constant rate. Finally, when the object has been brought to rest the distance stops increasing and the object has traveled 36.6 km, which has been determined solely by measurements with the accelerometer.

If we now put two accelerometers, one aligned along North–South and the other along East–West on a platform, and we know the starting position it is possible to determine the distance traveled along N–S and E–W independently and hence track the position on the Earth. A problem is that unless the platform is precisely perpendicular to the force of gravity, this will be measured as an acceleration and produce erroneous calculation of position. In the original inertial navigation systems, the platform on which the accelerators are mounted was kept perpendicular to Earth-vertical by mounting it in a gyroscope-controlled

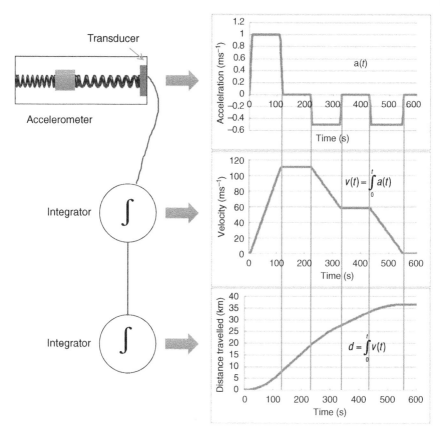

Figure 1.29 Principle of inertial navigation in one dimension. Integrating the measured acceleration gives velocity and integrating the velocity gives the distance traveled.

gimbal system that would maintain orientation while the aircraft pitched and rolled as illustrated in Figure 1.6. There are other complicating factors such as the aircraft traveling over a curved rotating Earth, which means that the platform has to be constantly adjusted to maintain itself perpendicular to gravity as described in detail in Chapter 9. In more modern systems the platform is fixed immovably to the aircraft and gyroscopes are used to determine the flight attitude and thus the component of Earth's gravity being measured so it can be deducted from the integration in software. These so-called *strap-down* systems are also described in detail in Chapter 9 and in this section the focus will be on the historical development.

The concept of inertial navigation was around before World War 2 but the first operational inertial guidance systems, albeit a partial version, was used in the German V2 (or A4) rockets in 1944. In the guidance system beneath the warhead was an accelerometer with a single integration stage that could determine velocity. The engineers would calculate what velocity was required to be reached after taking off from the starting position in order to reach London (and later other cities) ballistically. The system would then be set to shut off the engine at the required velocity and the missile would continue on a ballistic trajectory reaching its target if the calculations were correct. The system was crude and had an accuracy in the region of 5 km but this was later improved by using additional radio guidance.

Post war, the focus of development in INS was for use in missiles and the cold war spurred the design of smaller, lighter, and more accurate systems. They also played an important role in space exploration as probes that are sent into interplanetary space have no other navigation system to rely on apart from tracking by Earth stations. It was the military that initiated the use of INS in aircraft and the first flight test was in 1958. By the early 1960s many strike aircraft had an INS and a typical example is the Litton LN-3-2A system shown in Figure 1.30. These early platform-stabilized INSs were very expensive precision-engineered assemblies that could shield the accelerometers from Earth's gravity within an aircraft that could produce accelerations of between −5 g and +9 g in maneuvers. In order to attain the required accuracy, the gimbal assembly,

Figure 1.30 Litton LN-3-2A Platform-stabilized INS used in military strike aircraft in the early 1960s. *Source:* Reproduced with permission of https://en.wikipedia.org/wiki/LN-3InertialNavigationSystem. Licensed under CC BY-SA 3.0 [5].

accelerometers, and electronics all had to be kept in a temperature-controlled environment. In any INS system, the accuracy degrades with time and for the LN-3 devices this was quoted at 2 nautical miles error per hour of running.

The early systems were too expensive to use in commercial aviation but costs gradually reduced and in the late 1960s Delco introduced the popular Carousel INS system, which was installed on 707, 727, 737, 747, DC10, and Tristar airliners. This was also a gyro-stabilized platform INS and is shown in Figure 1.31a with the Earth-centered platform containing the accelerometers indicated. The control units in the cockpit were the Control Display Unit (CDU) and the Mode Selector Unit (MSU) (Figure 1.31b). When the INS was first switched on (MSU switched to STBY), the current position of the aircraft would be entered on the CDU and the MSU would then be switched to ALIGN. This would initiate a process taking approximately 10 minutes in which the gyro-stabilized platform set its attitude to obtain zero measured acceleration, during which the aircraft had to be at rest. Other procedures were also carried out during this initialization process to obtain the direction of true North and are described in Section 9.2.4. Waypoints would be entered into the CDU and when the system was ready it could be switched to NAV from which point any movements were recorded by the accelerometers. After takeoff, the unit could drive the autopilot to fly the aircraft to its destination via the entered waypoints. The availability of INS in aircraft from the late 1960s produced a step change in the ease of long-distance navigation, especially over oceans where radio navigation aids

(a)   (b)

Gyro stabilised platform
containing accelerometers

Figure 1.31 (a) Hardware and control electronics of the Delco Carousel with the case open to reveal the gyro-stabilized Earth-centered platform containing the accelerometers. *Source:* Reproduced with permission of National Air and Space Museum. (b) Lower part: Control Display Unit (CDU) used to enter the position at startup and waypoints. Upper part: Mode Selector Unit (MSU) to turn the unit on, initiate the align procedure, and select Navigation mode. *Source:* Screenshot of the CIVA simulator for X-Plane reproduced with permission from Philipp Münzel.

were not available. It remained the main tool of long-distance navigation till GNSS was available in the mid-1990s and thereafter was always used in combination with GNSS. Prior to INS, LORAN-C was available but did not have complete global coverage and was much more awkward to use than INS. A significant aspect of oceanic navigation prior to INS had been a combination of dead reckoning (see Section 6.5) and star sightings with some aircraft being fitted with periscopic sextants.

Despite mass production, INSs with gyro-stabilized platforms were still very costly due to the highly demanding engineering required to provide an Earth-centered platform for the accelerometers of sufficient stability to maintain acceptable navigation errors. This prompted an alternative approach in which the accelerometer platform is attached directly to the frame of the aircraft and gyroscopes mounted on the platform measure the pitch, roll, and yaw. If the attitude of the aircraft is known, a computer can then transform the measured output from the accelerometers to what the values would be if the platform was exactly perpendicular to gravity. This *strapdown* approach relies to a large extent on the availability of fast powerful processing since measurements from all accelerometers and gyroscopes have to be taken at 1000s of times a second and processed at that rate. Airborne computers in the 1960s built from discrete transistors were not up to the job but the microelectronic revolution starting in the late 1960s eventually made sufficiently powerful processors available.

Another technical requirement that needed to be met for the strapdown systems to work was sufficiently accurate gyro measurements of the attitude of the INS platform. The preferred option is to measure the rate of rotation of the platform by *rate gyros* (see Section 3.1.10) and from this compute the final angle relative to a given aircraft axis that the platform has rotated to. In the 1960s, strapdown rate gyros were not sufficiently accurate to achieve acceptable navigation performance. An alternative to rotating mechanical gyros is an optical method using a phenomenon known as the Sagnac effect, discovered in 1913 by the French physicist Georges Sagnac. He showed that when rays of light move in opposite directions around a circular cavity on a turntable, the light traveling with the rotation arrives at a target slightly after the light traveling against the rotation. The discovery of lasers in the 1950s made it possible to observe tiny changes in the interference pattern of two beams in circular paths and a practical Ring Laser Gyro (RLG) (Figure 1.32: see Section 3.2.4) first appeared in 1963 [12]. Since changes in interference patterns can be measured with high sensitivity, the RLG can detect small rotations of the table on which it is located and was a good candidate to act as rate gyro. Strapdown INS systems using RLGs were first tested in missiles in 1974, then in commercial aircraft in 1978, and were first used in passenger flights on Boeing 757 and 767 airliners [13]. From the start, at least two independent INSs were installed as redundant units so that if one developed a fault, navigation could continue with the other.

Further improvements came in the early twenty-first century with the development of MEMS (see Section 3.2.6) accelerometers used on the INS platform.

Figure 1.32 Single axis of a ring laser gyro. *Source:* Reproduced with permission of https://pl. wikipedia.org/wiki/%C5%BByroskoplaserowy#/media/File:RinglasergyroscopeatMAKS-2011airshow.jpg. Licensed under cc by sa 3.0 [5].

These are micromachined into Si and are exceedingly small and light with a three-axis accelerometer packaged onto a chip a few mm across. In principle, MEMS rate gyros could also be used on the INS platform but currently these have not achieved the required accuracy for INS. If this is achieved in the future, however, there is the prospect of an "INS on a chip" weighing a few grams that would fit into a matchbox. Current units have MEMS accelerometers and RLGs used as rate gyros on the INS platform.

One of the most significant changes that has occurred with INS in the last few years is a change in the philosophy of navigation from system-based (i.e. depending on a specific system like GNSS or INS) to performance-based. This involves combining all the available navigation sensors on an aircraft to achieve a specific navigation performance as discussed in the next section, thus current INS systems are combined with GNSS in a single navigation system. In this case the inertial measurement equipment is referred to as an Inertial Reference System (IRS) as its sensors have become part of a larger navigation system. In addition, the ADC can be combined with the IRS to produce an Air Data Inertial Reference System (ADIRS) in which the IRS has access to measurements of height, speed, Mach No, etc.

1.8.4 Combining Systems: Performance-Based Navigation (PBN) and Required Navigation Performance (RNP)

A modern airliner has a number of navigation sensors, including GPS, INS, and VOR, some of which may be triple redundant and all of which have different levels of accuracy of position in different flight phases. For example, the position error with INS always increases with time (see Chapter 9), whereas for GPS it depends on the number and location of satellites in view and whether augmentation is available. For VOR tracking the position error decreases with decreasing distance to the VOR beacon. Since the early 1980s, Flight Management Computers (FMCs) have been a feature on the flight deck to manage navigation and integrate it with performance management and other functions [14]. Partly, FMCs were introduced to decrease crew workload as airlines started to move to two-crew operations. Thus, the output from all navigational sensors along with those of the ADC, engine performance indicators, etc. are available to the FMS. The navigation sensors are combined using a process known as Kalman filtering (see Section 9.9), which essentially means taking a weighted average of the positions provided by the different sensors. The weighting for a specific sensor increases with the accuracy of that sensor in the relevant flight phase, so, for example, INS would have a greater weighting at the beginning of a flight than at the end. The data can all be combined to give an accuracy parameter known as Actual Navigation Performance (ANP), defined as the radius of a circle centered on the computed current position, where the probability of the aeroplane remaining continuously inside the circle is 95% per flight hour.

This facilitates a new concept in Air Traffic Management (ATM) in which routes are not specified by traditional airways or approaches via beacons but by optimized curved paths in three dimensions with a specified Required Navigational Performance (RNP) for aircraft to follow that path. In its simplest form this defines the maximum allowed deviation from the path so the radius of the ANP circle must be less than the deviation. For example, RNP 4 specifies that the deviation is less than 4 nautical miles, which may be appropriate for enroute navigation, while RNP 0.1 would be used for an approach. The RNP path would be between two points that may be at different heights so that in addition to Lateral Navigation (LNAV), Vertical Navigation (VNAV) is also important. GNSS and INS both provide altitude information and this can be combined with the height reading from the ADC. A good proportion of the current fleet of commercial airliners has the avionics necessary for this change and the ANP along with the RNP for a particular phase of flight is displayed on the PFD as shown in Figure 1.33a. The bars at the bottom of the display show the difference between the ANP and the RNP with reference to Figure 1.33b and the deviation from the ideal LNAV path is indicated by the small triangular pointer. The gap between the bars provides some indication of the leeway available for deviation to the flight crew, for example, if they needed to fly around a thunderstorm, deviations up to the edge of the bar are acceptable without calling ATC for a clearance. ANP–RNP bars are also displayed at the side of the PFD for VNAV.

Figure 1.34a and b illustrate the difference between a conventional approach using radio beacons and a curved three-dimensional approach RNP. In the latter case, there is a saving in distance traveled and fuel used while maintaining safe

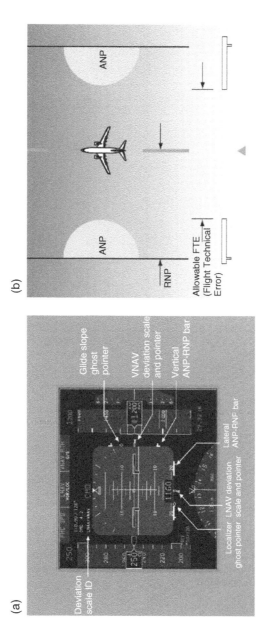

Figure 1.33 (a) PFD showing ANP–RNP for both LNAV (bottom of the display) and VNAV (right-hand side of the display. *Source:* Reproduced with permission of Aero Magazine, 2001. (b) Illustration of the information provided by the ANP–RNB bars on the PFD. The deviations from the ideal lateral and vertical paths are also shown providing some indication to the crew about how much deviation is permissible for the current RNP. *Source:* Reproduced with permission of Aero magazine 2001 [15].

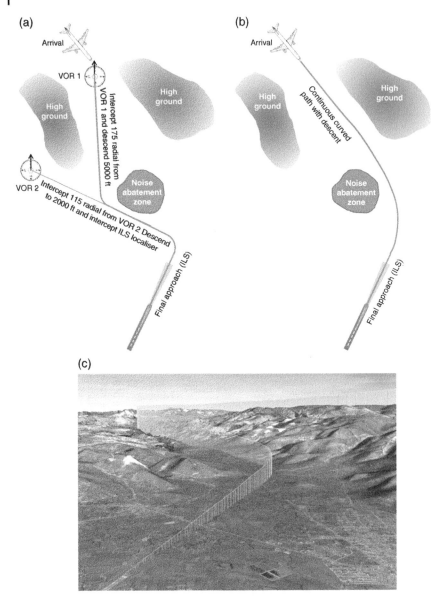

(a)

Arrival

VOR 1

High ground

High ground

High ground

Intercept 175 radial from VOR 1 and descend 5000 ft

VOR 2

Intercept 115 radial from VOR 2 to 2000 ft and intercept ILS localiser

Descend

Noise abatement zone

Final approach (ILS)

(b)

Arrival

High ground

High ground

Continuous curved path with descent

Noise abatement zone

Final approach (ILS)

(c)

Figure 1.34 (a) Example of a conventional approach procedure using specific radials from VORs to line up with a runway. (b) Curved three-dimensional RNP approach to line up on the same runway. (c) RNP approach to Cajamarca airport, Peru. *Source:* Reproduced with permission of https://en.wikipedia.org/wiki/Requirednavigationperformance#/media/File: RNPTrack3D.png. Licensed under CC BY-SA 3.0 [5].

separation from terrain and avoiding overflying noise abatement zones. RNP approaches are particularly suited to mountainous areas as curved tracks can be defined to safely navigate obstacles. The first RNP approach tested was in 1996 by Alaskan Airlines into Juneau airport, which has a challenging local terrain and an example of the track for an RNP approach to Cajamarca airport, Peru, is shown in Figure 1.34c. Such tracks will be stored on the navigation database and can be made active for the autopilot to follow.

1.9 Development of Auto Flight Control Systems

Like many of the topics in this chapter, the history of the autopilot begins in the nineteenth century before the first powered flight. As pointed out by McRuer and Graham [16], an important mind shift occurred with the Wright brothers aeroplane, which had well thought out controls for correcting changes in pitch, roll, and yaw. They effectively abandoned the search for a design with inherent stability in all axes and used feedback from the pilot to induce stability in the aeroplane. The pilot uses sensors to detect changes, calculates the response required to bring the aircraft back to level flight, applies the necessary force, which is amplified to move the control surfaces to produce the response. All these functions can be reproduced by a machine and the important point is that since the dawn of aviation someone or something needs to be in control of the aircraft. Since before the Wright brothers, inventors had experimented with the something, that is, a machine that could maintain an aircraft in stable flight. All early autopilots (although the term was not used till later) were based on the rigidity of the plane of rotation of gyroscopes, which, as pointed out in Section 1.4, must have been noted before recorded history.

The basic mechanism by which a gyroscope can be used to stabilize the flight attitude of an aircraft is illustrated in Figure 1.35 for just one axis (roll). The gyroscope cage is in a single gimbal, which is attached to the aircraft frame and the rotor is set spinning with its plane parallel to the wings (and to the horizon). Figure 1.35b shows what happens when the aircraft rolls by using a pointer attached to the gimbal and a scale attached to the rotor cage. The gyroscope maintains its original plane of rotation parallel to the horizon and the pointer indicates the roll angle, which is in fact one of the motions of an artificial horizon (see Section 3.1.9). Figure 1.35c shows the rolled attitude as seen from an observer in the aircraft, that is, the gimbal appears to have stayed stationary while the rotor has tilted. This movement in the rotor can be used to actuate a device that controls the ailerons and rotates the wings back to parallel to the horizon. Having two similar gyroscopes set up to rotate in orthogonal directions can also control the pitch and yaw to maintain straight and level flight.

(a)

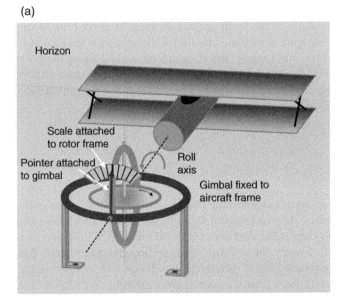

(b)

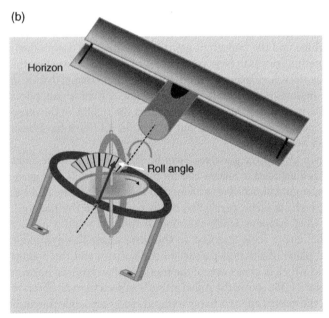

Figure 1.35 Gyroscope set up to control roll. (a) The caged rotor is in a single gimbal attached to the aircraft frame with a pointer attached to the gimbal and a scale attached to the cage. The gyro is set spinning with its plane of rotation parallel to the wings and the horizon (b) When the aircraft rolls the rotor plane stays parallel to the horizon and the pointer shows the roll angle (one axis of an artificial horizon). (c) Viewed from within the aircraft the plane of the rotor rotates as the aeroplane rolls and this movement can control an actuator that rolls the wings back to be parallel to the horizon. The pitch and yaw axes can similarly be controlled by orthogonal gyroscopes.

(c)

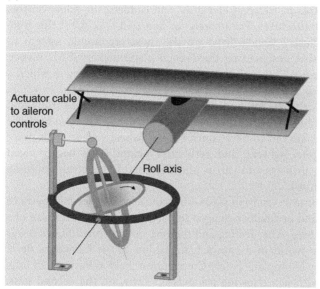

Figure 1.35 (Continued)

The first attempt to use this as method of control was by the US-born inventor Hiram Maxim in 1891, who used a gyroscope with a steam driven rotor to produce pitch stability in his heavier than air flying machine. The movement of the rotor relative to the aircraft longitudinal axis was used to operate a valve that allowed steam into a piston that controlled the elevators. The invention did not develop much beyond the drawing board due to the aeroplane being destroyed during early tests after which Maxim appeared to lose interest and moved on to other things. It was left to Lawrence Sperry to bring the gyroscopic control of flight to the attention of the world in 1914.

Lawrence, the son of Elmer Sperry, who founded the Sperry gyroscope company in 1910, was a talented inventor and by the age of 17 had built his first aeroplane. He was also a natural pilot and taught himself to fly but decided to enroll in the flying school run by Glenn Curtiss to obtain his Federal aeronautics license in 1913. This was the 11th to be issued and he was the youngest holder of a license in the United States. He had the free run of the workshop at the Curtiss school and immediately began work on a method to stabilize aircraft in flight based on gyroscopes. To the young Sperry the rigidity of gyroscopes to maintain a reference plane seemed a natural characteristic to exploit to maintain an aircraft at a constant flight attitude. He used an independent gyroscope to control each of the aircraft rotations, that is, roll, pitch, and yaw.

Although most aircraft flying in 1913 had adopted the control surfaces we are familiar with today, that is, ailerons, elevators, and rudders, the pilot controls depended on the manufacturer. Fortunately for Sperry it was about this time that the aviation industry standardized on the most ergonomic system to emerge, which was that designed by the French plane maker SPAD owned by Armand Deperdussin. This is the system that is still used today in which a stick or control yoke is moved backwards and forward to control the elevators (pitch), side to side (or wheel left and right) to control the ailerons (roll), and foot pedals for the rudder (yaw). Until recently the tendency has been for large aircraft and cabin GA aircraft to have control yokes (wheels) while sticks were used on open cockpits, fighters, and aerobatics planes. This has changed recently with Airbus airliners and Cirrus and Diamond GA aircraft adopting sticks for manual control.

With a universal control system in place it became a lot simpler for Sperry to design his autopilot and he finally managed to package it in a box the size of a cabin bag weighing 18 kg. After testing and refining in the United States it was first displayed to the world in a spectacular fashion at the Concours de la Securité en Aéroplane (Aeroplane Safety Competition) held in Paris on 18 June 1914. Not content with just showing a hands-off flyby, on the second pass, Sperry's French mechanic climbed out onto the wing while Sperry held his hands in the air (Figure 1.36a). On the final pass by the judges stand, both men climbed out onto the wings and waved to the ecstatic crowds as the plane flew on straight and level by itself. The slight rolls that developed as they climbed out were quickly corrected as the plane rolled back to wings level. It was a hard act to follow and Sperry won the prize of 50 000 francs and at the same time became a renowned inventor. The Sperry autopilot was used in the first cruise missile, the Kettering Bug built in 1918 (Figure 1.36b), of which

(a)   (b)

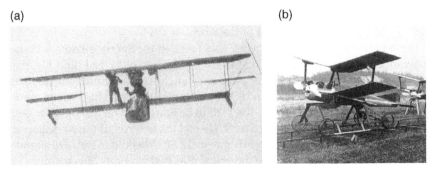

Figure 1.36 (a) Lawrence Sperry and his mechanic demonstrating the gyroscopic autopilot in Paris in 1914. (b) The Kettering bug cruise missile in 1918, which was fitted with the Sperry autopilot. *Source:* (a) HistoricWings.com and (b) HistoricWings.com.

many were built but never used in action. A gyroscopic stabilization mechanism was, however, used on the German V1 cruise missile in World War 2.

From 1914 to post World War 2, gyroscopes remained the basic sensing mechanism for control but they became more compact and the actuator systems that moved the control surfaces became increasingly sophisticated and included pneumatic servos. In addition, the autopilots started to take additional inputs from the compass and the pressure altimeter so that they could maintain an aircraft on a specific course and height. In 1930, a US Army Air Corps plane was kept at a constant altitude and heading for three hours [17]. The Royal Aircraft Establishment (RAE) in the United Kingdom were also very active in autopilot development, though much of it was kept secret and in 1930 they demonstrated automatic altitude and heading control of an aircraft for over 400 miles [18]. These developments were taking place when the length of commercial flights was increasing and it was recognized that pilots required assistance for flights that were several hours long, especially in nonvisual conditions.

In 1932, the prototype of what was to become the standard autopilot for commercial airliners, the Sperry A2, was developed. By then, airliners were already fitted with an extensive set of gyroscopic instruments (see Section 1.4) to cope with flight in low visibility conditions and one of the innovative steps in the A2 design was to use the instruments to supply the gyroscopic sensing. This saved weight as there was no need to have an independent set of gyros and the autopilots started to resemble modern systems in which sense inputs come from the FMS. The timing of the A2 coincided perfectly with the needs of commercial aviation and its profile was greatly enhanced by Wiley Post in his record-breaking round the world flight in a Lockheed Vega aircraft, the "Winnie Mae" in 1933 (Figure 1.37a). Post had seen the A2 prototype at the Sperry factory and insisted

(a)   (b)

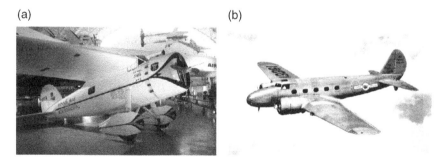

Figure 1.37 (a) Lockheed Vega "Winnie Mae" piloted by Wiley post on his record-breaking round the world flight in 1933 used the Sperry A2 autopilot to rest. Source: (a) https://en.wikipedia.org/wiki/WileyPost#/media/File:NASM-LockheedVega-WinnieMae.jpg. Licensed under CC BY-SA 3.0. (b) Boeing 247 airliner was the first to be fitted with an A2 autopilot as standard.

on having one fitted to his plane, which he used extensively to get some rest on his 7-day 19-hour flight. The A2 was first introduced into commercial service on the Boeing 247 (Figure 1.37b) in 1934 and was quickly in widespread use by the airlines. It used pneumatics to sense and amplify the indications of the sensors (gyroscopes) and hydraulics for the actuators on the control surfaces; however, the trend was to move to all-electric systems, that is, electrically driven gyroscopes and sensors, power amplification, and electric motors to drive the control surfaces. Other companies including Honeywell and Bendix in the United States and Siemens, Askania, and Anschütz in Germany entered the field. Similar rapid developments were also being made by the RAE in the United Kingdom though this continued to remain mostly secret and very little was published.

During World War 2, more development steps in autopilots emerged in Germany, including yaw damping to remove some fundamental aerodynamic instabilities of aircraft and the principle of rate-rate control. This is where rate gyros are used to measure the rate of rotation in roll, pitch, or yaw and the correcting control surface is moved at a rate that is proportional providing more positive control. In the United States the move to all electric autopilots was realized and also the emergence of a new level of control, that is, the ability of autopilots to maneuver aircraft rather than maintain a set attitude.

All the mechanisms presented so far would be described in modern terminology as *inner loop systems*, that is, they contain a single feedback loop that *maintains* a specific flight parameter (height, heading, etc.). It was becoming clear that it would be necessary to introduce a higher level of control, which on reception of some additional input could *change* one of the flight parameters and then let the inner loop system maintain it at the new value. An example is obtaining information from a bombsight, change to a new heading, and maintain it till new information is received. Such a control method is referred to as an *outer loop system* and is the basis of how the autopilot in a modern airliner can be made to follow a flight plan. Another example of an outer loop system is commands from the pilot controls for the autopilot to change to a new flight attitude, which forms the basis of fly-by-wire systems,

The sophistication that had been reached by autopilots was demonstrated convincingly in 1947 when a US Air Force C47 Skymaster flew from Stephenville, Newfoundland to Brize Norton in England without any pilot input from brake release at takeoff to brakes on after landing. This impressive feat was achieved with the latest technology of the time including a Sperry A-12 autopilot with approach coupler and Bendix automatic throttle control. The commands to the autopilot were input from punched cards interpreted by an IBM electronic controller so as a system it had the same capabilities as a modern autopilot commanded by a FMC. Considering that landing an aircraft is the most difficult task a human pilot performs, this flight seemed to indicate that there was nothing an autopilot could not achieve and indeed the majority of commercial flights today are carried out mainly by the autopilot.

Emerging from the war, the autopilot systems had clearly become more sophisticated but some deficiencies still remained including their tendency to over-control and induce oscillations. As pointed out by Mcruer, Graham, and Ashkensas [19], the theoretical basis on how to remove this kind of deficiency had been available for some time but the merging of theory and the practical application of autopilots did not occur till after World War 2. For example, the Proportional-Integral-Derivative (PID) control method used in a range of industrial control processes to achieve the optimum rate of convergence of a control parameter with a desired value followed by optimum stability was published in 1922 [20]. This merging of theory and practical application in the late 1940s and early 1950s coincided with the development of computers that could be used to solve the complex equations involved so the development of autopilots accelerated.

The next major development was routine landing by the automatic pilot (autoland) in commercial flights. This involved not only the development of equipment but rules and procedures such as safety minima. Although the principle of autoland had been demonstrated in 1947 by the US Air Force, this was with nonstandard equipment not generally available on airliners. In 1945, the British Government set up the Blind Landing Experimental Unit (BLEU) to carry out research and development on low visibility landings. This was an imperative for the nationalized airlines, BOAC set up in 1940 and BEA in 1946 since the large amount of air pollution created by coal fires coupled with damp weather could close London airports for days on end. Autoland was developed by BLEU initially for RAF aircraft and in the early 1960s for BEA's Trident fleet. It used a triple-redundant control with three independent processing channels and if one failed the other two would "out-vote" it and provide the output for the controls. As already described in Section 1.7.3, the technology came to maturity with the first fully automatic landing on a commercial flight with passengers at Heathrow airport in a British European Airways Trident in 1965.

Thus, the fundamental methods and hardware for autopilots to control the aircraft through every phase of flight were in place by the 1960s. From then on, the increasing sophistication of the FMS in the cockpit driven by ever-more powerful digital computers meant that new modes could be invoked, for example, terrain following and terrain avoidance. In 1972, the first digital fly by wire (FBW) system without a mechanical backup was tested by NASA on a Crusader F-8C. In this type of servo-mechanism there is no direct mechanical or hydraulic coupling between the flight controls and the control surfaces but the flight controls send electronic signals to the same servos that the autopilot uses to move the control surfaces. The 1970s also saw the development of full authority digital engine control (FADEC) where engine control is delegated

Table 1.1 Typical modes of the autopilot in a commercial aircraft.

No.	Mode	Action
1	Heading	Follows a selected heading (e.g. 280°).
2	LNAV	Follows the lateral route entered in the FMS.
3	VOR/LOC	Follows a selected track to or from a VOR or ILS localizer entered manually or selected by the FMS.
4	Altitude hold	Holds the altitude while pilot maintains lateral control.
5	Vertical speed	Maintains a specified vertical speed (climb or descent) till a selected altitude is intercepted.
6	Level change	Climbs or descends to the selected altitude while maintaining the selected speed.
7	VNAV	Follows the vertical component of the route entered into the FMS.
8	ILS/approach	Follows the glide slope and localizer to the runway and carries out autoland if the necessary systems are available.

to an electronic subsystem that takes care of all the details in demanding a given power from the engine. It has inputs from the ADC and FMS so that when a given power is requested, the system controls all the engine settings required, for example, fuel flow, intake configuration, propeller pitch, etc. to deliver that power. This makes engine control much simpler for both human pilots and autopilots as they just need to specify a single number – the percentage power required and the FADEC does the rest. It is an example of the distributed processing that is found on a modern aircraft.

The autopilot modes selectable by the pilot in a typical modern airliner are shown in Table 1.1 and it is evident that the pilot can demand all the necessary actions during the course of a normal flight to be carried out by the autopilot. If the autopilot is switched to LNAV and VNAV, then it will follow the flight pattern entered into the FMS (usually by airline operations) but the extra flexibility of the modes shown in Table 1.1 allow the crew to make changes easily in response to requests from ATC. The FMS menu system is also carefully designed so that modifications can be made to the flight plan with minimum workload.

The increase in sophistication of autopilots since the demonstration by Lawrence Sperry in Paris a century ago is truly impressive. Most commercial flights are conducted mainly by the autopilot and it is known that this saves fuel and increases passenger comfort. It must be borne in mind, however, that even today there are situations, either due to turbulence or malfunctioning sensors that an autopilot cannot deal with and it drops out leaving it to the human pilots to complete the flight.

References

1 Roger Ford (2000). The risks of travel. *Modern Railways* (October) (article cites figures based on UK Department of the Environment, Transport and the Regions (DETR) survey).

2 http://ethw.org/Milestones:First_Blind_Takeoff,_Flight_and_Landing,_1929 (accessed 11 June 2018).

3 Benniwitz, K. (1922). *Flugzeuginstrumente*. Berlin: Richard Carl Schmidt and Co.

4 Wing Cdr. Roderic Hill (2005). *The Baghdad Air Mail*. Nonsuch Publishing Ltd.

5 Creative commons (1953). http://creativecommons.org/licenses/by-sa/3.0/ legalcode (accessed 11 June 2018).

6 www.airbattle.co.uk/b_research_3.html (accessed 11 June 2018).

7 http://www.aps.org/publications/apsnews/200604/history.cfm (accessed 11 June 2018).

8 http://creativecommons.org/licenses/by/2.0/legalcode (accessed 11 June 2018).

9 Johnson, R. (2003). Blind flying on the beam: aeronautical communications, navigation and surveillance: its origins and the politics of technology, part III: emerging technologies, the radio range, the radio beacon and the visual indicator. *Journal of Air Transportation* **8**: 79–104.

10 https://timeandnavigation.si.edu/multimedia-asset/gee-chart-reims-chain-december-1944-11000000-scale (accessed 11 June 2018).

11 Helfrick, A. (2015). The centennial of avionics: our 100 year trek to performance-based navigation. *IEEE A & E Systems Magazine, Sept* **30** (9): 36–45.

12 Collinson, R.P.G. (2003). *Introduction to Avionics Systems*, 226. Springer.

13 Savage, P.G. (2013). Blazing gyros – the evolution of strapdown inertial navigation technology for aircraft. *AIAA Journal of Guidance, Control and Dynamics* **36**: 637–655.

14 Miller, S. (2009). Contribution of flight systems to performance-based navigation. *Boeing Aero Magazine* QTR_02.09, pp. 20–28.

15 Carriker, M., Hilby, D., Houck, D., and Rolan Shomber, H. (2001). Lateral and vertical navigation deviation displays. *Boeing Aero Magazine*, No. 16 (October), pp. 29–35.

16 McRuer, D. and Graham, D. (1981). Eighty years of flight control: triumphs and pitfalls of the systems approach. *Journal of Guidance and Control* **4**: 353.

17 Now – the automatic pilot (1930). *Popular Science Monthly* (February), p. 22.

18 Robot air pilot keeps plane on true course (1930). *Popular Mechanics* (December), p. 950.

19 Mcruer, D.T., Graham, D., and Ashkenas, I. (1973). *Aircraft Dynamics and Automatic Control*, 6. Princeton University Press.

20 Minosrsky, N. (1922). Directional stability of automatically steered bodies. *Journal of American Society of Naval Engineers* **34**: 280–309.

2

Pressure Instruments

In this chapter, the properties of the atmosphere will be described and the precise meanings of sea level, elevation, height, altitude, and flight level (FL) will be defined. The instruments that determine altitude, rate of climb or descent, airspeed, and Mach number by measuring static and dynamic air pressure will be described. Initially, the operation of traditional direct-reading instruments will be explained as these are still used in the majority of General Aviation aircraft and some commercial aircraft. In addition, the fundamental measurements that they make are the same as the ones performed by transducers used in glass cockpits. Finally, the operation of the Air Data Computer (ADC), used in the majority of commercial airliners, and how it processes and displays the relevant output data in the cockpit will be described.

2.1   Layers of the Atmosphere

The Earth's atmosphere is continuous from the ground to the border with space, but it can be described in terms of layers at different altitudes (Figure 2.1a), each having distinct properties. One of the properties that distinguishes the different layers is how the temperature varies with altitude, which is shown in Figure 2.1b for the International Standard Atmosphere (ISA: see the following section). In the *troposphere*, which is the lowest layer next to the ground, the temperature decreases linearly with height up to the *tropopause*, which marks the boundary between the troposphere and the next layer, the *stratosphere*. At the bottom of the stratosphere the temperature remains constant with altitude but above 20 km the temperature starts to decrease again and reaches a minimum at the *stratopause* at an altitude of 50 km, which marks the boundary between the stratosphere and the next layer, the *mesosphere*. Within the mesosphere the temperature starts to increase again. The layer above the mesosphere is known as the *thermosphere* beginning at 85 km and the boundary of space

Aircraft Systems: Instruments, Communications, Navigation, and Control,
First Edition. Chris Binns.
© 2019 John Wiley & Sons, Inc. Published 2019 by John Wiley & Sons, Inc.
Companion website: www.wiley.com/go/binns/aircraft_systems_instru_communi_Navi_control

(a) (b)

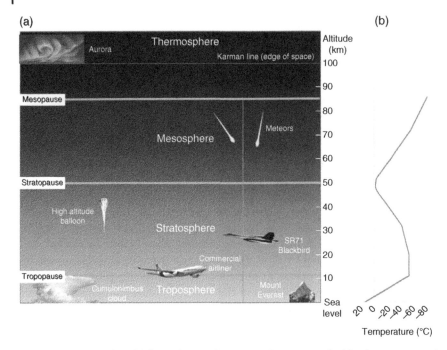

Figure 2.1 (a) Altitudes of different layers of the atmosphere as specified for the International Standard Atmosphere (ISA). (b) Variation of temperature with altitude in the ISA.

is generally accepted to be at the *Karman line* at an altitude of 100 km. This is named after the physicist Theodore Karman who was the first to demonstrate that above this line an aircraft would have to travel faster than orbital velocity to generate enough aerodynamic lift to maintain altitude. Thus, aerodynamic flight is impossible, even in principle, above this height, though in practice it is limited to below around 30 km (for more on flight in thin atmospheres see Interesting Diversion 2.1).

Aircraft are confined to the bottom two layers of the atmosphere, that is, the troposphere and the stratosphere. Commercial jet transport aircraft generally fly at the tropopause, supersonic aircraft in the stratosphere, and propeller aircraft in the troposphere. Virtually all weather is confined to the troposphere, which contains 75% of the mass of the atmosphere and 99% of the water vapor.

2.2 The International Standard Atmosphere (ISA)

Pressure instruments use atmospheric pressure to determine altitude and airspeed, but the pressure at a specific height varies with position on the

Table 2.1 Mean sea level conditions defined for the ISA.

Quantity	Symbol and value
Pressure	$P_0 = 101\,325\ \text{N m}^{-2} = 1013.25\ \text{hPa}$
Density	$\rho_0 = 1.225\ \text{kg m}^{-3}$
Temperature	$T_0 = 15\ °\text{C} = 288.15\ \text{K}$
Speed of sound	$a_0 = 340.291\ \text{m s}^{-1}$
Gravitational acceleration	$g_0 = 9.80665\ \text{m s}^{-2}$

Earth and also changes from day to day. So, in order to calibrate pressure instruments a model is used, which establishes an averaged universal set of conditions and is referred to as the ISA. This was introduced by the ICAO in 1952 and according to the model the atmosphere contains no dust or water vapor and is at rest with respect to the Earth's surface (i.e. no winds or turbulence). The conditions defined for the ISA at mean sea level (MSL) are shown in Table 2.1.

The altitude of the tropopause varies with latitude in the real atmosphere and is lower at the poles than at the equator but in the ISA, it is assumed to be 11 km above MSL all over the Earth. The temperature variation with height in the ISA up to an altitude of 20 km is shown in Figure 2.2 and is assumed to decrease

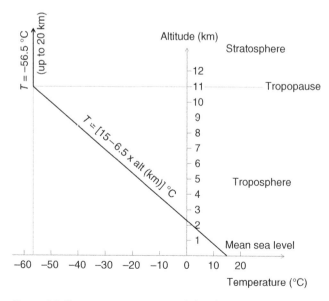

Figure 2.2 Temperature variation with height in the ISA up to 20 km.

linearly at a rate of 6.5 °C km⁻¹ (1.98 °C/1000 ft) from MSL up to an altitude of 11 km where it has reduced to −56.5 °C given by 15 − (11 × 6.5) °C. Into the stratosphere it is assumed to stay constant with height at −56.5 °C up to 20 km. There are also specifications for the temperature variation with height at higher altitudes [1], which are shown in Figure 2.1, but they are not relevant to commercial air transport and only a few aircraft designed to fly at very high altitudes are capable of climbing to above 20 km. For the rest of this section the calculation of the pressure and density as a function of height will be limited to below 20 km altitude.

Once the temperature is defined, the pressure variation with height in the ISA can be determined by assuming the air behaves as an ideal gas and is calculated using the hydrostatic equation and the specified temperature lapse rate. The pressure variation dp through an infinitesimal element of density ρ and height dh with unit area (Figure 2.3) is given by:

$$dp = -\rho g dh \qquad (2.1)$$

where g is the acceleration due to gravity. The equation of state for air is:

$$p = \rho RT \qquad (2.2)$$

where R is the gas constant and T is the absolute temperature.

From Equations (2.1) and (2.2):

$$\frac{dp}{p} = \frac{-g dh}{RT} \qquad (2.3)$$

So, the pressure, p, at an altitude, h, is given by integrating from sea level ($p = p_0$, $h = 0$) to h. The gravitational acceleration, g, varies with height but the variation between sea level and $h = 20$ km is only 0.3%, so for the

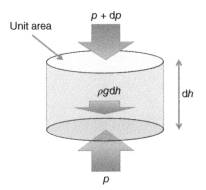

Unit area

$p + dp$

$\rho g dh$

dh

p

Figure 2.3 Variation of pressure across a height element of the atmosphere with unit area.

purpose of this book we will assume it is constant and take it outside the integral, thus:

$$\int_{p_0}^{p} \frac{dp}{p} = \frac{-g}{R} \int_{0}^{h} \frac{dh}{T} \tag{2.4}$$

The ISA model specifies that up to the tropopause (11 km) $T = T_0 - 0.0065h$, where $T_0 = 288.15$ K (15 °C) and h is in meters. So, Equation (2.4) becomes:

$$\int_{p_0}^{p} \frac{dp}{p} = \frac{-g}{R} \int_{0}^{h} \frac{dh}{T_0 - 0.0065h} \tag{2.5}$$

or

$$\ln\left(\frac{p}{p_0}\right) = \frac{g}{0.0065R} \ln\left(1 - \frac{0.0065h}{T_0}\right) \tag{2.6}$$

thus, for $h \leq 11$ km,

$$p = p_0 \exp\left(\frac{g}{0.0065R} \ln\left(1 - \frac{0.0065h}{T_0}\right)\right) \tag{2.7}$$

At higher than 11 km the temperature remains constant at $-56.5\,°C = 216.65$ K, so Equation (2.4) becomes:

$$\int_{p_0}^{p} \frac{dp}{p} = \frac{-g}{216.65R} \int_{0}^{h} dh = \frac{-gh}{216.65R} \tag{2.8}$$

that is, for $h > 11$ km,

$$p = p_0 \exp\left(\frac{-gh}{216.65R}\right) \tag{2.9}$$

The gas constant for air is 286.9 J kg^{-1} K^{-1}, so the pressure variation with altitude predicted by Equations (2.7) and (2.9) can be plotted and is shown by the blue curve in Figure 2.4. Aircraft altimeters are calibrated so that the pressure measurement in hPa produces a display of height according to the relationship plotted in Figure 2.4. It is emphasized that this relationship and thus the altimeter calibration assumes ISA conditions and that the true height in non-ISA conditions will depart from the altimeter reading.

The corresponding variation in density can easily be obtained from

$$\rho \propto \frac{p}{T} \tag{2.10}$$

i.e.

$$\frac{\rho}{\rho_0} = \frac{pT_0}{p_0 T} \tag{2.11}$$

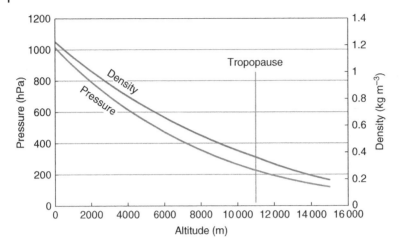

Figure 2.4 Variation of pressure (blue curve) and density (red curve) with altitude in the ISA.

So, up to 11 km (from Equation 2.7):

$$\frac{\rho}{\rho_0} = \frac{pT_0}{p_0(T_0 - 0.0065h)} = \left(\frac{T_0}{(T_0 - 0.0065h)}\right) \exp\left(\frac{g}{0.0065R} \ln\left(1 - \frac{0.0065h}{T_0}\right)\right)$$

$$(2.12)$$

and for $11\,\text{km} < h < 20\,\text{km}$,

$$\frac{\rho}{\rho_{11}} = \frac{p}{p_{11}} = \exp\left(\frac{-gh}{216.65R}\right) \qquad (2.13)$$

where in Equation (2.13) ρ_{11} and p_{11} are the density and pressure values at $h = 11\,\text{km}$. The density variation with height is displayed by the red curve in Figure 2.4.

2.3 Nonstandard Atmospheres

The real atmosphere can vary significantly from ISA conditions at a given height and in addition, atmospheric motion, which is assumed to be absent in the ISA, can have a significant effect on aircraft performance. A full discussion of the properties of the real atmosphere are beyond the scope of this book, but one correction to the properties of the ISA for real conditions that is useful and simple to apply is the modification to the air density required for non-ISA temperatures. Up to 11 km the density in the ISA is given by Equation (2.12), but if the temperature at the flight altitude is not the ISA value then the density will be modified. Aircraft normally have an outside air temperature (OAT) probe (see Section 2.12) and if the temperature outside the aircraft is T, measured in Kelvin, we can determine the deviation from ISA conditions (ΔT) and thus from the equation of state ($p = \rho R T$) we can write:

$$\frac{p}{R} = \rho_{ISA} T_{ISA} = \rho_{ACT} T_{ACT} = \rho_{ACT}(T_{ISA} + \Delta T) \tag{2.14}$$

or

$$\rho_{ACT} = \frac{\rho_{ISA} T_{ISA}}{T_{ISA} + \Delta T} \tag{2.15}$$

where ρ_{ACT} and T_{ACT} are the actual density and temperature of the air outside the aircraft that correspond to the temperature correction, ΔT, to the ISA temperature, T_{ISA}, measured in Kelvin. Thus, Equation (2.15) predicts that if the measured temperature at a given altitude is lower than T_{ISA} ($\Delta T < 0$) then the actual air density, ρ_{ACT}, will be higher than that predicted by the ISA, ρ_{ISA}, and vice-versa. This is intuitively obvious but Equation (2.15) allows a quick calculation of the difference knowing only the OAT.

Beermat Calculation 2.1

An aircraft is flying at an altitude of 10 000 ft and the OAT probe records a temperature of −10 °C. What is the density of the air compared to ISA density?
    10 000 ft = 3048 km, so the ISA temperature is: $T_{ISA} = 15 - 3.048 \times 6.5 = -4.81\,°C = 268.35\,K$
    Thus, the temperature deviation from ISA is $\Delta T = -5.19\,K$.
    Hence, from Equation (2.15), $\dfrac{\rho_{ACT}}{\rho_{ISA}} = \dfrac{T_{ISA}}{T_{ISA} + \Delta T} = \dfrac{268.35}{268.35 - 5.19} = 1.02$
    Hence, the lower temperature has produced a 2% increase in density.

2.4 Dynamic Pressure and the Bernoulli Equation

The ISA describes how the static pressure varies with altitude but for providing lift, thrust, and operation of some of the pressure instruments, it is the dynamic pressure produced by air moving relative to a surface that is the important quantity. We will assume in this section that the fluid (air) is incompressible, that is, its density is constant with pressure. To relate the dynamic pressure to the

Figure 2.5 Constant rate airflow in a tube of varying cross-section.

airspeed, consider an element of air of mass Δm flowing in a tube with varying cross-section (Figure 2.5) at a constant rate of flow, that is constant mass/unit time.

It is assumed that the tube is horizontal so there is no change in gravitational potential energy of the element. The change in energy of the element is just the change in kinetic energy between point 1 and 2, that is:

$$E_2 - E_1 = \frac{1}{2}\Delta m \left(v_2^2 - v_1^2 \right) \tag{2.16}$$

This change in energy is given by the change in the work done on the element between points 1 and 2, that is:

$$W_2 - W_1 = \frac{\Delta m}{\rho}(p_2 - p_1) \tag{2.17}$$

thus, $\quad \dfrac{\Delta m}{\rho}(p_2 - p_1) = \dfrac{1}{2}\Delta m \left(v_2^2 - v_1^2 \right) \tag{2.18}$

or $\quad (p_2 - p_1) = \dfrac{1}{2}\rho \left(v_2^2 - v_1^2 \right). \tag{2.19}$

This is normally expressed as:

$$\frac{\rho v^2}{2} + p_S = \text{constant} = p_t \tag{2.20}$$

where p_S is the static pressure, p_t is the total pressure, and $\dfrac{\rho v^2}{2}$ is the dynamic pressure. The assumption that air is incompressible is reasonable for speeds below around 300 knots. As the airspeed increases beyond this into the transonic and supersonic regime, Equation (2.20) has to be modified to take into account the changes of density with pressure. It is also worth remembering that Equation (2.20) is a special case of the Bernoulli equation where the gravitational potential energy has been neglected.

Interesting Diversion 2.1: Aviation on Mars

The dependence of dynamic pressure on the square of the airspeed has some interesting consequences, for example, to generate ten times the lift from a given wing only requires just over three times the airspeed, so with this in mind is it possible to fly an aircraft on Mars? At first glance, it seems an insane idea given that the surface pressure on Mars at the surface datum (there is no sea level on Mars) is around 8 hPa compared with 1013 hPa at ISA sea level on Earth. There are several factors that work in our favor, however, the biggest one being the squared dependence of dynamic pressure (or lift) on air density. In addition, the Martian atmosphere is mostly CO_2 giving it an average molecular weight of 43.34 compared to 28.96 for the Earth and the surface temperature at the equator averages $-63\,°C$, which pushes up the density further. Thus, the surface density of $0.02\ \mathrm{kg\ m^{-3}}$ is 1.6% that of Earth compared with only 0.8% for the ratio of pressures. Finally, the Martian gravity is 38% that of the Earth meaning that to fly we only need to generate 38% of the lift.

Suppose we took a Cessna 152, which, on Earth, cruises at 90 knots to Mars – how fast would it have to go to fly in the thin atmosphere? Let:

ρ_1 = density of the Martian atmosphere ($0.02\ \mathrm{kg\ m^{-3}}$)
ρ_2 = density of Earth's atmosphere ($1.225\ \mathrm{kg\ m^{-3}}$)
v_1 = cruising speed on Mars (unknown)
v_2 = cruising speed on earth (90 knots)

then

$$\rho_1 v_1^2 = 0.38 \times \rho_2 v_2^2$$

or

$$v_1 = \sqrt{\frac{0.38 \times \rho_2\, v_2^2}{\rho_1}}$$

putting in the numbers gives $v_1 = 434$ knots, which does not look so bad. In other words, if we strap rockets on to a Cessna 152 and accelerated it to 434 knots, which is not hard in the thin atmosphere (drag also scales with the dynamic pressure) it would apparently fly as normal.

The situation, however, requires more careful thought than this back of the envelope calculation. For example, the relatively low speed of sound of about 475 knots on the Martian surface due to the low temperature would mean that our rocket-powered Cessna would be traveling at Mach 0.91 – well into the transonic region with its attendant problems of high drag, Mach tuck, etc. Its subsonic design would result in disaster.

The prospect of flying planes on Mars for exploration is taken seriously by NASA, however, and several designs for aircraft optimized to fly on Mars have

(Continued)

Interesting Diversion 2.1: (Continued)

Figure ID2.1.1 Impression of flying wing concept, under development by NASA for Mars planes, flying over the surface of Mars. *Source:* NASA.

been developed, the most recent one being a flying wing concept (Figure ID2.1.1). Also flight in very thin atmospheres has already been achieved on Earth, for example, the SR71 Blackbird reconnaissance aircraft can fly at an altitude of 85 000 ft at which the Earth's atmosphere has a pressure of 10 hPa, which is not much different to that on the surface of Mars. There are two big advantages to doing these high flights on Earth as opposed to low altitude on Mars however. One is that on Earth the aircraft can use normal oxygen breathing jet engines and the other is that the SR71 can descend into the thick pea soup of an atmosphere at low level and land or takeoff at a sensible speed.

These issues have also been addressed for Mars planes. An engine running on hydrazine, which does not require external oxygen, was designed and tested by NASA in a very high altitude drone in 1982 but the landing and takeoff problem is more challenging. One idea is to set up a deep stall so that the aircraft descends at a very high angle relatively slowly (still >100 knots ROD) and use rockets at the appropriate height to bring the vehicle to rest just above the surface. Takeoff could also be achieved with rockets that produced both lift and forward thrust to get the aircraft moving fast enough for the wings to generate the required lift. It is likely that there will also be challenges that have not yet been conceived and despite all the testing that can be done by drones, it is clear the first aviators on Mars will be brave souls.

2.5 Definition of Sea Level and Elevation

To define the height that an aircraft is flying requires a reference at the Earth's surface and a natural choice is sea level since gravity pulls water toward the center of the Earth, so it should all settle to a common level. By definition, the height of the ground topography or man-made object above sea level is known as its *elevation*. Thus, we are told that the elevation of the top of Mount Everest is 29 002 ft (8848 m) or the elevation of my local airfield is 469 ft (143 m). Defining accurate heights in global navigation, however, requires a more careful consideration of what sea level actually means and how it is determined.

On a perfectly smooth and spherical planet that is completely flooded (Figure 2.6a), the concept is straightforward as the depth of the water is the same everywhere and the surface of the water after averaging out the waves is always the same distance from the center of the planet. The Earth, however, is not perfectly spherical, it is an oblate spheroid in which the distance from pole to pole (12 714 km) is 42 km less than the diameter at the equator (12 756 km). On a perfect oblate spheroid, the depth of the water will smoothly change between the poles and the equator due to the slight variation in gravity, but it is still possible to calculate precisely the depth and thus know the position of sea level at every point on the Earth (Figure 2.6b). The situation is more complicated because the density of the Earth varies and in denser regions the gravity is locally stronger making sea level higher than in less dense parts (Figure 2.6c). The density of the Earth as a function of position is well mapped and so again, the variations in sea level over the Earth can be calculated.

The final complication is that the Earth's surface has a topography that varies between 11 034 m below sea level (bottom of the Mariana trench) and 8848 m above sea level (top of Mount Everest) and it is only partially covered with water. It is thus necessary to calculate what sea level would be in a land region that is remote from the sea. The variation in gravity caused by topography is specified in a detailed model known as the World Geodetic System (WGS: see Section 6.4) and this can be used to calculate the local gravity and determine what sea level would be if the land was removed but its gravity remained (Figure 2.6d). Thus, defining sea level for every point on the Earth relies on having a detailed knowledge of the variations in density and the topography, but these are available and sea level is known to be within 1 m for every point on the Earth.

2.6 Definition of Height, Altitude, and Flight Level

Before describing the pressure instruments it is important to tighten up on the definitions of height, altitude, and elevation, all of which refer to vertical distances but have definite meanings in aviation (see Figure 2.7). As already

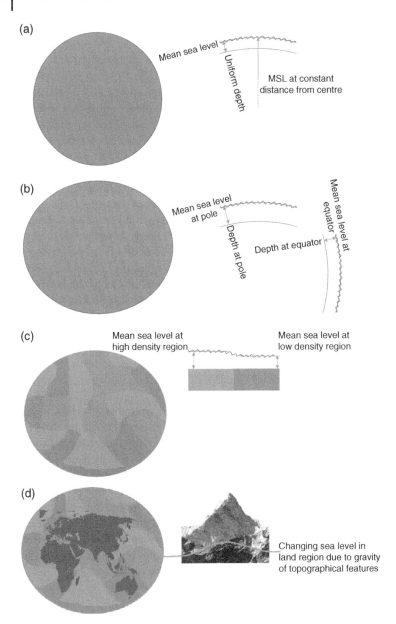

Figure 2.6 (a) On a flooded spherical planet, sea level is at the same distance from the center everywhere. (b) On a flooded spheroidal planet, sea level varies smoothly between the poles and the equator. (c) On a planet with a nonuniform density, sea level is higher in regions of low density. (d) On a partially flooded planet with continents and topography, sea level in land areas is calculated from the gravity of the topographical features assuming they are absent but their gravity remains.

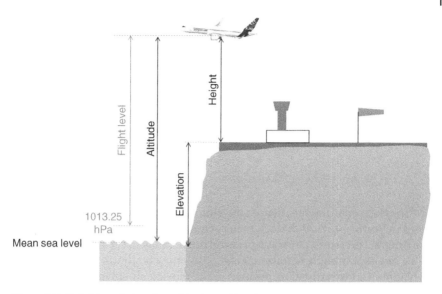

Figure 2.7 Definition of elevation, altitude, and height.

described in the previous section, elevation refers to the vertical distance above sea level of a fixed ground feature, which can either be natural terrain or an artificial structure such as a mast. Altitude is defined as the vertical distance of a flying aircraft above sea level, whereas height is the vertical distance above a user-defined datum such as a runway surface. Height is necessarily a local measure since as we move away from the datum at a constant altitude, the elevation of the ground underneath the aircraft and thus the height will change.

An additional measure used in aviation is FL, which is the height in hundreds of feet above the fixed pressure datum of 1013.25 mbar (i.e. sea level pressure in the ISA), so, for example, FL250 is 25 000 ft above the 1013.25 hPa level. Aircraft altimeters always have an adjustable barometric setting that sets a pressure for the base level that the height, altitude, or FL is displayed relative to and the values used are labeled by the codes, QFE, QNH, and SPS. The code QFE denotes the measured pressure at an aerodrome runway threshold and if the barometric setting is adjusted to this value the altimeter will read the height above the runway. The code QNH is the local sea level pressure determined from the known elevation of the aerodrome and the measured value of QFE assuming ISA conditions without taking into account temperature variations from ISA. Setting this value causes the altimeter to display altitude above the local sea level as defined in Section 2.5. The code SPS simply denotes 1013 hPa and if the barometric setting is this, then the altimeter reads the altitude relative to ISA sea level pressure, that is, FL.

Knowing the altitude is important for navigating to avoid terrain as the elevation of the ground is always given relative to sea level, but for each region there is an altitude above which terrain is no longer a problem and avoidance of other aircraft becomes the main priority. Thus, above a region-dependent "transition altitude," all aircraft use the same reference, that is 1013 hPa and adhere to FLs given to them by air traffic control. For example, over most of the United Kingdom, the transition altitude is set to 3000 ft, which is the internationally agreed minimum.

2.7 Pitot and Static Sources

The pressure instruments, that is, the altimeter, vertical speed indicator, airspeed indicator (ASI), and Mach meter require probes that measure the static and dynamic pressure. The static pressure probe can be, in principle, just a hole to the outside air but it requires careful positioning on the airframe. It cannot be placed in a forward-facing position as the moving air entering the port will produce an increase in pressure over static pressure. That is, the probe will also be measuring dynamic pressure just as in the pitot probe described below. Neither can it face rearward as the structure it is on will produce a partial void in its wake as it moves through the air and the measured pressure will be below ambient. The only option is to have the hole perpendicular to the airflow as shown in Figure 2.9a, but it is then possible to get a reduction in pressure due to the Bernoulli effect by the air flowing past the opening. This is minimized by the static port being flush with the surface and thus in the boundary layer (slow moving layer in contact with the airframe) and making the holes small encourages a stagnation point at the opening. Also, wind tunnel tests that measure the static pressure distribution over the airframe indicate positions where the pressure is equal to ambient as shown in Figure 2.8 [2]. In practice, for a typical airliner the position chosen is usually the one indicated in the figure as it has the smallest pressure variation as a function of position in front of the engines. Examples of static ports on a light aircraft and on an airliner are shown in Figure 2.9b and c.

The pitot source in contrast needs to be placed well outside the boundary layer and the opening needs to face the airflow as shown in Figure 2.9d. Air is brought to rest in the main chamber in which the pressure is thus the sum of static and dynamic pressure (also known as pitot pressure). The tube contains a heater to prevent icing and there is sometimes a drain at the bottom of the main chamber so that accumulated water can be removed. Figure 2.9e shows the pitot source under the wing of a light aircraft and Figure 2.9f is a photograph of a combined angle of attack vane and pitot source on a rotatable table on an Airbus A380. Combining the detectors in this way ensures

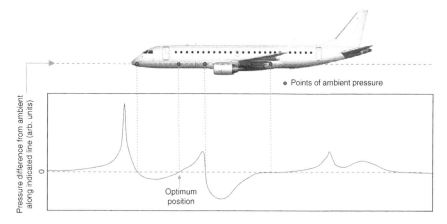

Figure 2.8 Typical static pressure distribution over an aircraft fuselage [2] showing the positions where it is equal to the ambient pressure and the optimum position for static ports. *Source:* Aircraft schematic from www.norebbo.com.

that the pitot tube is always presented parallel to the airflow. Pitot and static sources can also be combined into a single head as shown in Figure 2.9g under the wing of a light aircraft. As indicated in Figure 2.9a and d, the pitot and static pressures are passed into a pitot/static system (see Section 2.13), which either feeds the pressures via tubes directly into the instruments or uses transducers to convert the pressures into electrical signals that are processed by the ADC.

2.8 Pressure Altimeter

2.8.1 Basic Principles of the Pressure Altimeter

Although aircraft altitude can be obtained by a number of sensors including GPS, IRS, and radio/radar methods, the primary direct determination of altitude is by measuring the air pressure, which is supplied by the static source. The altimeter is a barometer that converts the measured pressure into an altitude assuming the aircraft is flying in the ISA, thus from the measured pressure it has to reproduce precisely the blue curve shown in Figure 2.4. Digitally this is straightforward as the relationship between pressure and height in the ISA is known and specified by Equations (2.7) and (2.9). In this section, however, the measurement system and mechanically driven displays in traditional pressure altimeters will be described. The detailed design varies

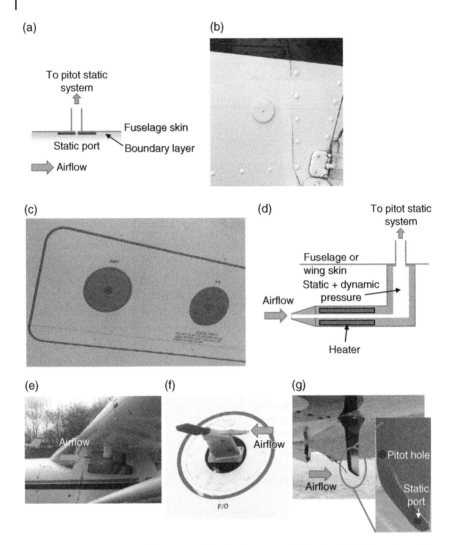

Figure 2.9 (a) Positioning of static port. (b) Static port on a light aircraft. (c) Captain and First Officer static ports on an Airbus A330. *Source:* Reprinted under Creative Commons license 3.0 [7]. (d) Schematic of pitot source. (e) Pitot source on a light aircraft. (f) Pitot source and angle of attack vane on an airbus A380. *Source:* Reprinted under Creative Commons license 3.0 [7]. (g) Combined Pitot and static source on a light aircraft.

with manufacturer but the general methodologies of measurement and display methods are presented below.

The basic sensing elements within the altimeter are aneroid capsules (Figure 2.10a), which are flexible sealed metal containers that are partially

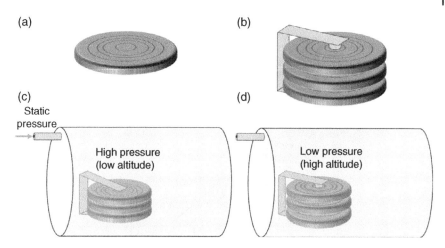

Figure 2.10 (a) Aneroid capsule. (b) Aneroid capsule stack to increase sensitivity and U-spring to prevent stack collapsing so that both decreases and increases in pressure are measured. (c) Aneroid stack in sealed housing at high static pressure. (d) Aneroid stack in sealed housing at low static pressure.

evacuated internally, so they would collapse under atmospheric pressure. Within an altimeter, several aneroid capsules are stacked to increase the sensitivity and are attached to a U-spring that prevents them from collapsing (Figure 2.10b) and allows increases and decreases in pressure to be measured. The aneroid stack is placed within a sealed housing into which is fed the pressure from the static probe and the detector expands and contracts as the pressure changes as illustrated in Figure 2.10c and d.

This relatively small movement must now be amplified and linked to the indicator, which we will assume is a three-pointer display (Figure 1.3c). Since the display needs to be linear with height, the linkage must somehow compensate for the nonlinear relationship between height and static pressure. The first step in the process is to convert the expansion and contraction of the aneroid capsules to a rotary motion, which can be done with the mechanism shown in Figure 2.11a. This is known as a tangent bar since the angle through which the rocking shaft rotates is given by an arctan function as shown below. The rotation angle, θ, produced by a capsule extension, x, depends on the starting angle of the rocking shaft, ϕ, relative to the horizontal. Let l be the original length of the rocking shaft bar as shown in Figure 2.11b, then we can write, using the law of sines:

$$\frac{l}{\sin\left[90 + (\phi - \theta)\right]} = \frac{x}{\sin\theta} \tag{2.21}$$

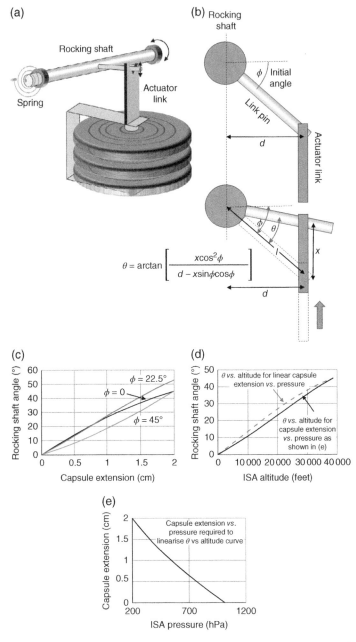

(a)

Rocking shaft

Actuator link

Spring

(b) Rocking shaft

ϕ / Initial angle

Link pin

Actuator link

d

ϕ θ

l

x

d

$$\theta = \arctan\left[\dfrac{x\cos^2\phi}{d - x\sin\phi\cos\phi}\right]$$

(c)

Rocking shaft angle (°)

$\phi = 22.5°$

$\phi = 0$

$\phi = 45°$

Capsule extension (cm)

(d)

Rocking shaft angle (°)

θ vs. altitude for linear capsule extension vs. pressure

θ vs. altitude for capsule extension vs. pressure as shown in (e)

ISA altitude (feet)

(e)

Capsule extension (cm)

Capsule extension vs. pressure required to linearise θ vs altitude curve

ISA pressure (hPa)

Figure 2.11 (a) Conversion from the linear extension of an aneroid capsule to rocking shaft rotation. (b) Arctan relationship between capsule extension and rocking shaft rotation angle for different initial starting angles, ϕ. (c) Plot of Arctan relationship for different initial starting angles, ϕ. (d) Rocking shaft rotation angle vs. altitude assuming linear capsule extension vs. pressure (dashed line) and for nonlinear capsule extension vs. pressure as in (e) (solid line). (e) Nonlinear capsule extension vs. pressure required to produce linear relationship between rocking shaft rotation angle and altitude.

or

$$\frac{l}{\cos(\phi-\theta)} = \frac{x}{\sin\theta} \tag{2.22}$$

thus, $l\sin\theta = x\cos\phi\cos\theta + x\sin\phi\sin\theta$ (2.23)

dividing through by cos θ gives:

$$\tan\theta = \frac{x\cos\phi}{l - x\sin\phi} \tag{2.24}$$

but $l = \dfrac{d}{\cos\phi}$ (2.25)

where d is the horizontal distance between the center of the rocking shaft and the actuator link. So, the relationship between the rotation angle of the rocking shaft, θ, and the capsule extension, x, is:

$$\theta = \arctan\left[\frac{x\cos^2\phi}{d - x\sin\phi\cos\phi}\right] \tag{2.26}$$

This first stage of the mechanism is designed to produce an angle of rotation of the rocking shaft that is proportional to the altitude, assuming ISA conditions so that an amplification of this rotation by a simple gearbox will maintain the linear relationship. Figure 2.11c shows x vs. θ curves assuming $d = 2$ cm and it is seen that varying the starting angle ϕ from 45° to 0 can produce a curve that has an increasing gradient ($\phi = 45°$), through almost linear ($\phi = 22.5°$) to a curve that has a decreasing gradient ($\phi = 0$). The pressure vs. altitude curve (Figure 2.4) has an increasing gradient so, assuming the capsule extension is proportional to pressure, choosing a value of ϕ that gives an x vs. θ curve with an increasing gradient (e.g. $\phi = 45°$) will help to linearize the altitude vs. θ curve.

As a specific example suppose, initially, that the extension of the capsule is proportional to pressure and that a change in pressure from 1013.25 to 200 hPa, corresponding to an altitude change of 0–38 600 ft in the ISA, causes x to change from 0 to 2 cm, as above. Combining the pressure vs. altitude curve (Figure 2.4 and Equations (2.7) and (2.9)) with the θ vs. x curve for a starting angle ϕ of 45° (Figure 2.11c) gives the θ vs. altitude curve shown by the dashed curve in Figure 2.11d. This still retains some curvature and making it linear can be achieved by designing the aneroid capsule and spring arrangement so that its extension vs. pressure curve is nonlinear (see Section 2.10) and as shown in Figure 2.11e. An adjustment is incorporated, for example, a screw that varies the distance d so that individual instruments can be calibrated.

The instrument also needs to incorporate a mechanism to compensate for the variation in the elasticity of the aneroid capsules with changes in temperature. This can be achieved by incorporating a bimetallic strip into the U-spring that

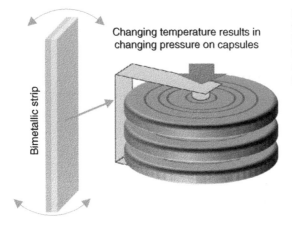

Changing temperature results in changing pressure on capsules

Bimetallic strip

Figure 2.12 Making part of the aneroid U-spring out of a bimetallic strip compensates for the variation of the elasticity of the capsules with temperature.

changes the spring force on the capsules in an opposite sense to that produced by the changing elasticity as illustrated in Figure 2.12.

2.8.2 Altimeter Display

Having produced a rocking shaft rotation that is linear with altitude and compensated for temperature, it is straightforward to amplify this and pass it on to the three-pointer indicator. As shown in Figure 2.13, the rocking shaft is attached to a sector gear that meshes with a pinion, which engages with a gearbox that drives three concentric shafts connected to the three pointers. The innermost pointer, which indicates hundreds of feet completes one revolution every 1000 ft. In the example given above a pressure change from 1013.25 to 200 hPa, which corresponds to an altitude change of ISA sea level to 38 600 ft produces a rocking shaft rotation of 45°. Thus, the gearing from the rocking shaft to the innermost indicator shaft must amplify the rotation by a factor 38.6 × 360/45 or about 310. The sector gear and first pinion shown in Figure 2.13 produce a 1 : 31 gearing and a further 1 : 10 is given by the next stage in the gearbox, which drives the innermost indicator shaft attached to the hundreds of feet pointer. A reduction gear of 10 : 1 then drives a concentric intermediate shaft attached to the 1000s of feet pointer and a further 10 : 1 reduction gear drives the outermost indicator shaft attached to the 10 000s of feet pointer.

Two additional mechanisms are incorporated, one to alter the barometric reference and the other to operate a window that shows a crosshatch pattern when the altitude is below 10 000 ft to help prevent misreading of the indicators. Changing the barometric reference requires a mechanism that rotates the pointer assembly without changing the extension of the aneroid capsules. This is achieved by having the capsules and first stage of gearing on a common mount

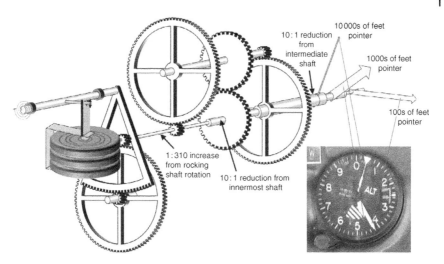

Figure 2.13 Gearing to amplify the rotation of the rocking shaft and drive three-pointer altimeter display.

and rotating the entire assembly as shown in Figure 2.14. The mechanism also requires a balance to compensate for the changing gravitational force on the capsule and gearing to drive the scale in the barometric reference window (see Ref. [2] for details). A disk attached to the outermost shaft driving the 10 000 ft pointer displays the crosshatch through a window in the main dial when the altitude is below 10 000 ft.

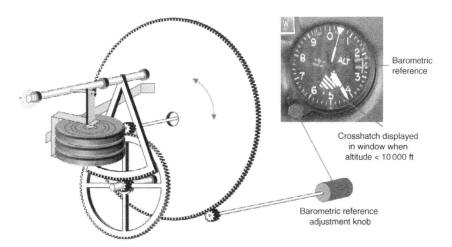

Figure 2.14 Mechanism to adjust barometric reference.

Beermat Calculation 2.2

In a specific altimeter, a change in the static pressure from the ISA sea level value (1013.25 hPa) to 100 hPa causes the width of the aneroid stack to change by 2.5 cm. If the distance between the actuator link and the rocking shaft center is 1.5 cm and the starting angle ϕ is 0, calculate the gear ratio required between the rocking shaft and the pointer indicating hundreds of feet (assume that the mechanism has been designed to produce a linear relationship between altitude and rocking shaft angle).

The change in pressure will produce a rocking shaft rotation of Arctan $(2.5/1.5) = 59.04°$.

100 hPa corresponds to an altitude greater than 11 km (see Figure 2.4), so Equation (2.9) should be used to obtain the altitude, i.e. $p = p_0 \exp\left(\dfrac{-gh}{216.65R}\right)$.

This can be rearranged to give: $h = \dfrac{216.65R}{g}\log_e\left(\dfrac{p_0}{p}\right)$, where $R = 286.9$ J kg^{-1} K^{-1} and $g = 9.807$ m s^{-2}. Thus, $h = 6338\log_e(10.1325) = 14\,677$ m $= 48\,154$ ft.

Hence the rocking shaft rotation of 59.04° must drive the hundreds of feet pointer through 48.154 revolutions, so the gearing ratio is 59.04 : (48.154 × 360) = 1 : 294.

The gearbox from the innermost pointer can also be used to drive the digital barrel type of display shown in Figure 2.15 though this is usually incorporated

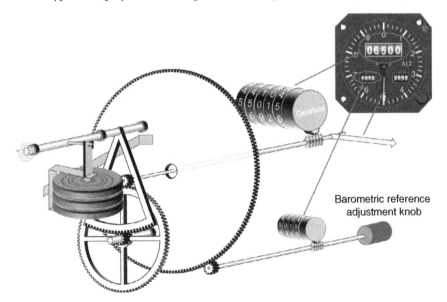

Barometric reference
adjustment knob

Figure 2.15 Mechanism in a digital barrel-type display.

into the servo altimeter described below and the simple direct drive mechanical system is normally used with a three-pointer display. In the barrel-type display, the barometric reference is also indicated on digital barrels driven by the barometric reference adjustment knob. The drive and gearing for the hundreds of feet pointer is the same in both types of display.

2.8.3 Servo Altimeter

A potential problem in purely mechanical altimeters is that very small movements in the aneroid capsules corresponding to small pressure changes may not be registered due to stiction in the gear assembly. This led to the development of the servo altimeter in which the movement of the capsules is sensed inductively as illustrated in Figure 2.16. The Inductive pick-up is known as an E–I bar due to the shape of the two components and the E-bar consists of a laminated soft magnetic material with a coil on each prong. An excitation signal is applied to the coil on the central prong, which magnetizes the core and induces an electrical signal in the outermost prongs given by Faradays law:

$$V = -N\frac{d\phi}{dt}$$ (2.27)

where V is the voltage induced across a coil of N turns by a time-varying magnetic flux ϕ. Since the signal in the coils around the outermost prongs is the gradient of the magnetic flux, if the applied signal is a sine wave, the induced signal is 90° out of phase with it ($d/d\theta$ [$\sin\theta$] = $\cos\theta$). The flux threading the pick-up coils depends sensitively on the air gap between the outermost prongs and the I-bar, which is also a magnetically soft laminated material. The pick-up coils are wired in series so that the total output signal is the difference in voltages induced in the pair, thus if the air gap is exactly the same for both then the combined signal goes to zero as shown in Figure 2.16a. As the aneroid capsules expand and contract, the air gap at the two pick-up coils is different and their combined signal is a sine wave that is +90° or −90° out of phase with the excitation signal. These signals are used to drive the altitude display as described below and illustrated in Figure 2.17.

The excitation and combined pick-up signals are amplified and passed in to the two coil sets of a two-phase induction motor whose direction of travel depends on the phase difference between the two waveforms. Thus, an expanded aneroid stack will drive the motor one way and a contracted aneroid stack will drive it the other way. The motor drives a magnet rotating inside a drag cup that uses the drag generated by eddy currents induced in the cup to transfer the rotation to a gearbox, which drives a digital barrel display and the hundreds of feet pointer. The gearbox also drives a mechanism that resets the E–I bar to equalize the air gaps and turns off the motor after the appropriate number of turns of the altimeter pointer.

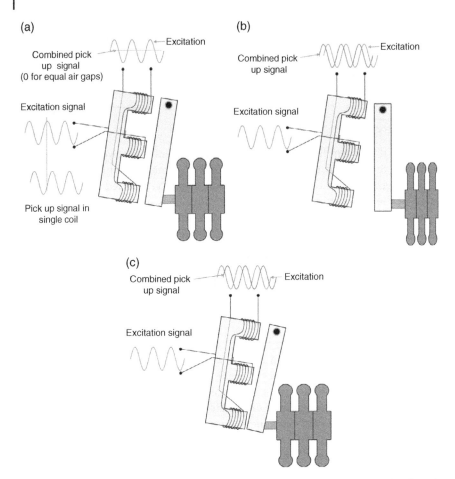

(a)

Combined pick up signal (0 for equal air gaps) ← Excitation

Excitation signal

Pick up signal in single coil

(b)

Combined pick up signal ← Excitation

Excitation signal

(c)

Combined pick up signal ← Excitation

Excitation signal

Figure 2.16 (a) E–I bar inductive pick up where the E and the I sections are made of a soft laminated magnetic material. The signal applied to the central prong of the E-bar produces an induced signal in the pick-up coils on the end prongs that is 90° out of phase and depends sensitively on the air gap between the E and I bar. The coils are wired in series and when the air gaps are the same the combined pick-up signal is zero. (b) and (c) If the aneroid expands or contracts there is a difference in the air gaps and the combined pick-up signal is ±90° out of phase with the excitation signal.

Suppose, for example, the altitude decreased by 1000 ft, which caused the aneroid stack to be compressed by 0.5 mm (shown exaggerated in Figure 2.17). The uneven air gaps in the E–I inductive pick up would drive the motor to show a decreasing altitude and the gearbox and reset mechanism are designed to precisely reset the E–I bar to equal air gaps after a single rotation of the hundreds of feet pointer, which turns off the motor. The barometric

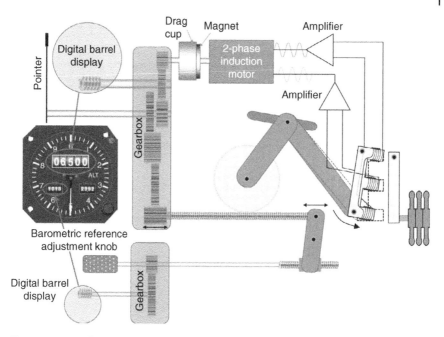

Figure 2.17 Mechanism to drive the altitude and barometric reference in a servo altimeter.

reference knob shifts the E–I bar, which forces the motor to reset it with a new altitude displayed.

2.8.4 Altimeter with Digital Encoder

Aircraft transponders (see Section 5.4.3) transmit a binary number to a radar ground station that reports the altitude of the aircraft. This is necessary because primary radar accurately measures line of sight distance to a target but not the altitude. In older aircraft, encoders to generate a binary number that corresponds to the altitude were incorporated within pressure altimeters and these are described below. In modern aircraft that incorporate digital data buses carrying data to the ADC, the correct binary number can be generated digitally.

First, a little detail is needed on the binary sequence itself, which is a modified form of Gray code, named after its inventor Frank Gray. This has the property that in a sequence of binary numbers only one bit changes as we progress from one number in the sequence to the next. The specific 11-bit variant used in aviation to transmit altitude data is referred to as a Gillham code and it is illustrated in Figure 2.18. By convention, the bits in the binary number are labeled D1, D2, D4, A1, A2, A4, B1, B2, B4, C1, C2, and C4, with

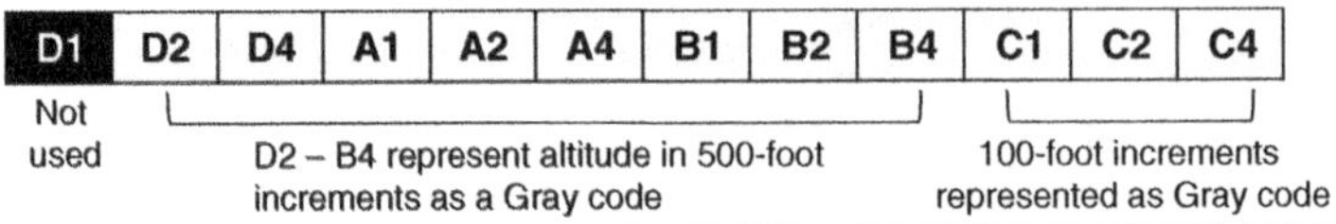

D1	D2	D4	A1	A2	A4	B1	B2	B4	C1	C2	C4
Not used	D2 – B4 represent altitude in 500-foot increments as a Gray code								100-foot increments represented as Gray code		

Altitude (feet)	Flight Level	D2	D4	A1	A2	A4	B1	B2	B4	Parity	C1	C2	C4	Octal	Construction
−1200	-	0	0	0	0	0	0	0	0	even	0	0	1	0001	$= 0 \times 500 - 1000 - 200$
−1100	-	0	0	0	0	0	0	0	0	even	0	1	1	0003	$= 0 \times 500 - 1000 - 100$
−1000	-	0	0	0	0	0	0	0	0	even	0	1	0	0004	$= 0 \times 500 - 1000 - 0$
−900	-	0	0	0	0	0	0	0	0	even	1	1	0	0006	$= 0 \times 500 - 1000 + 100$
−800	-	0	0	0	0	0	0	0	0	even	1	0	0	0004	$= 0 \times 500 - 1000 + 200$
−700	-	0	0	0	0	0	0	0	1	odd	1	0	0	0014	$= 1 \times 500 - 1000 - 200$
−600	-	0	0	0	0	0	0	0	1	odd	1	1	0	0016	$= 1 \times 500 - 1000 - 100$
−500	-	0	0	0	0	0	0	0	1	odd	0	1	0	0012	$= 1 \times 500 - 1000 - 0$
−400	-	0	0	0	0	0	0	0	1	odd	0	1	1	0013	$= 1 \times 500 - 1000 + 100$
−300	-	0	0	0	0	0	0	0	1	odd	0	0	1	0011	$= 1 \times 500 - 1000 + 200$
−200	-	0	0	0	0	0	0	1	1	even	0	0	1	0031	$= 2 \times 500 - 1000 - 200$
−100	-	0	0	0	0	0	0	1	1	even	0	1	1	0033	$= 2 \times 500 - 1000 - 100$
0	FL000	0	0	0	0	0	0	1	1	even	0	1	0	0032	$= 2 \times 500 - 1000 - 0$
100	FL001	0	0	0	0	0	0	1	1	even	1	1	0	0036	$= 2 \times 500 - 1000 + 100$
200	FL002	0	0	0	0	0	0	1	1	even	1	0	0	0034	$= 2 \times 500 - 1000 + 200$
300	FL003	0	0	0	0	0	0	1	0	odd	1	0	0	0024	$= 3 \times 500 - 1000 - 200$
400	FL004	0	0	0	0	0	0	1	0	odd	1	1	0	0026	$= 3 \times 500 - 1000 - 100$
500	FL005	0	0	0	0	0	0	1	0	odd	0	1	0	0022	$= 3 \times 500 - 1000 - 0$
600	FL006	0	0	0	0	0	0	1	0	odd	0	1	1	0023	$= 3 \times 500 - 1000 + 100$
700	FL007	0	0	0	0	0	0	1	0	odd	0	0	1	0021	$= 3 \times 500 - 1000 + 200$
Sequence continues up to the maximum altitude available with 11 bits shown below															
126,700	FL1267	1	0	0	0	0	0	0	0	odd	0	0	1	4001	$= 255 \times 500 - 1000 + 200$

Figure 2.18 Construction of 11-bit binary Gillham code number to represent altitude/ Flight Level.

D1 never used so there are 11 bits in the number. The altitude is reported in increments of 100 ft, so each binary number in the sequence corresponds to a FL. Bits D2–B4 are sequenced as a normal Gray code and represent altitude in 500-ft increments with a base value of −1000 ft. In the remaining bits C1–C4, five values out of the possible eight encode hundreds of feet relative to the bigger number. The scheme used is for the C1–C4 sequences: 001, 011, 010, 110, and 100 to represent −200, −100, 0, +100, +200 ft relative to the 500-ft number if the bigger number (D2–B4) is odd parity while if it is even parity, the same C1–C4 binary sequence represents +200 to −200 ft relative to the 500-ft number. Thus, for example, 00000010 001 represents 700 ft because 00000010 has magnitude 3 in Gray code so the big number represents $3 \times 500 - 1000 = 500$ ft and since the parity of the 8-bit binary number is odd, the C1–C4 bits 001 give +200 ft yielding a total altitude of 700 ft. The first few altitudes/FLs and their construction are shown in Figure 2.18. Students familiar with binary numbers will recognize these four groups of 3-bit numbers (e.g. A1, A2, and A4) as a 4-digit octal number, which is also shown in Figure 2.18. Note that the Gillham code, despite consisting of a construction of two Gray code numbers preserves the basic property of Gray codes in its entire 11-bit length, that is, only one bit changes each time we increment from one number to the next.

Using the full 11-bit number allows altitudes up to 126 700 ft to be represented, which is much higher than any aircraft can fly in practice. Discarding the D2 bit allows altitudes up to 62 700 to be represented, which is more than adequate for commercial transport aircraft and discarding both D1 and D4 bits allows altitudes up to 30 700 ft to be represented, which is more than adequate for general aviation aircraft. The lowest altitude that can be represented is below sea level to allow for the remote possibility that an aircraft is flying at low level in one of the few places on the globe where the surface is below sea level, the lowest lying being the shore of the Red Sea at an altitude of −1355 ft.

In modern aircraft with an ADC where the altitude information is digitized, it is straightforward to derive the Gillham code and pass it on to the transponder via the digital data bus. This can be done analytically, though to speed up processing it is better to derive the code using a look-up table. In older systems, the code was derived using a mechanical/optical system that was either stand-alone or incorporated into the altimeter. Figure 2.19 shows a typical optical encoder built into an altimeter where Figure 2.19a shows the transparent encoder disk containing the 11 concentric rings with the bit pattern for each binary number written radially in which the dark areas represent "0" and the transparent areas "1." For clarity, the encoder has been restricted to 7-bits, that is, bits D2, D4, A1, and A2 have been set permanently to zero. The remaining bits can represent altitudes up to 6700 ft = FL67. The disk is driven by the aneroid capsule and rocking shaft assembly and geared so that one complete revolution of the disk corresponds to the total altitude range of the instrument, in this case, altitudes −1200 to 6700 ft. The electronic read out is produced by shining a lamp through the disk onto a bank of 11 photosensors aligned with the concentric circles representing D2–C4 on the disk as illustrated in Figure 2.19b. A solid-state stand-alone encoder is shown in Figure 2.19c and this type of unit measures pressure using a piezoresistive transducer that generates an output voltage proportional to pressure. This is then converted to a digital signal by an analog-to-digital converter and the digital signal is further processed to output Gillham code that is passed to the transponder.

2.9 Vertical Speed Indicator (VSI)

A pilot needs to know the rate of climb or descent of an aircraft and while the altimeter indicates whether an aircraft is ascending or descending, it is often important to have a quantitative value for the rate of climb (ROC) or rate of descent (ROD). In an aircraft with an ADC and digital processing this can be determined by numerically differentiating the altitude and displaying the rate

(a)

(b)

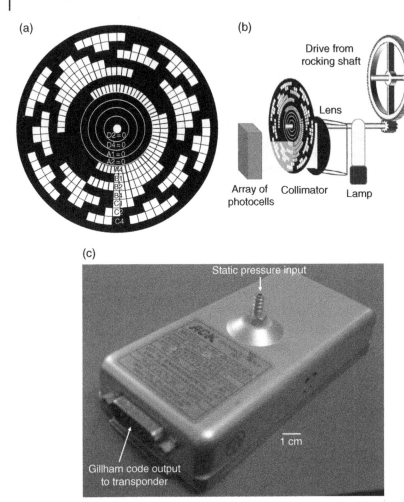

Drive from rocking shaft

Lens

D2=0
D4=0
A1=0
A2=0
A4
B1
B2
B4
C1
C2
C4

Array of photocells   Collimator   Lamp

(c)

Static pressure input

1 cm

Gillham code output to transponder

Figure 2.19 (a) Optical Gillham code encoder disk with encoding reduced to 7 bits for clarity (bits D2, D4, A1, and A2 set to zero). (b) Production of digital signal from encoder disk. (c) Stand-alone solid-state encoder using a piezoresistive transducer. *Source:* Reprinted under Creative Commons license 3.0 [7].

of change digitally (see Figure 1.4c). In this section, the method of measuring ROC and ROD directly and displaying the absolute value on a direct-reading instrument will be described.

The basic principle of a direct-reading VSI is illustrated in Figure 2.20a. The instrument has a single pressure input from the static source, which is fed via a capillary into an aneroid capsule in a sealed case. The capillary is included to prevent damage to the aneroid capsule in case of pressure bursts and this type

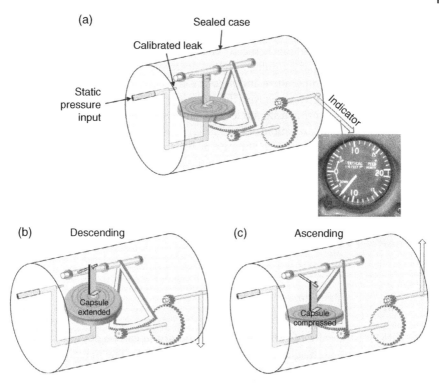

Figure 2.20 (a) Principle of operation of a direct-reading VSI. The static pressure is fed into the capsule and also the case via a calibrated leak, which introduces a time delay for pressures to equalize. (b) During a descent, the static pressure is rising and the pressure in the capsule is maintained higher than the case. (c) During a climb, the static pressure is falling and the pressure in the capsule is maintained lower than the case.

of protection must always be included when a pressure source is connected directly into the capsule. The static pressure is also passed directly into the case via a calibrated leak and so there is a significant time delay required for the pressures in the capsule and the case to equalize after a change. If the altitude is decreasing, the static pressure is rising so the delay in equalizing pressures produces a higher pressure in the capsule than the case causing the capsule to extend and drive the display to indicate a descent (Figure 2.20b). During a climb, the pressure is falling so that the pressure in the case is higher than the capsule causing it to compress and drive the display accordingly (Figure 2.20c). The principle is straightforward but to obtain quantitative information that is independent of altitude and air temperature and also indicates the ROD/ROC on a linear scale requires careful design, much of which is focused on the calibrated leak, also known as the metering unit.

The metering unit can either be a passive design that is naturally self-compensating for pressure and temperature changes or a system that detects temperature changes and mechanically controls the flow. The passive type consists of a combination of a capillary and an orifice through which the flow rate of air has different temperature and pressure characteristics (Figure 2.21). The aneroid capsule directly measures the pressure differential, ΔP, maintained either side of the metering unit and let us assume that the mechanism has been designed to produce a rotation of the display that is proportional to ΔP. This pressure difference depends on the rate at which the pressure is changing at the input and the rate of flow, Q, through the metering unit so that a high rate of flow will reduce the pressure difference and a low rate of flow will increase it. In a steady climb or descent, assuming a small rate of change of pressure so that responses are linear, for a given air density and temperature, ΔP is given by:

$$\Delta P \propto \frac{1}{Q} \times \frac{dP}{dt} \tag{2.28}$$

or

$$\Delta P \propto \frac{1}{Q} \times \frac{dP}{dh} \times \frac{dh}{dt} \tag{2.29}$$

where dh/dt is the rate of climb or descent and dp/dh can be obtained directly from the pressure vs. height data for the ISA (Figure 2.4). The magnitude of dp/dh is shown in Figure 2.21 (its absolute value is negative) and it is evident that at 30 000 ft (Figure 2.21a) the gradient is smaller than at sea level (Figure 2.21b). Thus, the reduction in the value of dp/dh would produce, for the same ROC or ROD, a smaller value of ΔP and thus a lower indication of vertical speed at high altitude than at low altitude.

As summarized on the right side of Figure 2.21a, at low air density and temperature (i.e. high altitude) the capillary produces a smaller Q and thus a higher ΔP while the aperture produces the opposite characteristic. At high air density and temperature (i.e. low altitude: Figure 2.21b) the difference in flow rates is opposite. The different behaviors are mostly due to the fact that flow through the capillary is laminar while it is turbulent through the aperture. With careful design of the dimensions of the capillary and aperture, it is possible to compensate for the change in the value of dp/dh with altitude in Equation (2.29) and produce a true indication of the ROC or ROD that is independent of the altitude at which it is measured.

The active type of metering unit is illustrated in Figure 2.22 and consists of two porous ceramic leaks with a valve between them that opens directly into the instrument case. With changing altitude, the temperature of the air fed into the metering unit varies, which controls the opening of the valve by the bimetallic strip. If the valve is fully open there is only one ceramic leak in line giving the highest possible flow rate, while if it is fully closed both ceramic leaks

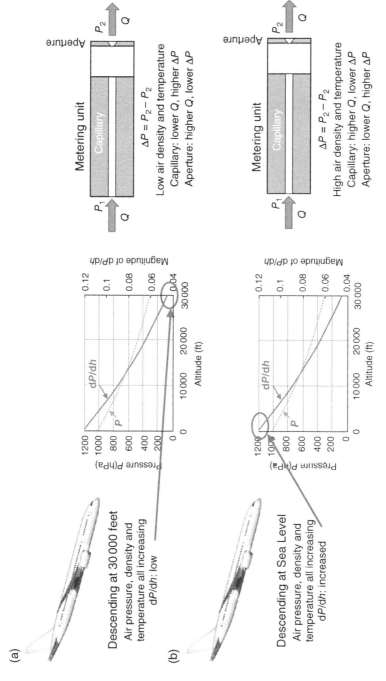

Figure 2.21 Changes in the properties of air and dp/dh in a descent at 30 000 ft (a) and at sea level (b) (left). Pressure and dp/dh vs. altitude (middle). Schematic of capillary plus aperture metering unit and the changes in the flow rate, Q, and pressure differential, ΔP, at high and low altitude (right). *Source:* Aircraft schematic from www.norebbo.com.

(a)

Descending at 30 000 feet

Air pressure, density and temperature all increasing
dP/dh: low

Pressure P(hPa)

Magnitude of dP/dh

dP/dh

P

Altitude (ft)

Metering unit

Aperture

P_1

Q

P_2

Q

Capillary

$\Delta P = P_2 - P_2$

Low air density and temperature
Capillary: lower Q, higher ΔP
Aperture: higher Q, lower ΔP

(b)

Descending at Sea Level

Air pressure, density and temperature all increasing
dP/dh: increased

Pressure P(hPa)

Magnitude of dP/dh

dP/dh

P

Altitude (ft)

Metering unit

Aperture

P_1

Q

P_2

Q

Capillary

$\Delta P = P_2 - P_2$

High air density and temperature
Capillary: higher Q, lower ΔP
Aperture: lower Q, higher ΔP

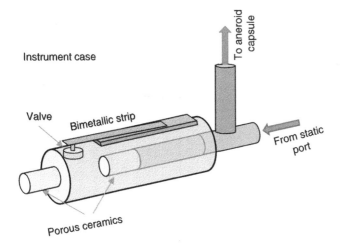

Instrument case

To aneroid capsule

Valve

Bimetallic strip

From static port

Porous ceramics

Figure 2.22 Active type of metering unit consisting of two porous ceramic leaks in which the changing air temperature controls the opening of a valve into the case via a bimetallic strip. Thus, the flow rate into the case can be controlled at any value between the maximum rate (one leak only) or the minimum rate (both leaks).

come into play producing the lowest possible flow rate. The valve can also be partially open so that an intermediate flow rate can be achieved. Thus, ΔP can be controlled to compensate for the changing value of dp/dh in Equation (2.29) to produce a ROD or ROC indication that is independent of altitude.

2.9.1 Instantaneous Vertical Speed Indicator (IVSI)

One operational problem with the simple VSI described above is that it takes up to 10 seconds for the pressure differential to stabilize and produce an accurate reading on the instrument face. Thus, for a given ROC or ROD the pilot would have to set a pitch attitude and power setting learned by training and then wait to see if the desired climb or descent rate was achieved followed by adjustment of power or pitch attitude to achieve the target rate. Right at the start of the climb or descent, however, there is a vertical acceleration of the aircraft and in the Instantaneous Vertical Speed Indicator (IVSI) this is detected by an accelerometer that then sets an appropriate pressure differential. When the steady vertical rate has been achieved, the acceleration goes to zero but by then the appropriate pressure differential has been achieved. The system used is illustrated schematically in Figure 2.23.

The accelerometer, also called a dashpot, is a spring-loaded piston within a vertical cylinder across the two pressure lines connected to the aneroid capsule and the metering unit. Suppose the aircraft initiated a climb from level flight.

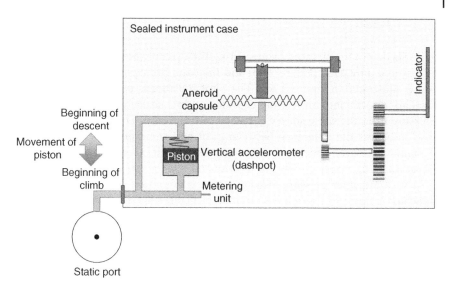

Figure 2.23 Dashpot accelerometer in an IVSI. At the start of a climb, the vertical acceleration will move the piston downward immediately establishing the correct pressure differential for the expected climb rate. The opposite motion does the same at the start of a descent.

During the climb, the pressure at the static port is decreasing, so the internal pressure in the aneroid capsule is lower than the instrument case causing the capsule to compress and drive the indicator to display the climb rate. As explained above, however, it takes time to establish the steady-state pressure differential while the vertical acceleration at the start of the climb causes the piston to move downward immediately reducing the pressure in the capsule and increasing it in the case. As the steady climb is achieved, the acceleration goes to zero and the piston returns to the start position by which time the metering unit has set up the required pressure differential. With careful design of the dimensions of the piston and spring constant, it is possible for the accelerometer to set up the required pressure differential corresponding to the climb rate that will be achieved. It is easy to see that the opposite pressure differential is established by the accelerometer in a descent.

One disadvantage of the system is that in a steep level turn, the load factor (*g*-force) required to maintain height will also move the piston and initially show a climb. When the turn is established the instrument reading will return to zero as the metering unit equalizes the pressures with the new position of the piston. When rolling out of the turn the removal of the load factor will return the piston to its original position and the instrument will temporarily display a descent.

2.10 Airspeed Indicator

All ASIs determine airspeed by measuring the dynamic pressure of air brought to rest in a pitot tube (Figure 2.9d) and the basic principle of a direct-reading ASI is illustrated in Figure 2.24a. The pitot pressure is passed via a damping capillary directly into an aneroid capsule within a sealed case while the static pressure from the static source is passed into the case. The capsule extension thus measures the difference between pitot and static pressures, that is, the dynamic pressure, which, for noncompressible air, is equal to $\rho v^2/2$ (see Equation 2.20), where ρ is the air density and v is the airspeed.

As with the other pressure instruments, it is important that the scale on the indicator is linear to make it easy to read in flight. This is achieved in the first stage of the instrument, that is, the capsule expansion and tangent bar mechanism (Figure 2.24b), so that the rocking shaft rotation angle has a linear dependence on airspeed. Assuming the capsule extension is proportional to the difference between the inside and outside pressure, it will vary with airspeed according to the quadratic function shown in Figure 2.24d. As was shown in Section 2.8.1, varying the parameters, d, the distance between the actuator link and the rocking shaft, and ϕ, the initial starting angle of the link pin (Figure 2.24c), enables significant control over the relationship between the rocking shaft rotation angle and the parameter to be displayed, in this case, airspeed. Figure 2.24e shows the rocking shaft rotation angle vs. airspeed if d and ϕ are optimized to produce a linear response and it is seen that this is achieved to a close approximation over the range 100–400 knots.

The remaining nonlinearity can be removed by controlling the capsule extension vs. pressure, for example, by using a variable spring loading on the aneroid capsule as illustrated in Figure 2.25. The length of the screws can be adjusted thus introducing an extra three parameters that can be varied. The number of screws can be increased and their positions varied to produce any desired variability in the spring stiffness. Thus, the relationship between the pressure differential across the capsule and its extension can be given any desired characteristic, which is useful for building in compensation for the "compressibility error." Equation (2.20) for the dynamic pressure assumes air is incompressible and this is a reasonable assumption below 300 knots but for higher speeds, it introduces a significant error. Thus, high-speed ASIs are calibrated assuming the dynamic pressure is given by [3]:

$$q = \frac{\rho v^2}{2}\left(1 + \frac{v^2}{4\,a_0^2}\right) \tag{2.30}$$

where q is the dynamic pressure and a_0 is the speed of sound at sea level.

Whichever formula applies (2.20 or 2.30) the conversion from dynamic pressure to airspeed can only be carried out for a specific air density and the ASI is

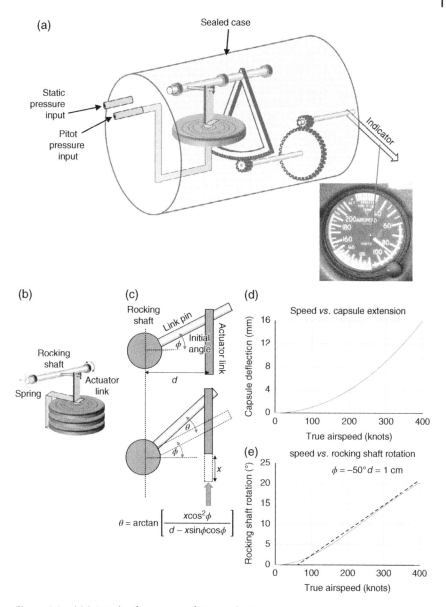

Figure 2.24 (a) Principle of operation of Airspeed Indicator (ASI). The pitot pressure is passed into the aneroid capsule and the static pressure is passed into the case so the capsule measures the difference, that is, the dynamic pressure (Equation 2.20). (b) Linkage between the aneroid capsule and the rocking shaft. (c) Relationship between capsule extension and rocking shaft rotation as a function of starting angle and d. (d) Capsule extension vs. airspeed assuming extension is proportional to the pressure difference and air is incompressible. (e) Rocking shaft rotation vs. airspeed for the optimum choice of ϕ and d showing an almost linear relationship above 100 knots.

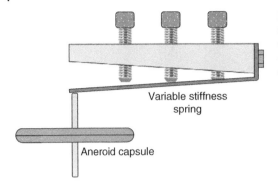

Figure 2.25 Controlling the relationship between the extension of the aneroid capsule and the dynamic pressure using a variable stiffness spring.

Variable stiffness spring

Aneroid capsule

calibrated to read the true airspeed (TAS) in the ISA at sea level, where $\rho =$ 1.225 kg m^{-3}. At any other density, the indicated airspeed (IAS) $\neq$ TAS and as the aircraft height increases the IAS under-reads the TAS by an increasing factor, so a conversion must be carried out. In a large aircraft, this is taken care of by the ADC. As an example, if an aircraft flying at 10 000 ft under ISA conditions has an IAS of 250 knots, its TAS is 290 knots.

One advantage of using the dynamic pressure to indicate speed, however, is that all the important flight parameters, such as lift, stall conditions, critical speeds for lowering landing gear and flaps, etc. depend on the dynamic pressure, irrespective of the TAS. Thus, the IAS can be used by the pilot to maintain safe flying conditions and determine whether it is safe to lower flaps and landing gear without having to worry about converting to TAS. The latter is only important to determine the Mach number (see Section 2.11) and the ground speed (after applying a correction for wind), which is required for navigation.

A number of critical speeds (the so-called "V-speeds") have a standard terminology and are marked on the ASI in a standardized color format shown in Figure 2.26a for a twin-engine propeller plane. These are:

V_{S0} – Stall speed in the landing configuration.

V_{S1} – Stall speed in a given configuration (normally "clean," that is, flaps and landing gear raised).

V_{FE} – Maximum flap extension speed (or maximum speed with flaps in a given position).

V_{NO} – Maximum normal operating speed (only exceed in smooth air and with gentle maneuvers).

V_{NE} – Never exceed speed.

V_{YSE} – Maximum rate of climb speed on one engine for a twin-engine aircraft.

V_{MCA} – Minimum speed at which control can be maintained on one engine for a twin-engine aircraft at maximum takeoff weight.

The V-speeds are indicated either by radial lines at the appropriate value (V_{MCA}, V_{YSE}, and V_{NE}) or by the ends of white green and yellow arcs as shown

(a)   (b)

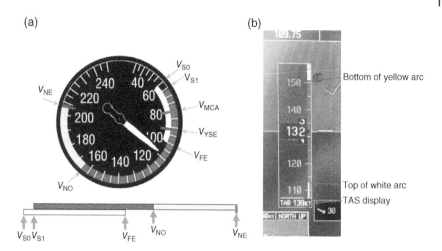

Figure 2.26 (a) Clockface of an airspeed indicator (ASI) on a twin-engine propeller aircraft showing color-coded arcs and radial lines indicating V-speeds. (b) Moving tape display on a glass cockpit showing the same color-coded bands and also a TAS indicator calculated the air data computer.

in Figure 2.26a. The ergonomic design allows quick recognition if any particular action is safe. For example, if the speed is below the top end of the white arc it is safe to lower flaps, the speed should only be higher than the end of the green arc in smooth air with gentle maneuvers to avoid overstressing the airframe, etc. For the twin-engine aircraft, an important safety issue is the ability to fly and climb on one engine. The blue radial line shows the speed (V_{YSE}) that achieves the maximum climb rate on one engine at full power (important if engine fails just after takeoff) and the lower speed red radial line (V_{MCA}) shows the minimum speed at which the rudder is able to counteract the yaw produced by the asymmetric thrust from only one engine. If an engine fails on a twin it is very important to maintain an airspeed above this value, otherwise the aircraft will yaw and roll (secondary effect of yaw) with insufficient dynamic pressure available on the rudder to maintain control even at full deflection.

This standardized system of colored arcs and lines is maintained on moving tape-type displays used in glass cockpits (Figure 2.26b) on large aircraft and some modern light aircraft. By convention, two V-speeds not shown on the ASI are:

V_{LO} – Maximum landing gear operation speed (up or down).
V_{LE} – Maximum speed with landing gear extended.

Conversion from IAS to TAS requires a correction for the air density and to determine this requires a knowledge of the altitude and OAT. In aircraft

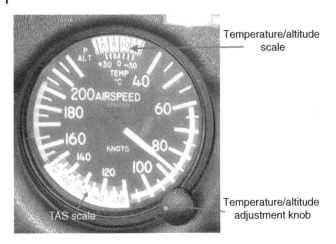

Temperature/altitude scale

Temperature/altitude adjustment knob

TAS scale

Figure 2.27 Direct-reading ASI incorporating a mechanism to indicate TAS.

with an ADC, the TAS can be calculated from the measured pitot pressure, static pressure, and OAT as shown in Section 2.14.2 and the value is displayed on the PFD as illustrated in Figure 2.26b. Some direct-reading ASIs incorporate a mechanism that enables them to display TAS in addition to IAS as shown in Figure 2.27. The pilot turns the knob at the bottom left of the instrument so that on the scale at the top, the measured OAT is aligned with the measured altitude. This then sets up the second scale at the bottom right of the instrument (in a similar manner to the rotary slide rule used by trainee pilots – see Section 6.5.1) to read the TAS over a limited range representing the normal range of cruise speeds. In the example shown, 10 000 ft has been set against –10 °C and under these conditions, 130 knots IAS would be a TAS of 150 knots.

During training, pilots are also presented with two other airspeeds, that is calibrated airspeed (CAS) and equivalent airspeed (EAS). CAS is the IAS corrected for instrument errors including any nonlinearity in the curve shown in Figure 2.24e and known errors produced by aircraft maneuvers. These include high-pitch angles used during a climb and high yaw rates, which both disturb the airflow into the pitot head. In addition, the CAS incorporates adjustments for compressibility errors if these have not been removed by the ASI mechanism as described above. In a glass cockpit, these corrections are all tabulated in the ADC, while for traditional cockpits the relationship between the IAS and CAS is tabulated in the aircraft flight manual (AFM). EAS is defined as the speed at sea level that would produce the same *incompressible* dynamic pressure as the TAS at the altitude at which the vehicle is flying. Essentially, it can be thought of as the IAS with no compressibility correction.

2.11 Mach Meter

2.11.1 Critical Mach Number

As explained below, it is important to know the Mach number as well as the TAS for any jet aircraft. The speed of sound, a, for an ideal gas is given by:

$$a = \sqrt{\frac{\gamma RT}{M}} \qquad (2.31)$$

where γ is the adiabatic index (1.4 for diatomic molecules), R is the gas constant ($=8.314 \, \text{J K}^{-1}$ mol), M is the molar mass (in kg mol^{-1}), and T is the absolute temperature in Kelvin. Thus, apart from a small correction to M due to the mixture of gases changing, the speed of sound in air depends only on temperature irrespective of height. Substituting appropriate values for the constants gives:

$$a \, (\text{knots}) = 38.967 \sqrt{T(\text{K})} \qquad (2.32)$$

The *local* speed of sound (LSS) is defined as the speed of sound in the undisturbed airflow around the aircraft, that is, at a sufficient distance for the air to remain undisturbed but close enough that the density and temperature can be taken to be the same. Thus, the LSS is 661 knots at ISA sea level ($T = 288$ K) and 588 knots at 30 000 ft ($T = 228$ K). The Mach number, Ma, for an aircraft is defined by:

$$Ma = \text{TAS}/\text{LSS} = \text{TAS}/a \qquad (2.33)$$

This is known as the free-stream Mach number, that is, the speed of the aircraft relative to the undisturbed air. Jet airliners (apart from concord) operate at speeds below the speed of sound but since air is accelerated over some parts (for example, the top of the wing), locally around the airframe the airflow can exceed the LSS. For $Ma > 0.8$, the speed is referred to as transonic, that is, the airflow around certain structures can exceed the LSS and shock waves will develop. The Mach number at which the airflow reaches the LSS somewhere on the airframe is referred to as the critical Mach number M_{CRIT}. Turbulence from the shock waves can overstress the airframe and also a rapid onset of drag and a significant movement of the center of pressure (wing chord position at which the lift acts) create other hazards such as a large pitching down moment (so-called Mach tuck). Thus, aircraft operating at transonic speeds have a maximum Mach number, M_{MO}, which may be slightly higher than M_{CRIT}, which should not be exceeded irrespective of the TAS. For example, M_{MO} for a Boeing 737 is 0.84, which according to Equations (2.32) and (2.33) is reached at a TAS of 494 knots at 30 000 ft under ISA conditions but on a cold day could occur at a significantly slower TAS.

Beermat Calculation 2.3

What is the Mach number of an aircraft flying in the ISA at 36 500 ft under ISA conditions at a TAS of 450 knots?

36 500 ft = 11 125 m, which is above the tropopause, so in the ISA the OAT is constant at −56.5 °C. Thus, from Equation (2.32), the LSS is:

$a(\text{knots}) = 38.967\sqrt{216.65} = 573.6 \text{ knots}$

So, the Mach number is $Ma = 450/573.6 = 0.78$

When climbing at a constant IAS, the Mach number increases because TAS increases and in addition the decreasing temperature reduces the LSS. At some height (depending on the OAT) the crew must stop climbing at a constant IAS and climb at constant Ma to avoid exceeding M_{MO}. Thus, since the advent of jet-powered commercial air travel, Mach meters that directly read the Mach number have been included in the cockpit instrumentation.

Interesting Diversion 2.2: World War 2 Aircraft at Transonic Speeds

During World War 2, there were several claims that propeller aircraft had exceeded the speed of sound in a dive but these can mostly be put down to errors in direct-reading ASIs. The compressibility of air at high Mach numbers was not properly compensated for and as shown in Figure 2.35 this can produce a significant airspeed over-read above 350 knots. In general, it can be regarded as impossible for the TAS of a propeller aircraft to exceed Mach 1 as, even at the high service ceilings achievable in military aircraft (for example, 36 500 ft in a Spitfire), this would require a TAS in excess of 570 knots (see Beermat Calculation 2.3). However, there is significant documentation that aircraft achieved their critical Mach number in dives. That is, the airflow at a specific point on the airframe (generally on the thickest part of the wing) exceeded Mach 1.

The value of the critical Mach number varies from wing to wing and, in general, a thicker wing will have a lower critical Mach number since it accelerates the airflow to a faster speed than a thinner one. Thus, the fairly thick wing on the P-38 Lightning (Figure ID2.2.1a) gave it a low critical Mach number of 0.69. The aircraft had an impressive service ceiling of 44 000 ft at which, in the ISA, $a = 552$ knots, thus the critical Mach No at 44 000 ft for a P38 corresponds to 381 knots, which it could reach fairly easily in a dive. The much thinner wing on the Supermarine Spitfire produced a critical Mach number of about 0.89 for this aircraft, so reaching its critical Mach number required a much higher TAS.

Forcing aircraft with a subsonic airframe design into the transonic region was extremely hazardous as the effect on the controls was not well understood and effects like "Mach tuck" could generate control forces that could not be overcome by the pilot and could maintain dives that were impossible to recover from. This killed a number of Mitsubishi Zero pilots, several of whom flew full-power dives into terrain. The highest substantiated Mach number achieved during the war was 0.92 in 1944 in an Mk11 Spitfire during tests at the Royal Aircraft Establishment, Farnborough, though the flight resulted in a damaged engine due to propeller overspeed.

(a)   (b)

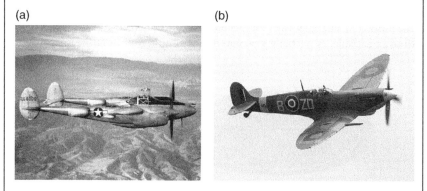

Figure ID2.2.1 (a) P38 lightning, critical *Ma* = 0.69. *Source:* Reprinted from Wikipedia. org: public domain image. (b) Spitfire, critical *Ma* = 0.89. *Source:* Reprinted from Wikipedia.org under Creative Commons license 3.0 [7].

2.11.2 Direct-Reading Mach Meter

Originally, Mach meters were direct-reading instruments driven purely by air pressure supplied by the pitot and static sources. From the equation of state of air (Equation 2.2), $p_s = \rho RT$, we can write, from Equation (2.31)

$$a = \sqrt{\frac{\gamma RT}{M}} \propto \sqrt{\frac{p_s}{\rho}} \tag{2.34}$$

where p_s is the static pressure while from Equation (2.20), we can write:

$$TAS \propto \sqrt{\frac{q}{\rho}} \tag{2.35}$$

where q is the dynamic pressure. Thus, from Equation (2.33):

$$Ma = \frac{q}{p_s} = \frac{p_t - p_s}{p_s} \tag{2.36}$$

where p_t is the total pressure measured in the pitot tube. So the Mach number can be obtained directly from the static and dynamic pressure feeds in the pitot/static system that is used to drive the other pressure instruments. Note that Equation (2.20) assumes incompressible flow and a full discussion of the effects of compressible flow on *Ma* is presented in Section 2.14 where it is shown that *Ma* still depends on the ratio p_t/p_s.

Figure 2.28 illustrates a mechanism using two aneroid capsules that will indicate the Mach number. The total pressure, p_t, is fed into an aneroid capsule and p_s is fed into the case, so this capsule measures q and is labeled the airspeed capsule. It drives the pointer via the ratio arm and ranging arm and provides an indication that changes with airspeed. A second aneroid capsule, labeled the altimeter capsule, is partially evacuated and simply monitors the surrounding

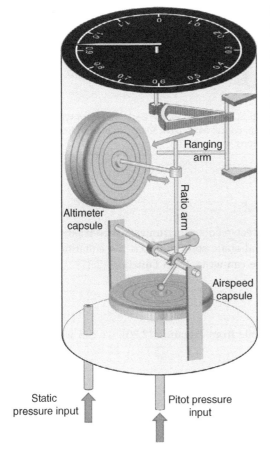

Figure 2.28 Principle of operation of the direct-reading Mach meter. The airspeed capsule drives the pointer via the ratio arm and ranging arm. The altimeter capsule sets the position of the ratio arm along the ranging arm so that it controls the movement of the scale by the ratio between q and p_s, which is proportional to *Ma*.

pressure p_s, as in an altimeter. This changes the position of the ratio arm rotated by the airspeed capsule so that it determines how much the movement of the airspeed capsule moves the scale. That is, it controls the movement of the scale by the ratio between q and p_s, which is proportional to Ma.

2.12 OAT Probe

2.12.1 Ram Rise and Total Air Temperature

A knowledge of the OAT is important to calculate the density of the air in which the aircraft is flying in order to derive the TAS from the IAS and to obtain the Mach number. The OAT is one of the inputs into the ADC to enable it to calculate all the required output parameters. The temperature required is the static air temperature (SAT), which is that for the undisturbed static air, which is not measurable directly. Air brought totally to rest from the moving airflow will be heated by the adiabatic compression to a value known as the total air temperature (TAT) and the relationship between this quantity and the SAT is described by [4]:

$$TAT = SAT\left(1 + \frac{\gamma - 1}{2}Ma^2\right) \tag{2.37}$$

where γ can be taken to be 1.4 for air, which is mainly composed of diatomic molecules. Thus, Equation (2.37) simplifies to:

$$TAT = SAT\left(1 + 0.2Ma^2\right) \tag{2.38}$$

In a real probe the temperature rise may not reach the full TAT, for example, the air may not be brought completely to rest, so Equation (2.38) is normally written:

$$TAT = SAT\left(1 + 0.2K_rMa^2\right) \tag{2.39}$$

where K_r is an empirical correction known as the *recovery factor* and the quantity $0.2K_rMa^2$ is known as the *ram rise*. The actual temperature recorded in the probe is known as the ram air temperature (RAT) where $RAT = K_r \times TAT$. Assuming $K_r = 1$, it is evident that the difference between SAT and TAT for Mach numbers below 0.2 is less than 1% and can be neglected. For an airliner operating at its maximum Mach number, however ($Ma \sim 0.8$), the ram rise is 13%.

Beermat Calculation 2.4

A Boeing 737 is cruising at its maximum Mach number (M_{MO}) of 0.84 at 32 000 ft under ISA conditions. What is the air temperature rise in the OAT probe measured in Kelvin if the recovery factor is 1?

32 000 ft = 9754 m, so under ISA conditions the OAT in K is given by 288–9754 × 0.0065 K = 224.6 K. This is the value of the SAT, thus from Equation (2.39):

TAT = 224.6(1 + 0.2 × 1 × 0.84²) = 256.3 K

So, the temperature rise is 256.3–224.6 = 31.7 K.

2.12.2 Direct-Reading Thermometer for Low Airspeeds

In low-speed general aviation aircraft, which travel at speeds below $Ma = 0.2$ (less than ~130 knots at sea level), the ram rise can be neglected and these aircraft are usually fitted with a simple direct-reading probe whose sensing element is a bimetallic coil illustrated in Figure 2.29a. The coil is exposed to the airflow within a housing and temperature changes cause the coil to wind or unwind directly driving a needle on a display. Figure 2.29b shows a typical probe installation on a light aircraft with the air inlet tube mounted through the front cockpit window.

2.12.3 Resistance Thermometer Probes

The most common temperature probes used in airliners are based on platinum or nickel resistance thermometers, which provide accurate temperature measurements over a wide range. Figure 2.30a shows the variation in the resistivity of Pt and Ni as a function of temperature in the range –80 to 1000 °C [5].

(a)   (b)

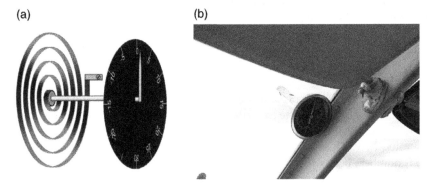

Figure 2.29 (a) Bi-metallic coil used as a temperature sensor in a direct reading OAT probe. (b) Typical installation in a light aircraft.

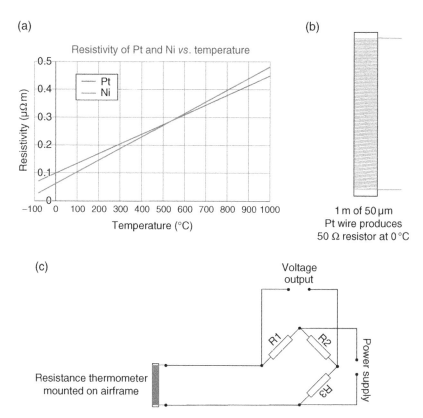

Figure 2.30 (a) Resistivity of Pt and Ni wire vs. temperature. (b) Typical dimensions of wire wound on a former to produce a temperature-sensing element. (c) Bridge circuit used to measure resistance changes.

These curves are very reproducible partly because both metals are easy to refine and high-purity material is readily available. Since both metals are quite good conductors (note the resistivity scale is μΩ m), producing an easily measurable resistance requires long lengths of very thin wires, which are wound on an insulating former (Figure 2.30b). For example, a 1 m length of 50 μm diameter Pt wire produces a resistance of about 50 Ω and the equivalent for Ni would be about 30 Ω. In older temperature probes, the resistance change drove a direct-reading indicator by changing the magnetic deflection of a coil in a magnetic field, a system known as a *ratiometer*. In more recent systems the resistance change is measured using a bridge-type circuit illustrated in Figure 2.30c and the signal passed to the ADC for processing.

The earliest airframe mountings for temperature-sensing elements (flush bulbs) were mounted against the aircraft skin and effectively measured the skin temperature. The main problem with these is that the recovery factor is uncertain and so obtaining SAT from the measured temperature was prone to error. Since the 1960s, the design of the housing has stabilized around the type shown as a simplified schematic in Figure 2.31a and installed on Boeing 737 as shown in Figure 2.31b. The housing resembles a pitot tube and brings the air

(a)

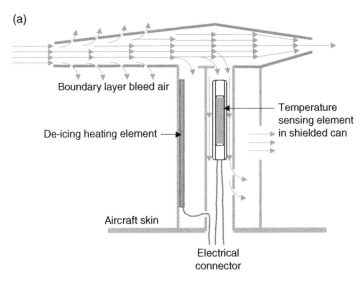

Boundary layer bleed air

Temperature
sensing element
in shielded can

De-icing heating element

Aircraft skin

Electrical
connector

(b)

Figure 2.31 (a) Simplified schematic of a total temperature probe. Air is brought to rest around the shielded temperature probe giving a recovery factor close to 1. The 90° deviation of the air tends to remove water droplets, which pass straight through. The air bleed holes in the outer casing remove boundary layer air heated by the de-icing element so that only the unheated core flow surrounds the temperature sensor. (b) Total temperature probe on a Boeing 737. *Source:* Reproduced from www.b737.org.uk.

around the sensing element to rest so that the recovery factor is close to 1. For this reason, the device is often known as a total temperature probe. To reach the temperature-sensing element, the airflow is diverted through 90° and so water droplets are separated and pass straight through. The pressure within the tube is higher than the surrounding air and the holes in the outer casing allow boundary air to escape. This is so that when the de-icing heater is on, only the unheated core flow can reach the temperature sensor.

2.13 Pitot–Static Systems

The pitot–static system supplies the data required for the altimeter, vertical speed indicator, ASI, and Mach meter to operate as well as information required for integrated systems (see Chapter 9). In smaller and older aircraft, the system consists of the pitot and static ports providing the respective pressures via tubing into the direct-reading instruments in the cockpit and Figure 2.32 shows a schematic of the system in a light aircraft.

In all but the simplest training aircraft there are two static ports, one on each side of the fuselage, which is useful for averaging out errors in pressure produced by maneuvers. For example, if the aircraft is yawing it will increase the static pressure on one side and decrease it on the other so that the common static line will stay at about the same pressure. The pitot and static lines are brought into the cockpit and passed to the various pressure instruments as shown. There will also be an alternate static source within the cockpit in case the external sources are blocked. In an unpressurized aircraft, the static pressure in the cockpit is slightly lower than that provided by the external ports due to the moving air past the cockpit. So, if the alternate static source is used, the altimeter will read a higher altitude than the correct one and the ASI will also read high, both of which are undesirable for safety.

With the move to glass cockpits, digital information is required for the PFD and so the static and pitot pressures are measured by transducers and digital signals are passed to the flight management system (FMS) without the pressure lines themselves passing to the instruments. An intermediate step was to connect the pressure lines into an ADC, as shown in Figure 2.33a. The ADC measures the pressures by internal transducers and with the OAT as an additional input carries out all the necessary computations to pass the required values of IAS, TAS, altitude, vertical speed, and Mach number to the FMS. This is the system found on a Boeing 757, which is illustrated schematically in Figure 2.33b. Redundancy is built in by having two ADCs driven by independent static lines and interconnected pitot lines so that three out of four pitot tubes and three out of four static ports can fail and the system will still provide valid inputs into one ADC. In addition, the pitot line and an alternate static line are connected to a traditional direct-reading altimeter and ASI so that these instruments are available in the case of total power failure.

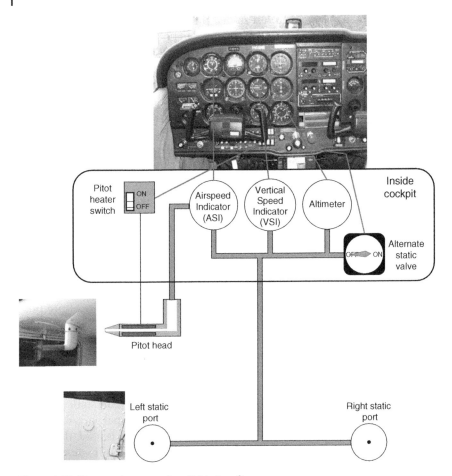

Figure 2.32 Pitot–static system in a light aircraft.

More modern pitot–static systems have transducers within the pitot and static port sources that pass digital information direct to an air data inertial reference unit (ADIRU) that processes the data along with the navigational information from the inertial reference system that drives the FMS (see Chapter 9). A pitot–static system of this type is found in the Airbus A320 series and is shown schematically in Figure 2.34. In this case there is a pitot and static line from the standby pitot head and static ports that are passed to a direct-reading altimeter and ASI in case of total power failure, but from the rest of the ports and probes only electrical signals are passed to the ADIRUs.

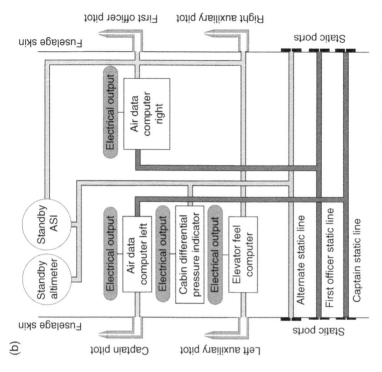

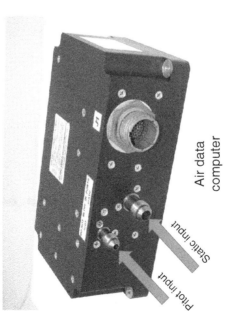

Figure 2.33 (a) Air data computer. *Source*: Reprinted from Wikipedia.org under Creative Commons license 3.0 [7]. (b) Schematic of a pitot-static system in a Boeing 757.

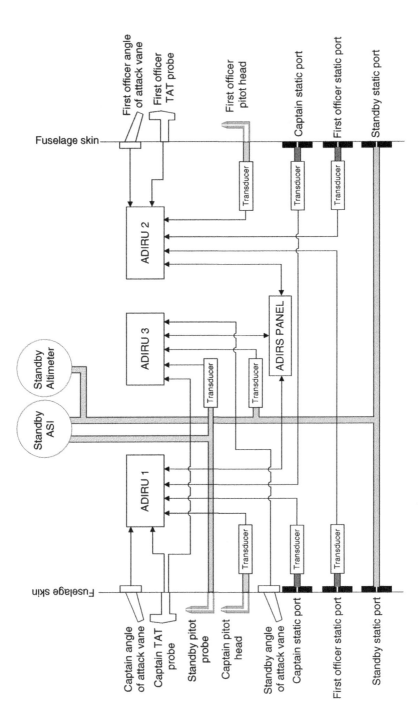

Figure 2.34 Pitot–static system in an Airbus A320 series.

2.14   Air Data Computer (ADC)

The ADC is either as a stand-alone system as shown in Figure 2.33a or is part of an ADIRU (Figure 2.34). The measured quantities are the total (or pitot) pressure in the pitot heads, p_t, the static pressure in the static ports, p_s, and the TATs in the OAT probes, T_t. From these quantities must be derived, as a minimum, the altitude, vertical speed, TAS (denoted by v), IAS, and Mach number (denoted by Ma). How these quantities are derived and indicated in direct-reading instruments was described in Sections 2.8–2.11, but in an aircraft with an ADC they are calculated directly from p_t, p_s, and T_t using the following procedures.

2.14.1   Altitude and Vertical Speed

As with direct-reading altimeters the altitude is determined only from the measured value of p_s assuming the ISA. Thus, Equation (2.7) (below 11 km) and Equation (2.9) (above 11 km) can be used to convert p_s to altitude or the data displayed in Figure 2.4 can be used in a look-up table and an interpolation algorithm used. Since the OAT is available to the ADC, it is possible to calculate a correction to the ISA-derived altitude for conditions on the day but this is not presented in the PFD so that all altitude displays in aircraft, whether traditional or glass cockpits are using the same scale. Once obstacle clearance is assured, the most important consideration is the altitude of aircraft relative to each other. The VSI is obtained simply by numerically differentiating the altitude with respect to time.

2.14.2   TAS and Mach number in Compressible Flow

In previous sections, Equation (2.20) was used as the relationship between dynamic pressure, given by $p_t - p_s$, and v, however, this assumes incompressible flow. In general, moving air tends to "get out of the way" rather than compress in response to an obstacle but as v approaches the LSS, this becomes increasingly difficult and compression occurs, resulting in a density that varies as a function of position. The relationship between p_s, p_t, and v in the case of compressible flow is given by [6]:

$$p_t = p_s \left[1 + \left(\frac{\gamma - 1}{\gamma} \right) \frac{\rho v^2}{2 p_s} \right]^{\frac{\gamma}{\gamma - 1}} \tag{2.40}$$

where ρ is the free-stream air density and γ is the adiabatic index that can be taken to be 1.4 for the Earth's atmosphere. The pitot (total) pressure as a

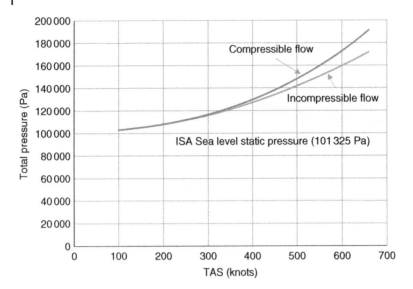

Figure 2.35 Total pressure as a function of TAS calculated for compressible flow (Equation 2.40) and incompressible flow (Equation 2.20).

function of v at ISA sea level ($p_s = 101\,325\,\text{Pa}$) given by Equation (2.20) for incompressible flow and Equation (2.40) for compressible flow is plotted in Figure 2.35. It is evident that the effect of compressibility starts to become significant above about 350 knots.

In principle, v could be obtained from the measured values of p_t and p_s using Equation (2.40) and by obtaining a value for ρ from the altitude, determined by the static air pressure measurement and Equations (2.12) and (2.13). This, however, assumes ISA conditions and it is preferable to obtain v entirely from the directly measured quantities p_t, p_s, and T_t, which is done as follows. The Mach number is given by [6]:

$$Ma = \sqrt{\left(\frac{2}{\gamma-1}\right)\left[\left(\frac{p_t}{p_s}\right)^{\frac{\gamma-1}{\gamma}} - 1\right]} \qquad (2.41)$$

which can be computed knowing only p_t and p_s and displayed on the PFD. The Mach number is defined by:

$$Ma = \frac{v}{a} \qquad (2.42)$$

where a is the LSS given by $a = \sqrt{\gamma R T_s/M}$ (see Equation (2.34)), R is the gas constant (= 8.314 J K^{-1} mol), M is the molar mass (in kg mol^{-1}), and T_s is the static temperature in Kelvin. From Equations (2.34) and (2.40):

$$v = Ma \times a = Ma\sqrt{\frac{\gamma R T_s}{M}} \tag{2.43}$$

For the mixture of gases that constitute air, $M = 0.028965$ kg mol^{-1} and converting v in Equation (2.41) from m s^{-1} to knots gives the following equation:

$$v = 38.967 Ma\sqrt{T_s} \tag{2.44}$$

The relationship between T_s and T_t (the measured quantity) was given in Equation (2.38), that is, $T_t = T_s(1 + 0.2K_rMa^2)$, where K_r is the probe recovery factor (close to 1.0 for modern probes). Thus, Equation (2.42) becomes:

$$v = \frac{38.967 Ma\sqrt{T_t}}{\sqrt{(1 + 0.2K_rMa^2)}} \tag{2.45}$$

Thus, v is determined entirely by the direct measurement of p_t, p_s, and T_t. This is the TAS and the IAS is obtained in the same way as for a direct-reading ASI, that is, converting the dynamic pressure, given by p_t-p_s, to a speed without considering the temperature so that IAS = TAS at sea level.

2.14.3 ADC Inputs and Outputs

As a minimum, the ADC has p_t, p_s, and T_t as inputs from the pitot and static sources and the TAT probe, but modern systems also have inputs from angle of attack vanes and are part of an ADIRU so that full three-dimensional navigation information is available from one unit. Older units have pressure lines passed into the ADC with the transducers converting the pressure to an electrical signal internal to the unit (Figure 2.33a), whereas on new aircraft the transducers are within the pitot and static sources so that only electrical signals are passed into the ADC. The transducers use piezoelectric or capacitive sensors and in either case the signal needs to be amplified and corrected for the nonlinearity of the transducer output as a function of pressure. This is done by a signal-conditioning unit within the ADC using a stored lookup calibration table. The output from the OAT probe is already an electrical signal as described in Section 2.12.3. A simplified schematic of the computations carried out in an ADC is shown in Figure 2.36. Note that in some aircraft the pitot head and angle of attack vane are combined as shown in Figure 2.9f.

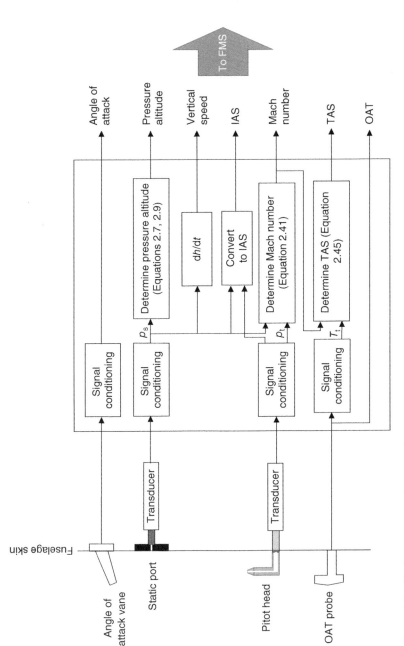

Figure 2.36 Air data computer inputs and outputs.

Problems

1 An aircraft is flying at an altitude of 36 000 ft and its OAT probe records a SAT of –50 °C. What is the density of the air compared to ISA density at the same altitude?

2 On a particular day, the Air Traffic Control at an aerodrome with an elevation of 440 ft reports a QFE value of 990 hPa to pilots. What will it give as the corresponding value of QNH?

3 Determine the 4-digit octal Gillham code allocated to a FL of 360 (36 000 ft above ISA sea level) by an encoding altimeter.

4 An aircraft flying at 32 000 ft under ISA conditions has an IAS of 400 knots. What is its TAS? (Assume incompressible flow).

5 If an aircraft takes off from an airport at sea level at a TAS of 100 knots, what TAS does it need to go to takeoff from an airport 2000 m above sea level (assume ISA conditions for both). What is the IAS in both situations?

6 A supersonic aircraft is flying at Mach 2.2 at an altitude of 55 000 ft under ISA conditions. What is the TAT measured by its temperature probe assuming a recovery factor of 1?

7 An airliner has a maximum Mach number (M_{MO}) of 0.81 and is climbing under ISA conditions at 440 knots. At what altitude does it reach the maximum Mach number? (Hint: this is best solved graphically by plotting the TAS and speed of $Ma = 0.81$ as a function of altitude).

8 A military jet is flying at low level under ISA conditions. Its static probe measures a pressure of 1005 hPa and its pitot head measures a total pressure of 1206 hPa. What is the TAS of the jet?

References

1 Clifford Matthews (2002). *The Aeronautical Engineer's Data Book*, 56. Butterworth-Heinemann.
2 Wuest, W. (1980). AGARD Flight Test Instrumentation Series. Volume 11 on Pressure and Flow Measurement, AGARD-AG-160.
3 Pallett, E.H.J. (1981). *Aircraft Instruments, Longman Scientific and Technical*. Pearson.

4 Lautrup, B. (2011). Chapter 14. In: *The Physics of Continuous Matter*. CRC Press.

5 Serway, R.A. (1998). *Principles of Physics*, 2e, 602. Fort Worth/London: Saunders College Pub.

6 Ames Research Staff. NACA Report 1135 (1953). http://www.grc.nasa.gov/WWW/K-12/airplane/Images/naca1135.pdf (accessed 12 June 2018).

7 Creative commons (1953). http://creativecommons.org/licenses/by-sa/3.0/legalcode (accessed 12 June 2018).

3

Gyroscopic and Magnetic Instruments

3.1 Mechanical Gyroscopes and Instruments

3.1.1 Basic Properties of Mechanical Gyroscopes

There are two properties of gyroscopes that are utilized in aviation, that is *rigidity* and *precession* (Figure 3.1). Rigidity is the tendency of the spin axis to remain fixed in inertial space and in practice the axis can be taken to be maintained relative to a distant star as illustrated in Figure 3.1a. The rigidity increases with the angular momentum of the gyroscope, which is given by:

$$L = I\omega \tag{3.1}$$

where I is the moment of inertia and ω is the angular velocity. The variables in bold (L, ω) are vector quantities, where the direction of the vectors is along the rotation axis, while I is a scalar.

Precession is the reaction of the gyroscope rotor to a torque, τ, applied perpendicular to the axis of rotation (i.e. perpendicular to L), which results in a rotation about an axis perpendicular to both τ and L (Figure 3.1b). A simple rule to determine which way the axis moves along the precession line is to rotate the applied torque by 90° in the direction of rotation of the rotor as illustrated in Figure 3.1c. If the angular rate of precession is denoted by Ω, then the relationship between, τ, L, and Ω is given by the vector cross product:

$$\tau = \Omega \times L = \Omega \times I\omega \tag{3.2}$$

or, considering only magnitudes

$$\Omega = \frac{\tau}{I\omega} \tag{3.3}$$

From Equation (3.3) it is clear that a high rigidity produces a low rate of precession, so in applications where a sensitive precession measurement is

Aircraft Systems: Instruments, Communications, Navigation, and Control,
First Edition. Chris Binns.
© 2019 John Wiley & Sons, Inc. Published 2019 by John Wiley & Sons, Inc.
Companion website: www.wiley.com/go/binns/aircraft_systems_instru_communi_Navi_control

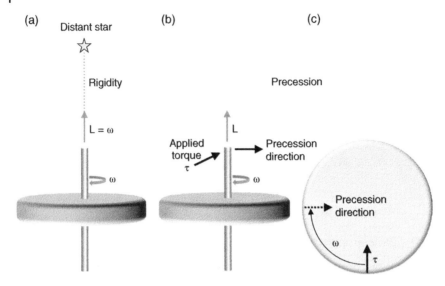

Figure 3.1 (a) The rigidity of a gyroscope axis with respect to inertial space can be represented by the tendency of the spin axis to remain aligned with a distant star. (b) Precession is the reaction to an applied torque and is orthogonal to the angular momentum, **L**, and the applied torque, τ. (c) The precession direction can be found by rotating τ through 90° in the direction of the rotor spin.

required, for example, a rate of turn indicator (Section 3.1.10), a lower rigidity gyro is used than in instruments that require the gyro to maintain a constant spin direction such as the direction indicator (DI) (Section 3.1.6). How rigidity and precession are used in practice to drive instrument displays is described in the following sections.

3.1.2 Gyroscope Wander

Only a mechanically perfect and frictionless gyroscope will maintain its spin axis relative to a distant star and in practice the direction of the axis will change over time due to friction in the bearings, rotor imbalance, etc. which all produce orthogonal torques on the rotor (Figure 3.2). This is described as *true wander*, that is, a real change in the alignment of the axis relative to inertial space. Even in a perfect gyroscope, however, if the gyro is set up to indicate a direction on the Earth such as in the DI (see Section 3.1.6), the fact that the vehicle is moving over a rotating sphere produces an *apparent wander* due to the changing frame of reference relative to inertial space. Apparent wander is also separated into Earth rate, due to the rotation of the Earth and transport wander due to a change

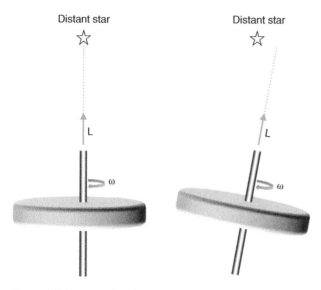

Figure 3.2 True wander of a gyroscope axis due to bearing friction, rotor imbalance, etc.

in the orientation relative to inertial space as the vehicle travels over a sphere. These are discussed in detail in Sections 3.1.7 and 3.1.8.

3.1.3 Labeling of Aircraft Axes and Rotations

In the instrument descriptions below, frequent reference is made to aircraft rotations and so these are defined in this section. There are various choices of Cartesian sets of axes, including an Earth-based system and an air-path system [1], but the set that will be used in this chapter is referred to as the body axis system and is illustrated in Figure 3.3. The axes all pass through the center of gravity of the aircraft and are fixed relative to the airframe thus they rotate with the aircraft. The longitudinal axis is parallel to the fuselage centerline, the lateral axis is orthogonal along the wings, and the vertical axis is orthogonal to both the others. It is assumed that the aircraft is a rigid body and does not buckle or bend during rotations. The rotations about the axes use the same terminology that has been used in ships since antiquity and are also illustrated in Figure 3.3. Roll is a rotation about the longitudinal axis, pitch is a rotation about the lateral axis, and yaw is a rotation about the vertical axis. A given set of pitch roll and yaw values is referred to as the *attitude* of the aircraft. The various instruments show the roll pitch and yaw so that the pilot has a picture of the aircraft attitude without any external reference.

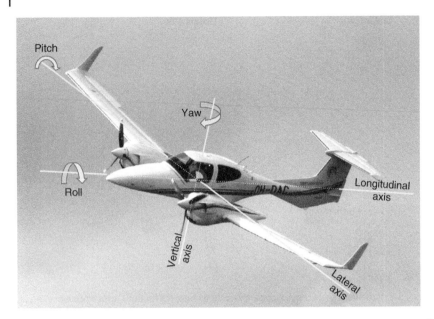

Figure 3.3 Labeling of the body axis system and rotations about the axes. *Source:* Aircraft image from https://en.wikipedia.org/wiki/Diamond_DA42 (public domain).

3.1.4 Types of Gyroscope

The generic types of gyroscope used in aircraft instruments are shown in Figure 3.4. The initial classification is into displacement gyros, which use the rigidity of the gyroscope axis to measure the change in angle of an aircraft axis and rate gyros that measure the rate of rotation about an aircraft axis. Displacement gyros are further classified into space gyros and tied gyros. The rotation axes in space gyros are not controlled in any way and maintain a constant direction relative to absolute space (i.e. the distant stars). These are used in Inertial Navigation Systems (INS – see Chapter 9). Tied gyros have the rotation axis controlled to maintain a specific direction, for example, normal to the surface of the Earth as in the case of an Earth gyro, which is used in the attitude indicator (AI) as described in Section 3.1.9.

3.1.5 Power for Gyroscopic Instruments

Power to spin a gyroscope is either produced by an electric motor or derived from air pressure directed against vanes on the circumference of the rotor. The most common sort of pressure drive uses an engine-driven vacuum pump to evacuate the case of the instrument so that air at atmospheric pressure,

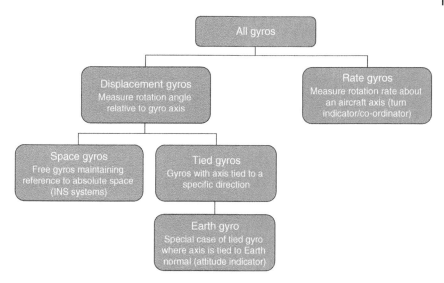

Figure 3.4 Generic types of gyroscope used in aircraft instruments.

supplied through an air filter, is drawn through a nozzle directed against the vanes as illustrated in Figure 3.5.

A typical system, depicted in Figure 3.6, also contains a simple regulator to maintain a roughly constant vacuum as the engine rpm is varied. Originally all gyro instruments were air-driven but modern light aircraft use a mixture of electrically-driven and air-driven instruments for redundancy and pilots are trained to fly safely using exclusively the air-driven or electrically-driven instruments. The normal setup has the AI (artificial horizon) and DI (or heading indicator, HI) driven by air pressure and the turn indicator electrically-driven. These individual instruments are described below. In a commercial airliner, the gyro instruments are all electrically-driven as these aircraft have a high level of redundancy of electrical power generation, including multiple generators, emergency battery supply, and ram air turbines that can be lowered into the air flow to supply power. More recently in commercial aircraft, mechanical rotating gyroscopes have been replaced by solid state MEMS and laser-based systems (see section 3.2).

3.1.6 Direction Indicator (DI)

The DI, also called a HI, illustrated in Figure 3.7, has a gyro with its axis in the horizontal plane in a double-gimbal assembly (two degrees of freedom) so that it can rotate freely around two axes both perpendicular to the spin axis. The cage can be locked (caged) in the horizontal plane and the outer gimbal rotated so that the card displays the indicated compass direction. The gimbal can then be unlocked and the horizontal axis will be maintained so that when the aircraft

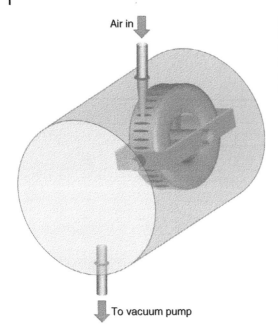

Air in

To vacuum pump

Figure 3.5 Air-driven gyroscopic instrument uses a vacuum pump to evacuate the sealed case so that air is drawn through a nozzle directed against vanes on the circumference of the gyroscope.

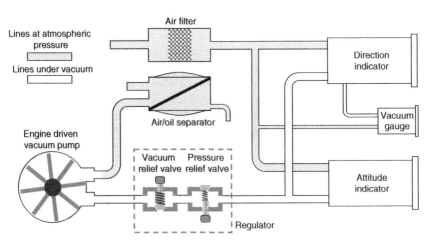

Figure 3.6 Typical system to drive air-powered gyroscopic instruments in a light aircraft.

turns, a geared drive moves a pointer to indicate the new direction on the card visible in the cockpit. The double-gimbal assembly allows the horizontal axis to be maintained as the aircraft pitches and banks. In Figure 3.7a, the rotor has been caged and set so that the display card indicates North and after the aircraft has yawed by 90°, the rigid rotor axis drives the gear train to rotate the display card by 90° to indicate East.

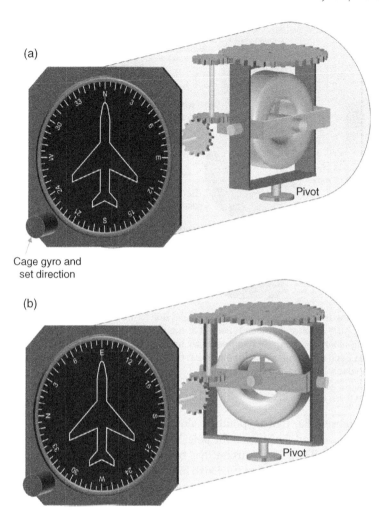

Figure 3.7 Schematic of direction indicator. The gyro axis is horizontal and can pivot inside the case with the two gimbals maintaining the axis direction as the aircraft pitches and rolls. The rotor can be caged and rotated to align it with the compass. (a) The rotor has been rotated so that the display card shows North. (b) After the aircraft has yawed by 90° the gyro axis is maintained and the gearing rotates the compass card to display the new direction.

3.1.7 Earth Rate

In the absence of true wander in a gyroscope, the axis is maintained relative to inertial space but even in a parked aircraft the rotation of the Earth will change the orientation of the aircraft relative to the original axis of the gyro so that the DI will show a steady change in direction. This is referred to as Earth rate and to illustrate the effect consider two parked aircraft with the gyro axis setup as

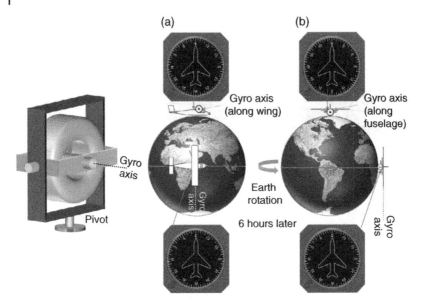

Figure 3.8 Earth rate: the apparent wander of a gyro axis due to the rotation of the Earth at all latitudes except the equator. (a) Aircraft parked at the equator and the pole with the gyro axis pointing along the wing. (b) Alignments six hours later. For the aircraft at the equator there has been no change while for the one at the pole the axis, which has remained fixed relative to space, is now at 90° to the wing.

illustrated in Figure 3.8a, that is, pointing along the wing, with one aircraft parked on the equator and the other at the pole. With this alignment of the gyroscope axis the DI display in both aircraft has been rotated so that it indicates East. Six hours later (Figure 3.8b) as the Earth has rotated through 90°, the gyro axis remains fixed relative to space and so for the aircraft at the pole the axis has rotated by 90° and is pointing along the fuselage while for the one at the equator it has remained along the wing. Thus, the DI on the aircraft at the pole rotates at the same rate as the Earth ($15°\,h^{-1}$) while the DI on the aircraft at the equator continues to point in the same direction. At intermediate latitudes, the rotation rate will depend on the sine of the latitude, so, in general:

$$\text{Earth rate} = \pm\,15° \times \sin\left(\text{latitude}\right)h^{-1} \qquad (3.4)$$

with minus for the northern hemisphere and plus for the southern hemisphere since the Earth rotates in an anticlockwise direction when viewed from above the North pole.

The Earth rate will also be apparent in a flying aircraft though its magnitude may vary due to a changing latitude during the flight. At a specific latitude, the Earth rate is like a precession and can easily be compensated by introducing a

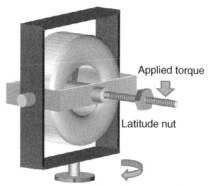

Applied torque

Latitude nut

Resulting precession
(opposite of Earth rate)

suitable torque perpendicular to the spin axis, whose magnitude is given by
Equation (3.2), which generates a precession opposite to the Earth rate.
A simple way to achieve this is by using a "latitude nut" illustrated in
Figure 3.9, which can be screwed in or out to change the torque on the spin axis.
This is used on small aircraft that tend to normally stay within a small range of
latitudes and the torque nut is set to counter the Earth rate at the airfield at
which they are hangared. On large aircraft that traverse the globe and also
on more complex small aircraft, the DI is slaved to a magnetic sensor that keeps
it aligned with the compass direction as described in Section 3.3.3.

3.1.8 Transport Wander

In addition to the Earth rate, if the aircraft is moving, the curvature of the Earth
will produce an apparent change in the axis leading to a shift in the DI, which is
known as transport wander and both these effects are known collectively as
apparent wander. The transport wander component of apparent wander is related
to the coordinate system used to describe the positions of points on the Earth's
surface, which is illustrated in Figure 3.10a and described in detail in Chapter 6.

A great circle is any maximum radius circumference centered at the Earth's
center and an example is shown in red in Figure 3.10a. Another example is the
equator, which is a special case of a great circle. Vertical great circles that pass
through the North and South poles are known as meridians and the N–S position on the Earth's surface is given by the angle along a meridian with 0 at the
equator and 90°N or 90°S at the poles. The E–W position is set to 0° for the
meridian passing through Greenwich and specified in degrees E or W from
the Greenwich, or prime, meridian along a minor circle drawn parallel to the
equator, which is known as a parallel. The red point at 40°N, 60°W is shown
as an example. A nautical mile is defined as one minute of arc around a great

circle so, for example, an aircraft traveling at 360 knots along a great circle is moving 6° of arc per hour.

Now consider two aircraft (Figure 3.10b), one at the equator and one at 60°N, both on the prime meridian, that set out heading directly E using the compass

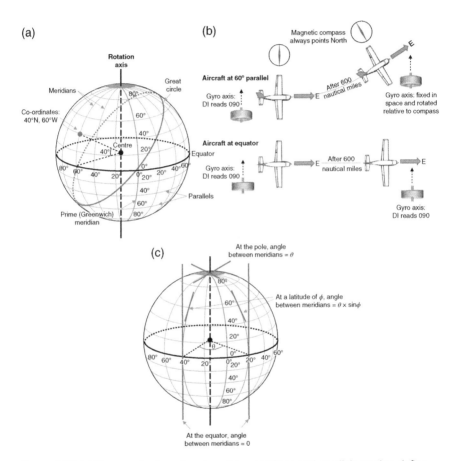

Figure 3.10 (a) The coordinate system based on meridians and parallels used to define points on the Earth's surface. A great circle is a maximum radius circle centered at the Earth's center. Meridians are great circles passing through the North and South poles and are used to define vertical angles (latitudes) from the equator. Parallels are minor circles drawn parallel to the equator and are used to define horizontal angles (longitudes) from the Prime Meridian, which passes through Greenwich. (b) In an aircraft setting out East along the equator, an un-slaved DI gyroscope axis maintains the same direction in space, which always coincides with the compass direction. For an aircraft at a latitude of 60°N, the gyroscope axis remaining fixed in space causes it to develop a misalignment with the compass direction. (c) Transport wander is the angle developed between the starting and current meridians. At the equator the angle between meridians is always zero and at the pole it is given by $\Delta(\text{longitude}) = \theta$. At a latitude ϕ it is given by $\theta \times \sin \phi$.

with an initial heading 090 on the DI, which in this example is not slaved to the compass. After traveling 600 nautical miles (10° of arc along a great circle) the aircraft at the equator will be at coordinates 00N, 010E and still traveling directly E according to the DI. The aircraft at 60°N, however, since the meridians are closer together, will have a greater longitude. Its E–W angle will have changed by 10/cos(60°) and its position will be 60N 020E. In addition, since the DI gyro axis has maintained the original direction in space, the DI will now be reading a heading greater than 090 due to transport wander. Note that in a calculation of DI readings both transport wander and Earth rate need to be taken into account (see Beermat Calculation 3.1 for the calculation of the actual reading for the example shown in Figure 3.10b).

Essentially, transport wander is due to the changing angle of the meridians relative to an axis that is fixed in space. So, the discrepancy between the compass, which always points to the North pole, and an un-slaved DI gyro axis is equal to the angle between the starting meridian and the current meridian on the journey. As shown in Figure 3.10c, the angle between meridians at the equator is always zero, at the pole it is the angular difference, θ ($=\Delta$(longitude)), between the meridians and at a latitude, ϕ, is given by:

$$\text{Angle between meridians} = \Delta(\text{longitude}) \times \sin(\text{latitude}) = \theta \times \sin\phi$$

$$(3.5)$$

For a given distance, d_E, traveled in nautical miles East or West, the change in longitude is given by:

$$\Delta(\text{longitude}) = \frac{d_E}{60} \times \frac{1}{\cos\phi} \tag{3.6}$$

So combining Equations (3.5) and (3.6) gives:

$$\text{Angle between meridians} (= \text{transport wander}) = \pm\frac{d_E \tan\phi}{60} \tag{3.7}$$

where plus is for traveling East and minus is for traveling West. If the aircraft is traveling with an Easterly ground velocity, v_E, (in knots) the rate of transport wander in degrees per hour is given by:

$$\text{Transport wander}(\text{degrees h}^{-1}) = \frac{v_E \tan\phi}{60} \tag{3.8}$$

Beermat Calculation 3.1

Calculate the reading on an un-slaved DI for the aircraft at a latitude of 60°N shown in Figure 3.10b after it has traveled due East, according to the compass, for 600 nautical miles assuming it has a ground speed of 300 knots.

In any calculation of DI readings, both transport wander and Earth rate need to be taken into account. Starting with transport wander, since we have the speed, Equation (3.8) can be used giving a transport wander in degrees per hour of:

$300 \times \tan(60°)/60 = +8.7° \, \text{h}^{-1}$ (plus for Easterly direction)

The Earth rate is given by Equation (3.4) and is:

$-15 \times \sin(60°) = -13.0° \, \text{h}^{-1}$ (minus for Northern hemisphere)

So the total wander is $-4.3° \, \text{h}^{-1}$ and after two hours the DI reading is $90 - 8.6 = 081$.

It is clear from Equation (3.7) that transport wander is zero at the equator and becomes infinite at the poles as, directly at the pole, any movement will cross all meridians at once. Transport wander is also zero for an aircraft traveling directly North or South. For an aircraft traveling in a general direction, the transport wander needs to be continuously calculated and integrated as is done in an inertial guidance system (see Section 9.2.3).

On a small training aircraft, the DI is only compensated for the Earth rate while the true wander, transport wander, and residual Earth rate produced by a change in latitude during flight will all cause the DI to drift with time relative to the compass heading. In this case the only remedy is to check the DI against the compass regularly and if necessary, cage the DI and reset it to the compass heading. Since all sources of wander are small this is accurate enough for most purposes. In more complex aircraft, a magnetic sensor is used to keep the DI synchronized to the compass and this is discussed after considering terrestrial magnetism and the magnetic compass in Section 3.3.3.

3.1.9 Attitude Indicator (AI)

The AI or artificial horizon shows the aircraft bank and pitch angles relative to the Earth's horizon and is particularly important to the pilot when flying without external visual references. It spins in a horizontal plane as indicated in Figure 3.11 and is actively maintained in that plane by feedback mechanisms so the rotor plane is always parallel to the Earth's surface, that is, the spin axis is always normal to the ground. It is thus a tied gyro, that is, it is tied to an external reference and since the external reference is the Earth's surface, it is a special case of a tied gyro known as an Earth gyro.

The feedback mechanism that maintains the external reference depends on how the gyro is powered but in the case of an air driven instrument, the method illustrated in Figure 3.11 can be used. The outer case is attached to the vacuum

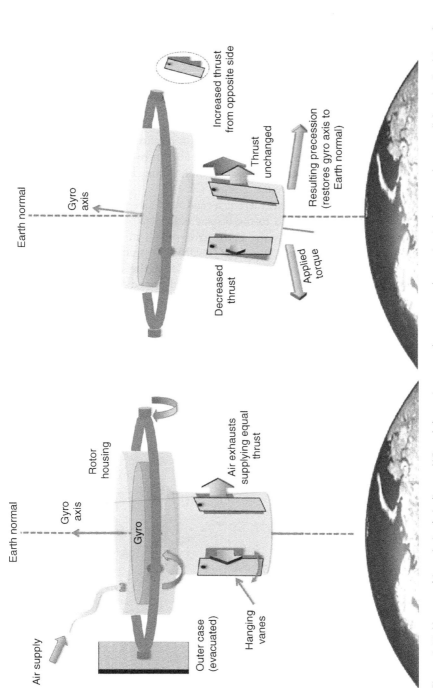

Figure 3.11 Vacuum-driven Attitude Indicator (AI) and the active control system used to maintain the gyro plane parallel to the Earth's surface, i.e. the gyro axis parallel to the Earth normal, providing an artificial horizon.

pump so that air is drawn through the gyroscope housing to spin the gyro as illustrated in Figure 3.5 and is exhausted through four ports in a chamber suspended under the gyro housing. This chamber acts as a pendulum providing coarse passive Earth alignment. The active Earth alignment is provided by hanging vanes that partially cover the exhaust ports. When the gyro is Earth aligned, air exits at the same rate from all four ports providing zero net torque on the gyro. If the gyroscope moves away from Earth alignment, the hanging vanes provide an uneven thrust from the exhaust ports and as shown in the diagram the net torque is naturally at 90° to the correction required so that the resulting precession is in the correct sense. The system also corrects for any true wander of the gyroscope axis and an elegant aspect of the design is that it is gravity itself that provides the Earth reference.

Thus, within the instrument there is a rigid reference to the actual horizon and the connectivity through to the display is illustrated in Figure 3.12a. As the aircraft pitches and rolls the Earth gyro, which in the example shown Figure 3.12 is air driven, uses the feedback system demonstrated in Figure 3.11 to maintain the horizon reference. If the aircraft pitches, the horizon bar moves a curved plate with blue (sky) and brown (Earth) areas above and below the horizon bar up and down relative to an aircraft model fixed to the instrument case. If the aircraft rolls, the roll indicator attached to the roll gimbal moves around a scale attached to the instrument case to indicate the angle of roll. A typical AI display, illustrated in Figure 3.12b, has roll angle marks at 10°, 20°, 30°, 45°, 60°, and 90° and pitch angle marks every 5°.

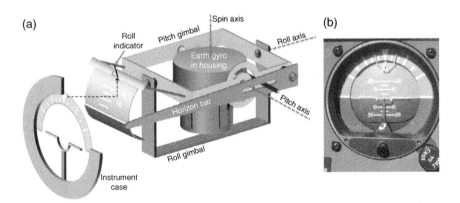

Figure 3.12 (a) System for transmitting the horizon reference provided by the gyro through to the display via the gimbals in an AI. (b) Typical AI display.

(a) 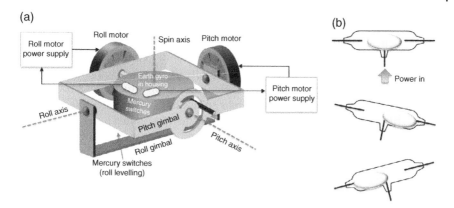 (b)

Figure 3.13 (a) Leveling system for electrically powered AI. The mercury switches detect any tilt and switch in torque motors to correct. The pitch torque motor acts on the roll axis so that precession corrects pitch while the roll torque motor acts on the pitch axis. (b) Three-pole mercury switch that can switch power to the torque motor to provide torque in either direction depending on the direction of tilt.

Air-powered AIs lend themselves to the elegant feedback mechanism to keep the gyro axis aligned to Earth normal but disadvantages include the lower rotation speed produced, which limits the rigidity of the gyro as shown by Equation (3.1). In addition, the drive sucks particles through the instrument, despite filtering, and the resulting wear on the bearings produces true wander. These disadvantages are overcome by using electrically-driven gyros, which can spin at rates up to 22 500 rpm or about twice as fast as air-driven instruments. The gyro still needs to be tied to the Earth reference and one scheme, illustrated in Figure 3.13, uses mercury level switches on the gyro housing to sense any movement of the axis away from Earth normal.

If the gyro axis moves away from Earth normal due to pitch, the corresponding mercury switch will provide power to the pitch motor, which will subject the gyro to a torque about the roll axis. The orthogonal precession will then cause the gyro to adjust its pitch till the plane of the gyro is again parallel to the Earth's surface and the mercury switch will disconnect the power. Similarly, movement away from the Earth normal due to roll is corrected by the roll motor producing a torque about the pitch axis. As shown in Figure 3.13b, a three-pole mercury switch can be used to switch power to produce torque in either direction. A complication with this control system is that acceleration and deceleration as well as unbalanced turns (see next section) can produce false correction signals. This is remedied by disabling the switches for detected accelerations >0.18 g and bank angles >10°. Note that the mechanical drive to the cockpit display (not shown for clarity) can be the same as that used in air-driven instruments and illustrated in Figure 3.12.

3.1.10 Turn and Slip Indicator and Turn Coordinator

The rate at which the aircraft is yawing (turning) is an important quantity for flight and this is determined by a turn indicator or turn coordinator using a *rate gyro*, which measures the rate of precession. The simplest instrument is the turn indicator, nowadays only found on vintage aircraft, in which the gyro is gimbaled as illustrated in Figure 3.14a. Apart from the spin axis, the gyro can only rotate about the longitudinal axis but the movement is limited by a spring. If the aircraft pitches, the gyro is rotated about the spin axis so there is no effect and if it rolls, then the precession would be about the vertical axis (Figure 3.14b), which is prevented by the gimbal, so again there is no effect. If the aircraft yaws,

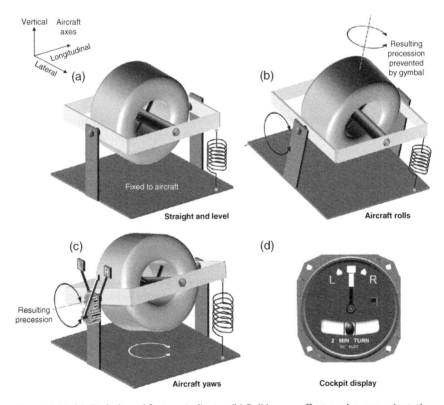

Figure 3.14 (a) Gimbal used for turn indicator. (b) Roll has no effect on the gyro orientation relative to the instrument case. (c) Yaw produces precession about the longitudinal axis, which, due to the spring, produces a deflection proportional to the rate of yaw. (d) Turn indicator display showing Rate 1 or standard turn marks and slip indicator (balance ball). *Source:* Reproduced with permission of http://aviation.stackexchange.com/questions/16534/what-is-the-difference-between-turn-coordinator-and-artificial-horizon. Licensed under CC BY-SA 3.0 [4].

however, the gyro precesses about the longitudinal axis, which due to the spring makes the gimbal adopt a fixed angle with respect to the instrument case (Figure 3.13c), which is proportional to the rate of yaw. A pointer driven by the moving gimbal has to be geared to rotate in the opposite sense as application of the rule to determine the direction of precession (Section 3.1.1) shows that a pointer attached directly to the moving gimbal would show the opposite direction of turn. The cockpit display of a turn indicator is shown in Figure 3.14d and has two marks either side of center. These indicate a turn rate of $3° \text{ s}^{-1}$, which is also known as a Rate 1 or standard turn. The majority of turns executed by commercial aircraft are at Rate 1.

When an aircraft rolls, the lift vector is rotated as shown in Figure 3.15b. The horizontal component of lift will cause a side-slip that produces an angle of attack of the airflow on the tail that will generate a horizontal lift force resulting in a rotation about the vertical axis (yaw). Although this is a secondary effect, it is the normal method used to steer the aircraft. In a *balanced* turn the aircraft speed, v, and radius of the turn, r, are such that they match the centripetal force provided by the horizontal component of the lift, i.e.:

$$L_H = \frac{mv^2}{r} \tag{3.9}$$

The balanced condition in a turn can be maintained with reference to the balance ball or slip indicator visible at the bottom of the instrument shown in Figure 3.14d. This is the simplest of all the aircraft instruments and consists

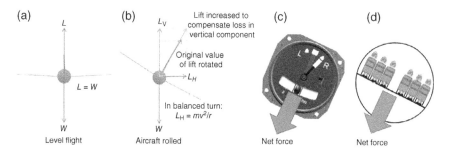

Figure 3.15 (a) In level flight, lift = weight. (b) Rolling rotates the lift vector providing a vertical and horizontal component. To maintain height, the overall lift must be increased to make the vertical component, L_V, equal to the weight. The horizontal component, L_H, provides the centripetal force for the turn. (c) In a balanced turn the vectorial sum of gravity and the centripetal force is towards the floor of the aircraft, which is indicated by the slip indicator (balance ball) staying between the marks on the glass tube. *Source:* Reproduced with permission of http://aviation.stackexchange.com/questions/16534/what-is-the-difference-between-turn-coordinator-and-artificial-horizon. Licensed under CC BY-SA 3.0. (d) In a balanced turn passengers feel a normal gravitational force with no sideways component.

of an agate or steel ball in a curved glass tube filled with a damping fluid. During straight and level flight, gravity makes the ball rest in the lowest part of the tube between the reference marks. During a turn gravity and centripetal force both act on the ball and if the turn is balanced, the vector sum of the forces will act along the normal axis, that is, directly towards the aircraft floor and the ball will remain between the reference marks (Figure 3.15c). Maintaining a balanced turn is important as it generates the least drag and seated or standing passengers feel a normal gravitational force with no sideways component (Figure 3.15d). The radius of the turn can be altered using the rudder and if the bank angle is such that L_H does not provide the appropriate centripetal force, then the turn is unbalanced and is referred to either as a skidding turn (bank angle too small) or a slipping turn (bank angle too large).

One problem with the simple turn indicator described above is that there is only a response after the yaw has developed and the instrument provides no indication of the initial roll before the yaw. A simple remedy is to angle the gimbal by 30° about the lateral axis as shown in Figure 3.16a. This produces a component of yaw during a roll and makes the gyro precess about the longitudinal axis in response to the initial roll rate as well as yaw. Thus, the instrument responds immediately a roll occurs to initiate a turn and maintains a reading as the roll rotation stops and the yaw sets in. If a suitable angle is used (~30°), then during a turn the initial roll produces a similar deflection to the final yaw so the pilot can set up a given rate of turn during the roll. The

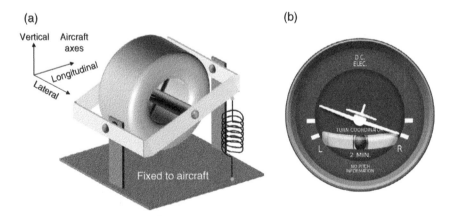

Figure 3.16 (a) In a turn coordinator the gimbal is rotated about the lateral axis by 30° so that both the initial roll rate and yaw produce precession about the longitudinal axis. (b) Turn coordinator display showing Rate 1 or standard turn marks and slip indicator (balance ball). *Source:* Reproduced with permission of http://aviation.stackexchange.com/questions/16534/ what-is-the-difference-between-turn-coordinator-and-artificial-horizon. Licensed under CC BY-SA 3.0 [4].

instrument set up this way is referred to as a turn coordinator and is usually distinguished by having a model aeroplane as the indicator as shown in Figure 3.16b, but still has the Rate 1 or standard rate turn marks. Note the instrument still also has the built-in slip indicator to facilitate balanced turns. The "No pitch information" warning is given to be sure the instrument is not confused with the AI.

3.2 Solid-State Gyroscopes

3.2.1 The Advantages of Solid-State Gyroscopes

Mechanical rotating gyroscopes all suffer from true wander but in the AI and turn coordinator, there is active control of the gyroscope plane of rotation so that it does not affect the operation of the instrument. True wander is much more problematic in INS where the gyroscope is not interfered with after it is set up and absolute rigidity is required for long periods. With high-precision engineering, it is possible to get true wander in a mechanical spinning gyroscope down to $<0.01°$ h^{-1} but only at great expense including environmentally controlled housings for the gyroscope. This has led to the development of solid-state gyro systems based on laser interferometry or micro-electro mechanical systems (MEMS) devices for use in guidance technology. These new types of gyro are smaller, cheaper, and much easier to integrate into glass cockpits. INS is dealt with in Chapter 9, however, the relevant gyroscopes are described here. In all cases, solid-state devices are rate gyros and in order to obtain a rotation angle, θ, from the precession rate, Ω, the output must be integrated with respect to time as in Equation (3.10):

$$\theta = \int \Omega \mathrm{d}t \qquad (3.10)$$

3.2.2 The Sagnac Effect

Essentially a mechanical gyroscope, by being rigid in inertial space, is able to detect rotation. The same property can be achieved optically using a phenomenon known as the *Sagnac Effect* discovered by the French physicist Georges Sagnac in 1913 and illustrated in Figure 3.17a. If two monochromatic light beams are sent in opposite directions round an optical table in a closed loop defined by mirrors as shown, they can both exit at the half-silvered mirror into a detector where the interference between them will produce fringes. If the table rotates, in the time taken for the beams to go around the loop, the half-silvered mirror will have moved and one beam will have its path slightly increased

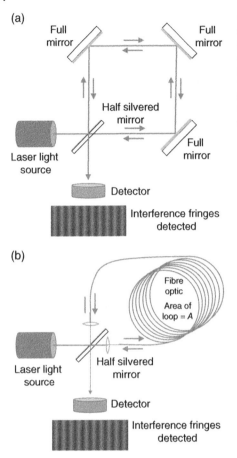

Figure 3.17 (a) Illustration of the Sagnac effect. Two monochromatic light beams propagating in opposite directions around a path defined by mirrors will recombine and interfere. If the optical table rotates, the interference pattern will shift due to the change in path length of the two beams. (b) Fiber-optic gyro based on the Sagnac effect.

relative to the static table and the other will have its path decreased. The phase difference of the two beams will thus change and the fringe pattern will shift. The fringe pattern is unaffected by uniform linear motion, which affects both beams equally and will only be changed by rotation of the table. The rigidity of the system relative to inertial space is given by the invariance of the speed of light for all inertial reference frames.

3.2.3 Fiber-Optic Gyroscope

A fiber-optic can define a loop for a light beam and if the loop is set up as shown in Figure 3.17b, then light beams propagating in both directions around the fiber-optic can be set up. Again, a rotation in the optical table generates a path length difference and thus a change in the relative phase of the two beams, which produces a shift in the interference fringes. It can be shown that the phase shift is given by [2]:

$$\Delta\phi = \frac{8\pi nA\omega}{\lambda c} \tag{3.11}$$

where A is the area of the loop, n is the number of loops of optic fiber, ω is the angular rate of rotation of the ring (with the angle units the same as $\Delta\phi$), λ is the wavelength, and c is the speed of light. From Equation (3.11) it is evident that the phase shift is proportional to the rate of rotation of the loop and that the instrument can be made more sensitive by increasing the number of fiber-optic loops.

3.2.4 Ring Laser Gyroscope

The ring laser gyro also uses the Sagnac effect but this time instead of passing the light into the system from the outside, the counter-propagating beams are set up inside a laser cavity that generates the light (Figure 3.18a). Normally the cavity is triangular and within a glass block that is filled with a mixture of He and Ne that generates red laser light with a wavelength of 633 nm. The gas is excited by setting up a discharge between a cathode and two anodes so that clockwise and anticlockwise propagating laser beams are set up and reflect around the three mirrors in the cavity as shown.

Within a laser cavity, there has to be an integral number of wavelengths and if the device is rotating about an axis perpendicular to its plane, the Sagnac effect will change the effective cavity lengths for the two beams. The result is that one beam will have its wavelength stretched relative to the one in the stationary cavity and the other beam will have its wavelength compressed. This is illustrated in Figure 3.18b for a circular cavity for clarity. As shown below these wavelength shifts are tiny compared to the natural width of the laser line, which is broadened by various processes including the Doppler shift due to the motion of the He and Ne gas atoms. However, the frequency shift of the two beams when they are combined at the detector produces a beat frequency, which can be thought of as the interference pattern in time rather than space. As demonstrated in Beermat Calculation 3.2, beat frequencies are typically a few kHz and easily detected. In order to determine the direction of rotation, one beam is reflected by a prism to produce a collinear beam with the counter-propagating one at the detector. Combining the two thus produces a spatial interference pattern as illustrated and the direction of movement of the fringes gives the sense of rotation of the cavity.

(a)

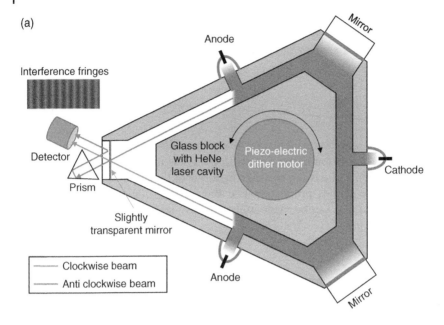

(b)

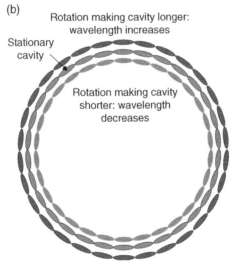

Figure 3.18 (a) In a ring laser gyro the counter-propagating beams are within a laser cavity. (b) Rotation of the cavity will produce a wavelength and frequency difference between the two beams giving a beat frequency at the detector.

Beermat Calculation 3.2

A ring laser gyro that produces laser light with a wavelength of 633 nm with an optical path length of 20 cm is rotated at a rate of 3° s⁻¹ in a plane parallel to the optical path. What is the beat frequency of the combined laser beams at the detector? For simplicity, assume the optical path is circular.

In a circular cavity, a rotation of 360° will produce a path length difference of 20 cm, so a rotation of $3°\ s^{-1}$ will produce a path length difference for one propagating beam that changes by $(3/360) \times$ path length $= 0.2/120\ m\ s^{-1} = 0.00167\ m\ s^{-1}$. So, the wavelength shift is $(0.00167/c) \times \lambda_0$, where c is the speed of light and λ_0 is the wavelength of the laser light in the stationary cavity. The factor $0.00167/c$ will also be the relative change in the frequency, f_0, given by $f_0 = c/\lambda_0 = 4.74 \times 10^{14}$ Hz.

So, the frequency shift is $(0.00167/c) \times 4.74 \times 10^{14} = 2633$ Hz, which is for one-beam only. The other beam will be shifted from f_0 by the same amount in the opposite sense so the beat frequency of the combined beams at the detector is 5.266 kHz.

The relationship between the beat frequency Δf, and the angular rate of rotation of the cavity, ω, is described by [3]:

$$\Delta f = \frac{4A\omega}{\lambda_0 p} \tag{3.12}$$

where A is the area and p is the path length of the cavity. The quantity $4A/\lambda_0 p$ is known as the scale factor, K, and since the quantities are all fixed for a specific cavity, ω is directly proportional to Δf.

One advantage of the ring laser gyro is that zero rotation corresponds to zero beat frequency so the instrument is self-calibrating. This is not so with the fiber-optic gyro, which gives a fringe pattern whatever the rotational state and a calibration procedure is required to determine what fringe pattern corresponds to zero rotation. On the other hand, a problem with the ring laser gyro in its simple form is that it suffers from a phenomenon known as "lock-in" at low rotation rates. This is a general feature of the behavior of coupled oscillators also known as "injection locking" where if the oscillators have only slightly different frequencies and the coupling is sufficiently strong, one oscillator captures the other and locks it to the same frequency. In the case of the ring laser gyro this forces the beat frequency to go to zero so that it does not indicate rotations below a certain rate. The solution is to rotationally dither the laser cavity through a small angle in the plane of the optical path by a piezoelectric motor as illustrated in Figure 3.18. Thus lock-in only occurs where the rotational velocity passes through zero and the induced errors cancel.

3.2.5   Micro-Electromechanical System (MEMS) Gyroscopes

Gyroscopes micromachined into Si known as MEMS are the most recent devices to be used in aircraft instruments and are likely to become dominant in new aircraft in the next few years. They are based on the rigidity of the plane of vibration of an oscillating system so, for example, a perfect frictionless pendulum gimbaled to swing freely will maintain its plane of oscillation relative to inertial space. The French physicist, Léon Foucault, used the effect in 1851 to demonstrate the rotation of the Earth.

Some insects use this characteristic of oscillators to sense rotation during flight by vibrating small appendages known as halteres, shown in Figure 3.19, on a crane fly. The same principle is used in MEMS instruments in which the oscillators are fabricated into Si and the precession caused by rotation is sensed by the change in capacitance of the oscillator relative to a fixed surface. As shown below it is also possible to machine onto the same piece of Si other devices so that an entire attitude and heading reference system (AHRS) can be included on a single chip. It is this sort of device that is normally used in aircraft instruments but the individual sensors on the chip are described below starting with the gyroscopes.

Figure 3.19 Vibrating Halteres on a crane fly that use the rigidity of the plane of oscillation to provide information on rotation during flight.

Figure 3.20 (a) Coriolis force observed acting on a mass, *m*, from within a rotating (non-inertial) reference frame. (b) The normally used "tuning fork" configuration in which the masses oscillate in opposition. Rotating the frame will produce an oscillating coriolis force on each mass with the forces 180° out of phase.

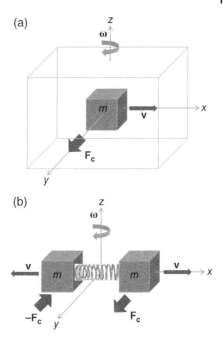

MEMS gyroscopes are based on the Coriolis force, which is often referred to as a pseudo force as it does not arise from any interaction between objects but is an apparent force that is present when a system is being observed from within a non-inertial reference frame. Figure 3.20a shows an example, involving a mass *m* moving with velocity **v** along a given direction. If the reference frame in which the mass moves is rotated at a rate ω then to an observer outside the system, who remains in a stationary (inertial) reference frame, the mass continues to move as before. However, an observer inside the rotating (non-inertial) reference frame sees the mass follow a curved path as if it was subject to a force, $\mathbf{F}_c$, given by:

$$\mathbf{F}_c = -2m\omega \times \mathbf{v} \tag{3.13}$$

The setup normally used is to have two masses oscillating in opposition as shown in Figure 3.20b, which is the so-called tuning fork configuration. In this case, when the frame is rotated about the axis shown, the Coriolis force on each mass will oscillate with the two forces in antiphase in the directions indicated by the arrows.

The movement caused by the Coriolis force is detected by the change in capacitance between the masses and fixed electrodes. A MEMS implementation of a device with an area of 0.24 × 0.37 mm is shown in Figure 3.21a. The masses, outlined in yellow, are held at the inner boundary so that they can flap up and

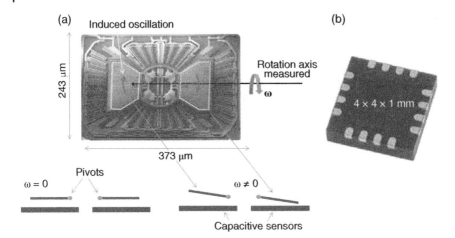

Figure 3.21 (a) MEMS implementation of a gyro to measure rotation about a single axis. (b) Chip package containing three-axis MEMS gyroscope. *Source:* © ST Microelectronics. Used with permission.

down. The direction of the driven oscillation is indicated by the red arrows and by comparison with Figure 3.20b it is clear that rotating the device about the axis shown will cause the two plates to flap as indicated in the bottom part of the figure for one half cycle of the driven oscillation. The motion is picked up by the capacitive sensors and the electrical signal is processed to yield the rate of rotation, ω. Three of these devices placed orthogonally on a chip are used to measure rotations about all three axes and Figure 3.21b shows a three-axis MEMS gyroscope packaged in a chip with dimensions $4 \times 4 \times 1$ mm.

3.2.6 MEMS Accelerometers

Miniature devices that measure linear acceleration can also be micromachined into Si and are usually combined with MEMS gyros on the same chip with the basic principle illustrated in Figure 3.23a. A sense mass attached to two springs is deflected under acceleration and the movement is detected by the change in capacitance between plates attached to the moving mass and fixed plates on the chip. The conversion of the movement into an output voltage signal (Figure 3.22b) is achieved by passing square waves onto the fixed plates with the signal inverted between adjacent plates. Thus, the moving plate picks up zero signal in the center but when deflected, it outputs a square wave whose amplitude is proportional to the deflection from center. The sense of the acceleration is determined by the phase of the detected square wave relative to the driver. The detected signal is amplified and passed into a demodulator that

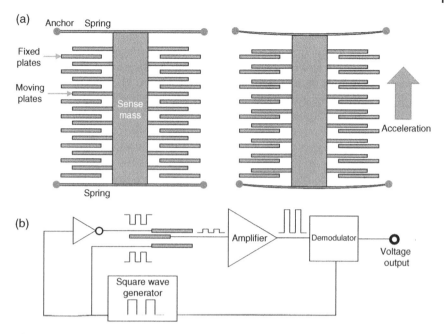

Figure 3.22 (a) MEMS implementation of a device to measure acceleration along a single axis. (b) Electronics used to convert an acceleration into an output voltage.

measures the amplitude at the driven frequency. The output from the demodulator is a voltage that is proportional to the plate deflection, which in turn is proportional to the acceleration. A single sense mass is sensitive to acceleration along one axis only so that three orthogonal devices are required to measure absolute acceleration but as with the other MEMS devices all three can be put onto a piece of Si a few mm across.

3.3 Magnetic Compass

3.3.1 Terrestrial Magnetism

The magnetism of the Earth is a result of the dynamo mechanism, in which an electrically conducting fluid that is rotating and convecting can maintain a magnetic field over astronomical timescales. In the case of the Earth, the conducting fluid is the molten Fe in the outer core of the planet in which the convection is driven by tidal heating and rotation is due to Coriolis forces produced by the Earth's rotation. The mechanism is dynamic and the magnetic field evolves with time so that the magnetic poles move and over geological timescales there are complete reversals of the polarity. At the current time the field is approximately dipolar with the magnetic axis tilted by about 10° relative to the spin axis. The wandering of the magnetic pole is sufficiently slow that magnetic compasses are useful for navigation and the poles of permanent magnets are defined by their behavior with respect to the magnetic poles of the Earth. Thus, North pole on a bar magnet actually means "North-seeking" pole (traditionally colored red) and similarly for the south pole (traditionally colored blue). This means that if the Earth's magnetic field is represented by a bar magnet at the core, it would be the South pole of the magnet that is oriented towards the North pole of the Earth.

Magnetic variation at a specific point on the Earth is the angle between the magnetic pole and true North, that is, the point where the meridians shown in Figure 3.10 converge, which is at the spin axis of the Earth. Variation ranges from 0° (currently in Western Europe) and can reach 180° on the line linking the magnetic and the geographical pole. At central latitudes, variation changes at a rate of about 1° per decade. The magnetic variation and its evolution with time are described in detail in Section 6.2.

Over most of the Earth the flux lines are not parallel to the ground but have a vertical component so that a freely suspended magnetic needle will make an angle to the horizontal known as the angle of dip (Figure 3.23). This depends on the relative strength of the horizontal component **H** (also known as the directive force) and the vertical component, **V**, of the total magnetic field, **M**, and is given by Tan(dip angle) = V/H. The direction-finding information comes

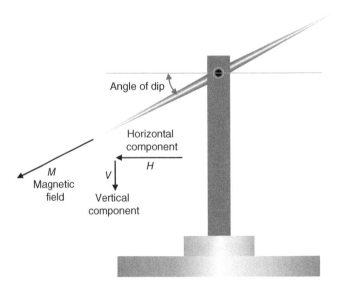

Figure 3.23 Angle of dip.

entirely from **H**, while **V** carries no directional information and also causes some technical problems discussed below. At the poles $H = 0$ and $V = M$ while at the equator $H = M$ and $V = 0$. The magnetic equator (also called the *aclinic* line) is the line joining all points with zero angle of dip and this is shown in Figure 3.24. Broadly, it follows the Earth's equator but there are significant deviations in latitude from zero. The minimum usable directive force to drive a compass is around 6 μT, which limits the use to latitudes below about 80°.

The magnetic field varies with time on different timescales, the most important evolution being the *secular change* in which the magnetic poles circle around the geographical poles with a cycle time of about 960 years, which is the cause of the steady movement in magnetic variation. Other less significant cycles are diurnal, due to the effect of the solar wind on the geomagnetic fields, annual due to the position of the Earth relative to the sun and 11 years corresponding to the sunspot cycle. In addition, there can be temporary changes in the variation for a few hours by up to a few degrees caused by magnetic storms from sunspots.

3.3.2 Direct Indicating Magnetic Compass

All aircraft, large and small, have a direct indicating magnetic compass as this provides a direction reference independent of all other aircraft systems and the internal construction of a direct indicating magnetic compass is shown schematically in Figure 3.25a. It consists of a magnetic sensor, which is usually two needle-shaped permanent magnets attached to a direction-labeled compass card with North being at the position of the North pole of the sensor. At all latitudes apart from the magnetic equator the sensor will try to adopt an angle relative to the horizontal due to the angle of dip. This is minimized by having a pendulous suspension of the magnet and card assembly by a low friction iridium pivot resting in a jeweled cup. The entire case is filled with a low-viscosity liquid (e.g. kerosene), which damps oscillations and lubricates the bearing while a float on top of the assembly reduces the friction further. The casing contains a compressible wall so that changes in pressure are accommodated.

This simple method of reducing the effect of the vertical component of the Earth's field is effective but it produces reading errors if an aircraft accelerates due to the weight vector of the card assembly acting to one side of the pivot axis as illustrated (exaggerated) in Figure 3.25b. Any acceleration or deceleration will cause the card assembly to swing but if the acceleration is to the North or South, since this is the plane of the angle of dip, the swing will not produce a change in reading. In the Northern hemisphere the North pole of the magnet is tilted towards the ground so that the weight is on the North pole side and as illustrated in Figure 3.25c an acceleration to either East or West will generate a torque on the card causing it to rotate so that the indication moves towards North. Similarly, a deceleration along either East or West will cause the compass

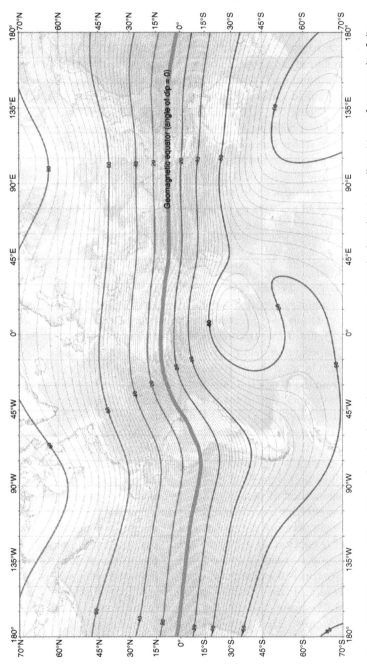

Figure 3.24 Map of spatial variation of angle of dip. The geomagnetic equator is the line that joins all positions of zero angle of dip.

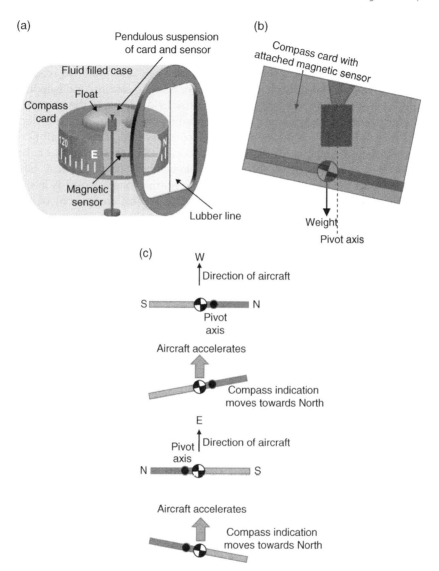

Figure 3.25 (a) Internal construction of a direct indicating magnetic compass.
(b) Pendulous suspension causes the weight of the suspended magnet assembly to be offset from its center of gravity. (c) For an aircraft traveling East or West in the Northern hemisphere an acceleration causes the compass indication to swing towards North and a deceleration produces a swing towards South. The reading errors are reversed in the Southern hemisphere.

reading to swing towards the South. These errors are reversed in the Southern hemisphere since in that case it is the South pole of the magnet that is tilted towards the ground so that the weight is on the South pole side.

The vertical component of the Earth's magnetic field also produces reading errors for an aircraft heading along the North or South quadrants and banking. Figure 3.26a shows the orientation of the Earth's field relative to the magnetic sensor in an aircraft flying straight and level heading directly North and for clarity the magnetic sensor is shown outside the compass card. If the aircraft performs a balanced turn, as described in Section 3.1.10, the resultant of the centripetal and gravitational forces acts along the aircraft vertical (see Figure 3.15) so the effective gravitational force on the compass card is in the same direction as in an aircraft flying level. The vertical component of the Earth's field, however, will have a component along the aircraft lateral axis, which will pull the North-seeking end of the sensor into the turn, that is, displace the reading away from North as illustrated for a bank to the left in Figure 3.26b. Clearly a bank to the right will cause a displacement in the reading away from North in the opposite direction so that for any turn that is moving towards North the reading will be "repelled" away from North and will lag behind the real heading. This is summarized in Figure 3.26c, which also shows the forces on the sensor during a bank when the aircraft is heading directly South. The sensor is still pulled away from reading 180 but the forces are in the opposite sense so that for an aircraft that is turning onto South the reading will be "attracted" towards South producing a value that will lie ahead of the real heading.

The upshot of this, when attempting to turn the aircraft onto a specific heading using only the compass, is that for turning towards the Southern quadrant, the pilot needs to overshoot the required heading by a specific amount before rolling the wings level and the compass will then settle back to the required value. When turning onto a heading in the Northern quadrant the wings need to be rolled level before the compass reading reaches the required heading after which the reading will continue to increase to the required value. For turning onto a heading in the East or West quadrant the roll out can occur when the compass reads the required heading. The turning errors are reversed in the Southern hemisphere since the V component of the Earth's magnetic field acts in the opposite direction on the sensor when the aircraft is banked.

The compass orientation is also sensitive to local aircraft magnetism, which is either classed as soft or hard. Soft magnetism arises from ferrous metal components becoming magnetized by electromagnetic fields from aircraft electrical systems while hard magnetism is any permanent magnetism in metal components. Any misalignment due to aircraft magnetism is termed deviation and in small aircraft is given by a table of corrections on a placard near the compass as shown in Figure 6.10. If any major work is done on an aircraft the compass deviation will be remeasured.

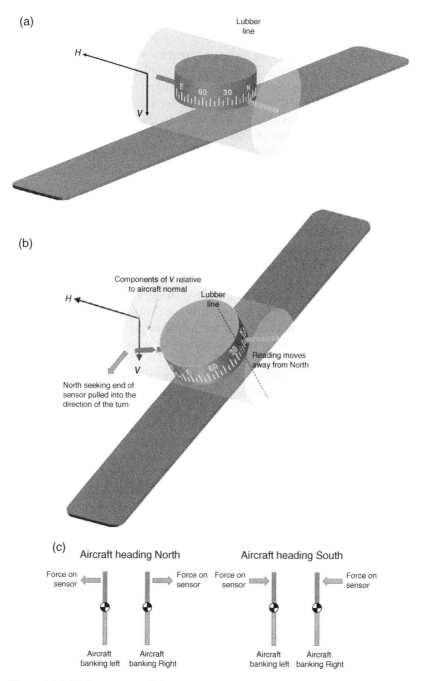

Figure 3.26 (a) Components of the Earth's magnetic field, *H* and *V*, relative to the compass sensor for an aircraft heading North. (b) Components of *V* relative to the aircraft normal axis in a banked aircraft. The component of *V* along the aircraft lateral axis pulls the North-seeking end of the magnetic sensor into the turn producing a shift in the reading away from North. (c) Summary of forces on the sensor in banked aircraft heading North or South.

The power of the direct reading compass is that for straight and level flight, a simple instrument will provide a heading reference independent of all other aircraft systems and will continue to provide a reading in the event of a complete systems failure. The errors introduced by accelerations and turns, however, mean that it is unsuitable as a reference for slaving the DI relative to magnetic North. For this purpose, magnetometers that measure the horizontal component of the Earth's magnetic field and provide an electrical signal proportional to H are used and these are described in the following two sections.

3.3.3 Flux Gate Sensor

Prior to the development of miniature magnetometers on Si chips that can be combined with MEMS gyroscopes (see next section), the most common type of magnetic sensor in use was the flux gate whose operating principle is illustrated in Figure 3.27. It is based on a circular soft iron or permalloy core that is magnetized by a drive coil whose field amplitude is strong enough to drive the core into magnetic saturation in both directions. Since the magnetization is around the loop the average magnetization in one half is in the opposite direction to that in the other half, which is indicated by the blue and red semi-loops in Figure 3.27a.

A sense coil is placed around the permalloy core and drive winding as shown in Figure 3.27b. The drive waveform for the core is shown in Figure 3.27c and the resulting magnetization of the core in the absence of an external field (Figure 3.27d) generates no magnetic flux and no voltage in the sense coil since the flux in the two halves of the core cancel at every point in the cycle. In the presence of an external field, H_{ext}, in the direction shown in Figure 3.27b, the red half of the core reaches saturation earlier than the blue half so for a short period in each cycle the flux is changing as shown in Figure 3.27e and a voltage is induced in the pick-up coil according to Faradays law:

$$\nabla \times E = -\frac{\partial B}{\partial t} \tag{3.14}$$

The flux changes in alternating directions twice per cycle so a series of pulses as shown by the black curve in Figure 3.27f are induced in the pick-up coil. To make this signal easier to amplify and process, it is usually converted to a sinusoidal signal (red curve in Figure 3.27f) by driving a tuned circuit with the pulses.

The system described in Figure 3.27 will only detect an applied field in one direction and the arrangement of core and pick-up coils normally used is shown in Figure 3.28a. The permalloy core is in the form of a ring with three arms connected to a central boss around which is wound the drive coil. Three sense coils are wound on the three arms spaced evenly around the core and comparing the three signals will determine the direction of the applied field. This is illustrated in Figure 3.28b for aircraft flying directly North and directly East. In the former

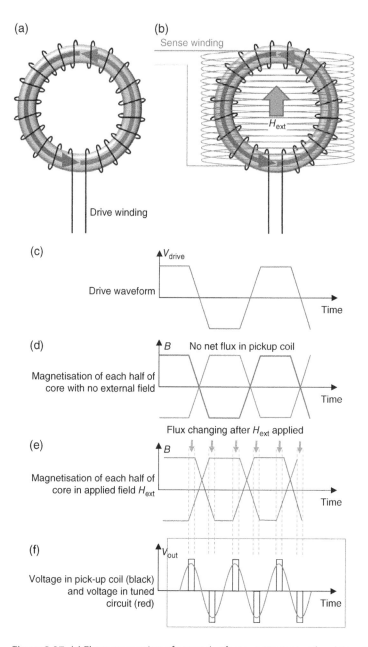

Figure 3.27 (a) Flux gate consists of a round soft magnetic core with a drive winding. (b) Flux changes are sensed with a pick-up coil. (c) The drive waveform (up to 10 kHz) has a sufficiently high amplitude to drive the soft magnetic core into magnetic saturation in both directions. (d) With no external field both halves of the core produce equal and opposite magnetic flux and thus no net flux in the pick-up coil. (e) In an external field (H_{ext}) one half of the core reaches saturation sooner than the other half so for a short period in each cycle the flux is changing. (f) The changing flux induces a voltage pulse in the pick-up coil twice per cycle (black curve). The pulsed output is converted to a sinusoidal waveform (red curve) by driving a tuned circuit.

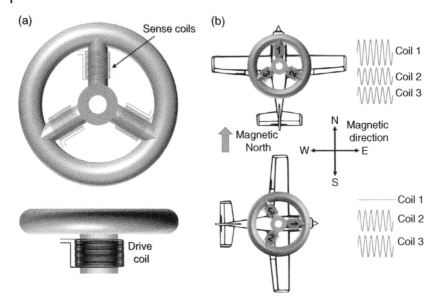

Figure 3.28 (a) Normal configuration of flux gate sensor with three pick-up coils to determine the direction of the applied field by comparing the relative signals. (b) Illustration of different signals on the pick-up coils for aircraft traveling in different directions.

case the applied field (Earth's field) is aligned with coil 1 and so this will be maximum amplitude while the signals on coil 2 and 3 will be equal and 50% of the amplitude of coil 1. For the aircraft flying East the applied field is perpendicular to coil 1 so the signal is zero while it is equal in coils 2 and 3 at 87% of maximum amplitude. The device is normally placed near one wing tip to be as far as possible from the magnetic fields generated by the aircraft.

The flux gate sensor is used to correct the DI and keep it aligned with the compass heading as shown in Figure 3.29. Any misalignment between the flux gate reading and the DI is transformed into the amplitude of an AC signal, which is rectified and drives a precession motor to achieve alignment. As explained in Section 3.1.9 to produce a change in the DI reading (yaw) the motor needs to rotate the gimbal about a perpendicular (roll) axis. In principle, the display could be a direct readout of the direction derived by the magnetometer but the system below utilizing a DI is used since the rigidity of the gyro will damp out fluctuations in the flux gate signal.

The control panel in the cockpit can be used to turn off the slaving by switching the DI gyro to "FREE" so that the DI continues to act as the basic uncontrolled instrument relying purely on the rigidity of the gyro. This is the recommended procedure in the presence of thunderstorms, which produce large fluctuations in the magnetometer output. In the free mode, the DI can be rotated manually at a slow or fast rate by driving the precession motor.

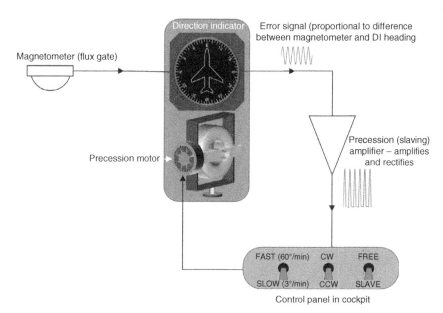

Figure 3.29 System used to slave a DI to a flux gate magnetometer.

3.3.4 Miniature Magnetometers

A more recent development is a miniature magnetometer on a chip that detects the direction of the Earth's magnetic field using asymmetric magneto-resistance (AMR), which was first discovered by William Thomson (Lord Kelvin) in 1851. When a current is passed through a magnetic material, it is found that the resistance depends on the angle between the current flow and the magnetization as illustrated in Figure 3.30a. The effect arises from the spin–orbit interaction leading to a higher probability of s-d scattering when the current is in the magnetization direction. This means that the resistance is highest when the current and magnetization are parallel ($\theta = 0$). Figure 3.30b shows data from permalloy, which demonstrates that the change in resistance is ~2% as the angle between the current and the magnetization varies between 0 and 90°.

To produce a magnetometer, a bar of magnetic material is magnetized along its easy direction by a local magnet and a current is passed through the bar. When an external field in the orthogonal direction (Figure 3.30c) is applied, it changes the direction of magnetization relative to the current flow and thus the resistance of the bar.

If the magnetization direction of the bar and the current in zero external field are parallel, however, any external field will reduce the resistance so the direction of the applied field cannot be determined. Also, there will be a highly non-linear response so that the detector is set up to initially have the current flowing

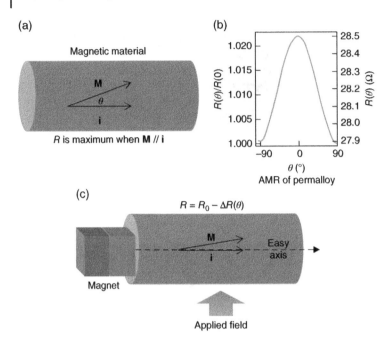

Figure 3.30 (a) Anisotropic magnetoresistance (AMR) is the dependence of the resistance of a magnetic material on the angle, θ, between the magnetization and the current flow. (b) Data from permalloy demonstrating a 2% change as θ is varied between 0 and 90°. (c) If the bar is magnetized along its easy axis and a bias current established, an applied field in the orthogonal direction will change the resistance.

at 45° to the magnetization by using a "barber pole" arrangement of electrodes as shown in Figure 3.31a. The aluminum "shorting bars" set up the electric field at 45° to the axis of the permalloy rod (the easy axis) so that the operating point is shifted to $\theta = 45$° (Figure 3.31b) giving a linear response and also a sensitivity to the direction of the applied field along the axis shown in Figure 3.31c.

The sensitivity is increased and the output is converted to a voltage by combining four detectors in a Wheatstone bridge arrangement illustrated in Figure 3.31c. An additional advantage of this setup is that with no applied field the bridge is in balance and produces zero voltage output as opposed to a background signal given by a single detector. The arrangement of Al shorting bars is such that an applied field along the sensitive axis, indicated by the arrow, changes the resistance of the two detectors in each arm in the opposite direction and the change is in the opposite sense in the two arms giving a fourfold increase in sensitivity. The only remaining issue is that a single Wheatstone bridge array will give the field component along a single axis so to produce a magnetometer, orthogonal detectors can be used to determine the absolute direction of the applied field as illustrated in Figure 3.31d. Figure 3.31e shows an entire magnetometer packaged into a chip a few mm across.

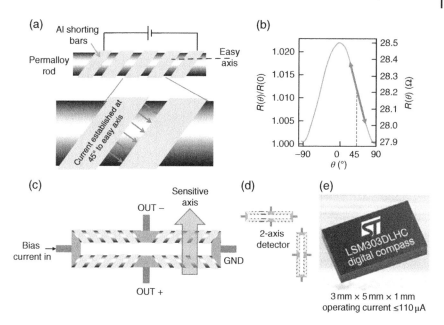

Figure 3.31 (a) Al "shorting bars" used to establish current flow at 45° to the magnetic easy axis. (b) Operating point is shifted to $\theta = 45°$ so that an applied field produces a linear response and the device is sensitive to the direction of the field along the sensitive axis. (c) Four bars are combined in a Wheatstone bridge configuration to increase the sensitivity and to provide a voltage output. (d) Orthogonal detectors are used to detect the absolute direction of the applied field. (e) The whole detector array is packaged onto a chip. *Source:* © ST Microelectronics. Used with permission.

3.4 Attitude Heading and Reference System (AHRS)

There is an increasing use of MEMS devices acting as detectors for aircraft instruments enabling a significant reduction in weight and improvement in reliability and stability in comparison with mechanical accelerometers and rotating gyroscopes. These detectors all provide voltage outputs suitable for displaying virtual instruments on glass cockpit displays. Also, several detectors can be integrated and the combination becomes an Attitude Heading Reference System (AHRS) that is, a single unit that can drive the artificial horizon, the DI, and the turn coordinator. In the MEMS industry, the miniaturization has become impressive with single chips available on the market for a few dollars that combine MEMS accelerometers, gyroscopes, miniature magnetometers, and Wi-Fi transmitters.

These cheap devices are used in applications such as mobile phones, hobby drones, etc. and are not suitable for commercial aviation because to achieve the required accuracy requires powerful onboard processing and Kalman

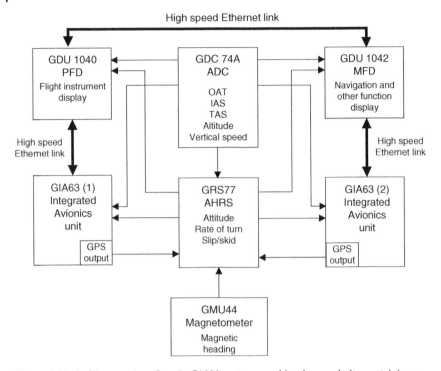

Figure 3.32 Architecture in a Garmin G1000 system used in glass cockpits containing a MEMS-based AHRS and a solid-state magnetometer.

filtering (see Section 9.9). In addition, the AHRS system needs to be strapped to the frame of the fuselage while the magnetometer needs to be at a remote location (usually the wing tip) well away from sources of magnetic field other than the Earth's field. Thus, units in the AHRS contain several electronics cards and are discrete rather than integrated, however, there is still a significant saving in weight and cost. The overall architecture of a modern glass-cockpit Garmin G1000 used extensively in general aviation and small commercial aircraft is shown in Figure 3.32.

3.5 Sensor Fusion

The AHRS system described in Section 3.4 is an example of combining inputs from many sensors that give complementary information to get an accurate picture of some aspect of flight, in that case, aircraft attitude. Increasingly, aircraft systems combine information from an array of sensors using software to evaluate different aspects of flight such as three-dimensional position, speed, rates of climb and descent, etc. This merging of data from multiple sensors is referred to as sensor fusion. This is important as individual sensors produce noise, they are unreliable in some aspects of flight, and may become faulty. If data from multiple sources are combined with a software-determined reliability factor, then despite all the problems with individual sensors, a robust determination can be made of the flight information required. As a simple example, altitude can be determined by the pressure altimeter, GPS positioning, or when close to the ground the radio altimeter. Rather than having three readings, these can be combined using software that assigns an accuracy and reliability to each to give a single reading with a high degree of confidence. Sensor fusion and the types of algorithm used will be discussed in more detail in Section 9.9.

Problems

1 The Earth can be considered to be a spinning gyroscope with its spin axis tilted at 23.5° to the orbital plane as illustrated in the figure. Because the planet is slightly nonspherical with an equatorial bulge (exaggerated in the figure), the Sun's gravity provides a torque in the direction shown that causes the spin axis to precess. It is estimated that the torque generated by the Sun is 4.48×10^{22} Nm and the moment of inertia of the Earth is 8.034×10^{37} kg m^2. Estimate the number of years taken for the Earth's axis to complete one revolution of the precession.

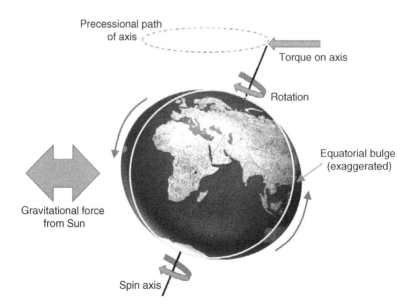

2 The figure below shows a simple mechanism to compensate for the Earth rate of a gyro using a so-called "latitude nut." If the aircraft's home airfield is at 52°N, at what distance, x, does the nut have to be positioned to remove the Earth rate precession? The nut weighs 10 g and the screw thread without the nut provides a torque of 4×10^{-5} Nm. The rotor has a moment of inertia of 2.4×10^{-3} kg m^2 and is spinning at 15 000 rpm.

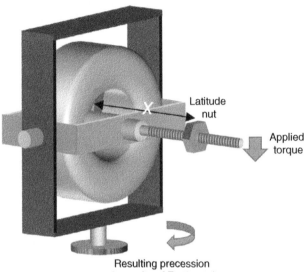

Latitude nut

Applied torque

Resulting precession
(opposite of Earth rate)

3 An aircraft weighing 6500 kg and traveling at 300 knots is performing a balanced rate 1 turn. What is the roll angle?

4 A vintage aircraft takes off from an airfield at a latitude 52°N and travels directly East according to the compass at 100 knots groundspeed for two hours. There is an un-slaved DI on board that also indicates East on takeoff but is then left free without adjustment by the pilot. Calculate the reading on the DI after two hours assuming there is no true wander in the gyro.

5 The detector in a ring laser gyro operating at a wavelength of 633 nm and an optical path of 15 cm is producing a beat frequency of 3 kHz. What is the rate of turn in the axis parallel to the plane of the optical path? You may assume the optical path is circular.

6 The data below shows the resistance change in the permalloy bar within a miniature magnetometer with the change in angle between the current flow

and the magnetization. From the data make an estimate of the resistance change in $\mu\Omega$ of the permalloy bar in a miniature magnetometer if it is rotated from position A to position B relative to the Earth's magnetic field as shown in the figure. The magnitude of the Earth's field is 50 μT and the saturation magnetization of permalloy is about 1 T.

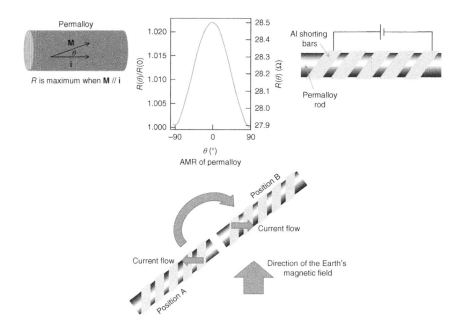

References

1 Ruijgrok, G.J.J. (2009). Chapter 1. In: *Elements of Airplane Performance*. Delft: VSSD Publishers.
2 Merlo, S., Norgia, M., and Donatii, S. (2002). Fibre gyroscope principles, chapter 16. In: *Handbook of Optical Fibre Sensing Technology* (ed. J.M. López-Higeura). Wiley.
3 Juang, J.-N. and Radharamanan, R. Evaluation of Ring Laser and Fiber Optic Gyroscope Technology. https://www.asee.org/documents/sections/middle-atlantic/fall-2009/01-Evaluation-Of-Ring-Laser-And-Fiber-Optic-Gyroscope-Technology.pdf (accessed 12 June 2018).
4 Creative commons (1953). http://creativecommons.org/licenses/by-sa/3.0/legalcode (accessed 12 June 2018).

4

Radio Propagation and Communication

4.1 Basic Properties of Radio Waves

As first verified by Heinrich Hertz in 1886, radio waves are electromagnetic waves. They travel at the speed of light and are the propagation of an oscillating electric and magnetic field oriented perpendicular to each other as illustrated in Figure 4.1a. This is an example of a *linearly polarized* wave, that is, the direction of electric field remains fixed and the direction of polarization is defined as being along the electric field. As shown in Section 4.4, this type of wave is emitted by a simple dipole antenna. Another type of polarization of radio waves used in aviation is shown in Figure 4.1b and is referred to as a *circularly polarized* wave. In this case, the electric field direction rotates once per cycle of amplitude variation so if an observer could detect the direction of the electric field it would rotate at the same frequency as the wave as it passed through the detector. For clarity, the magnetic field oscillation has been omitted from Figure 4.1b, but it is perpendicular to the electric field at all positions. The *helicity* of the wave is given by the direction of rotation of the electric field as it passes a fixed point, that is, right (clockwise) or left (anticlockwise). Different polarizations are utilized in aviation depending on the application and in the case of some radar applications the detected polarization is different to the emitted one to highlight specific features (see Chapter 5).

The relationship between frequency and wavelength of an electromagnetic wave is

$$f = \frac{c}{\lambda} \, (\text{Hz}) \tag{4.1}$$

where c is the speed of light and since this is very close to 3×10^8 m s^{-1}, two useful practical formulae for converting from frequency to wavelength and vice-versa are

$$f = \frac{300}{\lambda(\text{m})} \, \text{MHz} \tag{4.2}$$

Aircraft Systems: Instruments, Communications, Navigation, and Control,
First Edition. Chris Binns.
© 2019 John Wiley & Sons, Inc. Published 2019 by John Wiley & Sons, Inc.
Companion website: www.wiley.com/go/binns/aircraft_systems_instru_communi_Navi_control

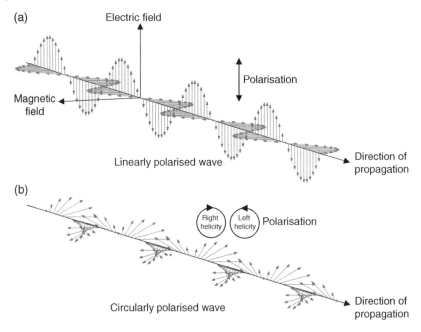

Figure 4.1 (a) Linearly polarized electromagnetic wave in which the electric and magnetic field directions are orthogonal and the polarization direction is defined as being along the electric field direction. (b) Circularly polarized electromagnetic wave in which the polarization direction rotates once every cycle of amplitude variation. The helicity can be right (clockwise) or left (anticlockwise). Only the electric field is shown but the magnetic field is perpendicular to the electric field at all positions.

$$\lambda = \frac{300}{f(\text{MHz})}\ \text{m} \qquad\qquad (4.3)$$

Radio waves used in aviation cover a huge frequency spectrum from 30 kHz to 40 GHz, which is six decades wide. This spectrum is subdivided into bands and the most commonly used classification is the one defined by the International Telecommunications Union (ITU), which is a specialized branch of the UN. Figure 4.2 shows the names of the frequency bands defined by the ITU and also indicates their role in civil aviation. Another classification scheme defined by the Institute of Electrical and Electronics Engineers (IEEE) is also used, especially for the highest frequencies. The IEEE scheme is the same as the one specified by the ITU up to the UHF band but further subdivides the highest frequencies into the C, L, S, X, and K bands as indicated in Figure 4.2. Other organizations, for example, the EU and NATO also specify frequency bands defined by specific applications but the names used in this book will conform to the ITU and IEEE schemes.

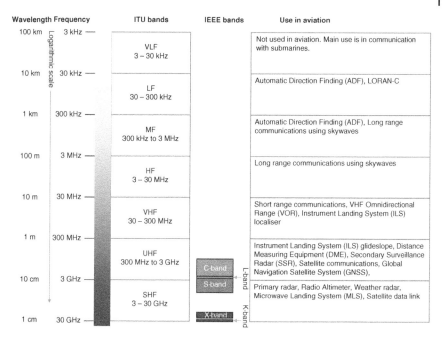

Figure 4.2 Names of frequency bands specified by the ITU and IEEE and their role in aviation.

4.2 Propagation of Radio Waves

4.2.1 Attenuation

Attenuation is the term given to the decrease in power density, that is, the power illuminating a unit area, as an electromagnetic wave propagates. In a vacuum, radio waves propagate in a straight line at the speed of light but the power density decays as the inverse square of the distance from the source simply due to the fact that the area illuminated by the same power increases as the square of the distance traveled as illustrated in Figure 4.3. The inverse square law applies equally to directional beams as long as the detector is smaller than the total wavefront.

When radio waves propagate through the atmosphere, extra attenuation processes come into play in addition to the inverse square law, that is, absorption and scattering. Absorption occurs when the radio energy is converted to another form within the absorbing body, for example, heat and is thus removed from the beam. For certain wavelengths, the ionosphere absorbs radio waves very strongly and the energy is converted into kinetic energy of the ions as described below. Scattering is where a particle reradiates the incident radio energy in all directions thus removing energy from the beam in the direction

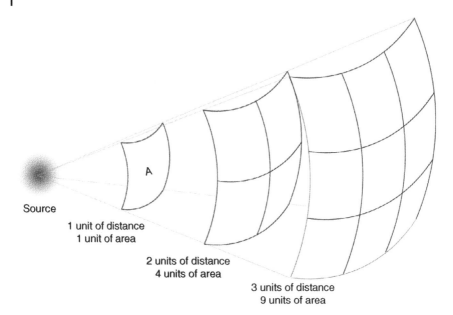

Figure 4.3 Illustration of the inverse square law. The power density illuminating a unit area decays as the square of the distance from the source due to the fact that the illuminated area increases with the square of distance traveled.

of propagation. Attenuation by absorption and scattering is mainly by water droplets and dust particles and increases with the frequency of the radio waves becoming very strong for frequencies above 1 GHz (Figure 4.4).

While propagating between the transmitter and receiver, radio waves will develop static noise, that is, random fluctuations of amplitude, due to atmospheric scattering and electrical interference from various sources including thunderstorms, solar activity, etc. For non-ionospheric propagation, the static

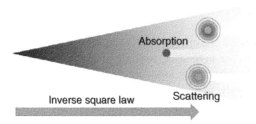

Figure 4.4 Attenuation processes for radio waves propagating through an atmosphere. In addition to the inverse square law, the radiation is attenuated by absorption and scattering.

interference is greatest at low frequency and at VHF and above becomes negligible compared with electronic noise in the receiver and man-made sources of electrical interference. Radio waves that interact with the ionosphere, however, will collect static noise at all frequencies. Propagation of radio waves is generally classified as ionospheric and non-ionospheric and these two modes are described below.

4.2.2 Non-Ionospheric Propagation

Non-ionospheric propagation can be subdivided according to frequency into surface (or ground) waves and space (or direct) waves.

4.2.2.1 Surface (or Ground) Wave: 20 kHz to 50 MHz (LF–HF)

Air has a refractive index, n, which is very close to that of vacuum ($n = 1$), however, the ground has a higher refractive index. At low wavelengths there is significant penetration of a radio wave into the ground and since the speed of propagation through a medium, v, is given by $v = c/n$, where c is the speed of light, the portion of the wave under the surface is retarded. This causes the wavefronts to bend (diffract) in the same direction as the curvature of the ground and the wave tends to follow the surface of the Earth as shown in Figure 4.5a. As the frequency increases, the surface attenuation increases and the range reduces. The losses are greater over land than sea due to the higher conductivity of sea water.

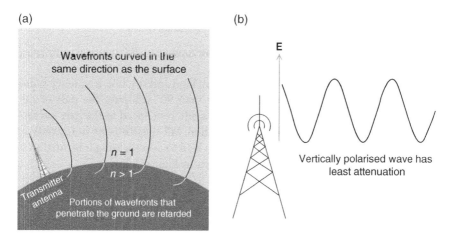

Figure 4.5 (a) Increased refractive index of the ground relative to air slows down radio waves that penetrate the surface, which causes the wavefronts to bend around the Earth's surface. (b) Attenuation is minimized for a wave with a vertical polarization.

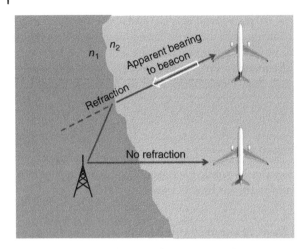

Figure 4.6 The different refractive indices of land and sea cause coastal refraction for non-normal incidence and possible bearing errors.

Generally, this propagation mode is utilized in the LF–MF frequency band (30 kHz to 2 MHz) and radio waves with a vertical polarization, that is, their E-field is vertical (Figure 4.5b), are used as this produces the minimum attenuation. The range of a transmitter is determined by its power and at 300 kHz, which is typical for a nondirectional beacon (NDB), the range is approximately given by

$$\text{Over sea}: \text{Range} = 3 \times \sqrt{\text{Power}(\text{W})} \quad \text{nautical miles} \tag{4.4}$$

$$\text{Over land}: \text{Range} = 2 \times \sqrt{\text{Power}(\text{W})} \quad \text{nautical miles} \tag{4.5}$$

For example, a 10 kW transmitter would have a range of 300 nautical miles over the sea and 200 nautical miles over land.

Because of the different refractive indices of land and sea, low-frequency radio waves crossing the coast at an angle other than 90° will have the part of the wavefront penetrating the land traveling at a different speed to the part in contact with the sea. The direction of propagation will thus be bent (refracted) as the radio wave crosses the coast and this can lead to false-bearing information if the aircraft is using a radio beacon as shown in Figure 4.6.

4.2.2.2 Space (or Direct) Wave: >50 MHz (VHF)

At frequencies above 50 MHz extending into the VHF band, there is negligible penetration of radio waves into the ground and waves are either absorbed or reflected when they encounter terrain. Thus, the propagation mode is a mixture of a direct wave from the transmitter and a ground-reflected wave incident at

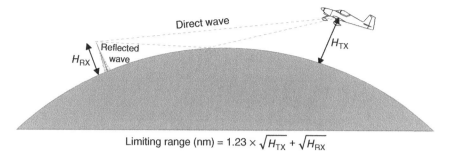

Limiting range (nm) = $1.23 \times \sqrt{H_{TX}} + \sqrt{H_{RX}}$

Figure 4.7 Space (direct) wave mode for VHF transmission.

the receiver (Figure 4.7). This mode is known as a space wave and it is present at all frequencies whatever other modes are present, but for frequencies above 50 MHz it is the only mode. VHF propagation is thus line of sight only and in principle is limited by the horizon, but there is usually some diffraction so the usable range extends to slightly beyond the horizon. The range was derived in Section 1.5 and shown to be given by

$$\text{Range} = 1.23 \left(\sqrt{H_{TX}(\text{ft})} + \sqrt{H_{RX}(\text{ft})} \right) \text{nautical miles} \qquad (4.6)$$

where H_{TX} and H_{RX} are the height of the transmitter and receiver in feet (Figure 4.7) and the pre-factor of 1.23 takes into account the additional 15% usually included for reception beyond the physical horizon. Note that the equation is independent of transmitter power since the line of sight distance is the limiting factor.

4.2.3 Ionospheric Propagation (Skywaves)

4.2.3.1 Origin of the Ionosphere

Ultraviolet photons from the Sun have sufficient energy to ionize atmospheric atoms and molecules and produce pairs of charged particles consisting of the ejected electrons and positive ions (Figure 4.8a). Free electrons and ions will recombine when they encounter each other, so to maintain a permanently ionized layer the atmospheric atoms have to be sufficiently dilute that the recombination rate is comparable to or less than the production rate. The variation in the ion density as a function of altitude is illustrated schematically in Figure 4.9b. Above about 600 km, the density of atmospheric atoms is too low for there to be a significant number of ions produced but the increasing density of the atmosphere with reducing height produces an increasing density of ions. Below a certain height the density of ions will start to decrease as the recombination rate starts to dominate and in addition, the production rate diminishes

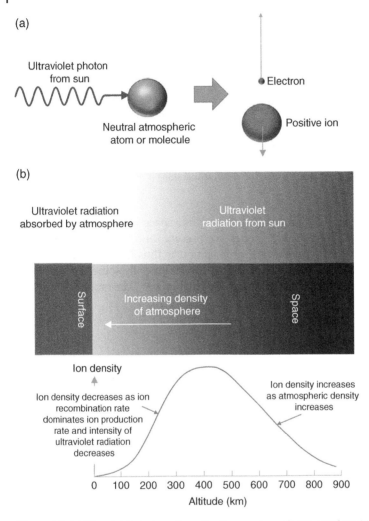

Figure 4.8 (a) Ultraviolet photons from the Sun produce electron and positive ion pairs. (b) At the upper end of the ionosphere the ion density is low due to the low density of atmospheric atoms. The ion density increases with decreasing altitude as the density of the atmosphere increases but reaches a maximum and starts to decrease again as the recombination rate starts to dominate the ion production rate and the ultraviolet intensity drops due to absorption.

due to absorption of the ultraviolet radiation by the atmosphere. Thus, there is a peak, which is typically at an altitude of about 300 km. The ion density becomes insignificant below about 60 km and above about 1000 km.

The structure of the ionosphere is more complex than the simple curve shown in Figure 4.8b and also goes through significant changes between day and night.

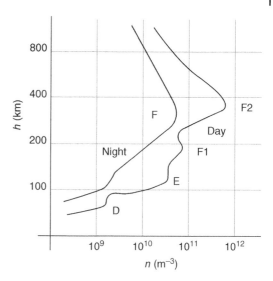

Figure 4.9 Diurnal and nocturnal variation of electron density with height in the ionosphere. *Source:* Reproduced with permission of Springer from Ref. [1], Fig. 2.21.

The upper atmosphere contains different chemical elements each with its own ionization potential and these tend to stratify at different heights so there is more than one ion density peak. In addition, other processes are involved in maintaining the ionospheric profile including ion diffusion and interactions with the solar wind (charged particles from the Sun) [1]. Typical electron densities as a function of altitude are plotted on a log scale in Figure 4.9 and it is evident that there is about a factor of 10 difference between day and night values. During the day, the different layers emerge clearly and are denoted by increasing altitude as the D, E, and F layer, with the latter further subdivided into the F1 and F2 layer. At night when the production of ions stops, the lowest lying D layer vanishes due to the rapid recombination of ions and electrons. In the higher layers the lifetime of the ions and electrons is sufficiently long that a significant electron density can be maintained until dawn when ionization starts up again though the F1 and F2 layers merge into a single F layer.

Thus, the D layer at a height of ~75 km forms at sunrise and disappears at sunset while the E layer at ~125 km is permanent though it reduces in altitude at sunrise and increases in altitude at sunset. The F layer is at ~225 km and splits into two at sunrise and rejoins at sunset. The splitting is greater in the summer and the F2 layer can reach an altitude of 400 km. Superimposed on the general behavior are fluctuations caused by high-energy radiation from the Sun. The ionosphere is stable during night and day but at dawn and dusk when it is changing there is interference with radio transmissions using the ionosphere.

4.2.3.2 Reflection and Absorption of Radio Waves by the Ionosphere

The ionosphere, which is a changing density of ions going through maxima is a medium with a changing refractive index that can reflect radio waves. Any radio wave that is reflected from the ionosphere back to the ground is referred to as a skywave. Above a critical frequency, radio waves will not be reflected by the ionosphere but will be absorbed and the critical frequency can be calculated as follows. The refractive index of a medium is given by

$$n = \sqrt{\epsilon} \tag{4.7}$$

where ϵ is the dielectric constant. For a wave of angular frequency $\omega = 2\pi f$, where f is the linear frequency in Hz, the dielectric constant is given by

$$\epsilon = 1 - \frac{\omega_P^2}{\omega^2} \tag{4.8}$$

where ω_P is the plasma frequency, which in a charged plasma like the ionosphere is given by

$$\omega_P = \sqrt{\frac{Ne^2}{\varepsilon_0 m}} \tag{4.9}$$

In Equation (4.9), N is the electron density, e is the electron charge, ε_0 is the permittivity of free space, and m is the electron mass. Note that the positively charged ions do not contribute significantly to the dielectric constant so the equation is written in terms of the electron density only. A typical value for N in the F-layer of the ionosphere (see Figure 4.9) is 10^{11} m^{-3} and putting in the values for e (1.6×10^{-19} C), ε_0 (8.85×10^{-12} Fm^{-1}), and m (9.1×10^{-31} kg) gives a typical plasma frequency $\omega_P \sim 18$ MHz, corresponding to a linear frequency, f of 2.8 MHz. This is the critical frequency, f_{crit}, since any radio wave whose frequency is above the plasma frequency will be absorbed by exciting plasma oscillations in the electron gas. Thus, Equation (4.9) can be rewritten as

$$f_{crit} = \frac{1}{2\pi} \sqrt{\frac{e^2}{\varepsilon_0 m}} \sqrt{N} \tag{4.10}$$

making it clear that f_{crit} depends only on the electron density, all other parameters in Equation (4.10) being constant. Putting in the values for all the constants yields, to a close approximation, the convenient expression:

$$f_{crit} = 9\sqrt{N} \tag{4.11}$$

For radio waves incident at an angle θ to the ionosphere (Figure 4.10) higher frequencies will be reflected and the maximum usable frequency, f_{muf}, for a specific value of θ is given by

$$f_{muf} = \frac{f_{crit}}{\sin\theta} \qquad (4.12)$$

It is clear from this equation that for any frequency greater than f_{crit}, there is a critical angle θ_{crit} above which radio waves will not be reflected. The D, E, and F layers have increasing electron densities so there is an increasing critical frequency with the altitude of the layer. For frequencies well above the plasma frequency, radio waves are not absorbed and the ionosphere becomes transparent, so it is possible to choose which layer to use for the skywave reflection by using a suitable frequency. In general, radio waves with frequencies up to 2 MHz in the MF band are reflected from the E-layer and those with frequencies 2–50 MHz in the HF band are reflected from the F-layer.

The distance traveled over land by a radio wave reflected from the ionosphere is known as the skip distance as illustrated in Figure 4.10 and clearly the skip distance depends on the altitude of the ionospheric layer used. There is also a ground wave propagating from the transmitter and in the HF band this has a much shorter range than in the LF band in which ground wave propagation is utilized. The region between the disappearance of the ground wave and the reflected skywave is unreachable by the radio wave and is known as the dead space.

As an example, if a transmission was required over a range of 1000 km using a skywave reflected from the F-layer, which has a typical height of 250 km, $\theta = 30°$, and for $N = 10^{11}$ m^{-3}, from Equations (4.11) and (4.12), f_{muf} would be 5.7 MHz. Due to uncertainties in conditions in the ionosphere, in practice, the optimum working frequency (OWF) is given by $0.85 \times f_{muf} = 4.8$ MHz.

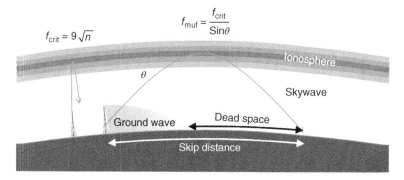

Figure 4.10 Skip distance and dead space for a skywave.

Generally, it is the E and F layers that are used for long-range HF communications. The maximum range for E-layer (height ~125 km) reflection is 1350 nm and for F-wave reflection (height ~ 225 km) is 2200 nm. It is possible to get even longer ranges, however, as returning skywaves can be reflected from the ground back to the ionosphere and multi-reflection waves can reach the opposite side of the globe.

Beermat Calculation 4.1

It is required to send a transmission using skywaves from the HF station at Shannon on the West coast of Ireland to an aircraft midway across the Atlantic 1300 nautical miles away. The seasonal average for the electron density and height of the F2 layer at the time of day in question are 5×10^{11} m^{-3} and 350 km, respectively. Determine the OWF for the transmission.

The distance from the transmitter to the aircraft is 2407.6 km and the geometry of the transmission is sketched in the figure. From this we find that $\theta = \tan^{-1}(350/1203.8) = 16.2°$.

From Equation (4.11), $f_{crit} = 9\sqrt{5 \times 10^{11}} = 6.36$ MHz. Thus, from Equation (4.12), the maximum usable frequency is 6.36/sin(16.2) = 22.78 MHz. The OWF is 85% of this value, that is, 19.36 MHz.

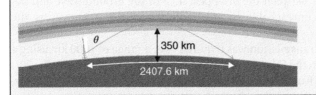

4.2.3.3 Ducting Propagation of Very Low Frequency (VLF) Waves

Finally, although it is not used in aviation it is worth highlighting another mode of ionospheric propagation in the case of VLF (3–30 kHz) waves, which is known as ducting propagation. The wavelength of 3 kHz waves is 100 km, which is similar to the height of the E-layer, so the ionosphere and the Earth's surface can act as a curved waveguide (see Section 4.4.8.1) to channel the waves around the Earth.

4.3 Transmitters, Receivers, and Signal Modulation

4.3.1 Basic Continuous Wave Morse Code Transmitter/Receiver

The most basic radio communication system that can be used to send messages in Morse code is shown schematically in Figure 4.11. The sinusoidal carrier wave, which in this example has a frequency of 100 kHz (LF band), is produced

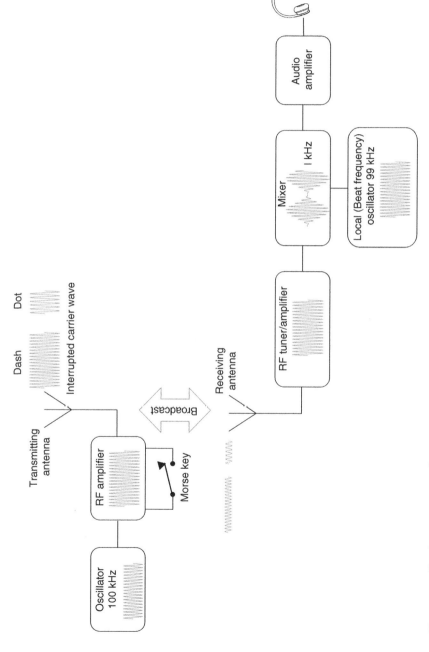

Figure 4.11 Simple transmitter and receiver for transmitting and receiving Morse code.

by an oscillator and then passed to a radio frequency (RF) amplifier. A Morse key connected to the initial stage of the amplifier turns the carrier on or off to produce transmitted dots and dashes. The receiver consists of a tuner, that is, a filter set to 100 kHz and an amplifier to boost the weak received carrier wave. The amplified signal is then converted to an audible tone by mixing it with a locally generated 99 kHz wave, which produces a 1 kHz beat frequency. This is heard at the headphones whenever the Morse key at the transmitter is pressed. The Morse code transmitter where the carrier wave is simply turned on and off illustrates the simplest possible form of modulation, which is referred to as keyed modulation. To transmit voice and data more sophisticated forms of modulation are required and these are described below.

4.3.2 Quadrature Amplitude Modulation of Carrier

Transmitted analog or digital information has to be imposed on the carrier wave somehow before transmission (modulation) so that it can be extracted at the receiver (demodulation). The signal required to be transmitted can be represented by a Fourier series up to a frequency that is much lower than the carrier frequency and is referred to as the baseband since, relative to the carrier frequency, it consists of a narrow range of frequencies close to zero. The time variation of the electric field $e_c(t)$ of the carrier wave is represented by the general expression:

$$e_c(t) = E_c \sin(2\pi f_c t + \phi) = E_c \sin(\omega_c t + \phi) \tag{4.13}$$

where E_c is the amplitude, f_c and ω_c are the linear and angular frequency, respectively, and ϕ is a phase shift that aligns the peaks and troughs of the wave with a specific time relative to $t = 0$. Note that applying a phase shift, $\phi = 90°$, converts the sine wave to a cosine wave. Information can be imposed onto the carrier by modulating the amplitude, E_c, the frequency, ω_c, or the phase, ϕ, and all these methods are used for different types of radio transmission in aviation. Each type of modulation will be looked at in more detail below but first a general method for modulation, referred to as *IQ* or quadrature amplitude modulation (QAM), will be described. There are many ways of imposing a modulation on a carrier but QAM is increasingly common as it is particularly suited to digital transmission.

In this method, the final modulated wave is constructed by amplitude modulating two waves with the carrier frequency whose phase is 90° apart and adding them as illustrated in Figure 4.12a. Since the waves are separated by a quarter cycle they are said to be in *quadrature*, hence the name of the method. A sine wave and cosine wave are in quadrature and conventionally the amplitudes of the cosine and sine waves are labeled "*I*" (for "in-phase") and "*Q*" (for quadrature), respectively, with the maximum amplitudes normalized to 1.

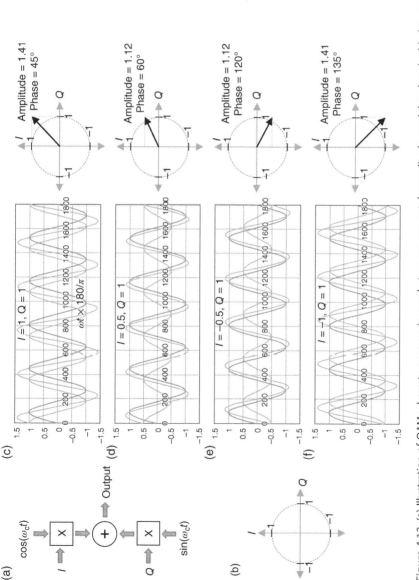

Figure 4.12 (a) Illustration of QAM where two waves in quadrature are separately amplitude modulated and added to produce the output. (b) Phasor diagram used to display instantaneous values of I and Q. (c) Cosine wave (blue line), sine wave (red line), and the output waveform for the simple case $I = Q = 1$ along with the corresponding phasor diagram. (d)–(f) Waveforms for different values of I with Q held at 1 showing changing amplitudes and phase. The x-axes of the waves have been converted to angle by multiplying the time t by $180\omega/\pi$.

The instantaneous values of I and Q can be represented on a *phasor* diagram illustrated in Figure 4.12b, which has I and Q represented on orthogonal axes and the combination at any instant is shown as a vector from the origin to the correct point in the I–Q plane. The vector is shown at one instant but in many types of modulation it is changing with time, for example, changing in length or angle or rotating about the origin.

The I and Q waveforms (blue and red lines, respectively) and their sum (green line) for the case when I and Q are held constant at the value 1 are shown in Figure 4.12c along with the corresponding phasor diagram. It is seen that the output waveform is phase shifted by 45° with respect to the I waveform and has an amplitude of $\sqrt{2}$. Changing I to 0.5 while maintaining $Q = 1$ produces an output waveform with a phase shift of 60° and an amplitude of 1.12 as shown in Figure 4.12d. Reducing I further to zero will clearly generate just the sine wave, that is the phase shift will be 90° and the amplitude will be 1 and continuing on to negative values of I will produce increasing phase shift and changing amplitudes as illustrated in Figure 4.12e and f. It is clear that changing I and Q through their entire range of values –1 to +1 will produce output waveforms whose amplitude varies from 0 to 1.41 and whose phase varies from 0 to 360°, thus the method can be used for amplitude and phase modulation. Frequency modulation (FM) can also be produced by a continuous change of phase as described below.

4.3.3 Superheterodyne Receivers and Demodulation of QAM Signals

At the receiver, the information modulated onto the carrier must be extracted and this is done using the superheterodyne principle, which was first patented in 1901. Originally, the idea lay dormant for a number of years but has become the most widely used method today. Figure 4.13a shows a block diagram for a superheterodyne receiver, which is drawn for an amplitude-modulated signal but the same principle applies to frequency and phase modulation. The first stage is an RF amplifier that provides some frequency selectivity to reduce noise and the output is fed into a mixer, which multiplies the modulated carrier by a sinusoidal wave at an intermediate frequency (IF) between the carrier and baseband frequency to produce a modulated IF waveform. The IF frequency is normally 455 kHz or 1.6 MHz for amplitude modulation and 10.7 MHz for FM. In fact, as shown in Section 4.3.4, the multiplication process at the mixer produces outputs at several frequencies not all of which contain the modulation signal but the unwanted frequencies can be filtered out. The IF signal is passed into a demodulator that extracts the baseband signal and this can be done by a number of techniques specialized to the application [2].

The simple block diagram in Figure 4.13a does not have the capability to extract separate I and Q signals from a broadcast waveform but it is straightforward to do this by including the front-end stage shown in Figure 4.13b. The total

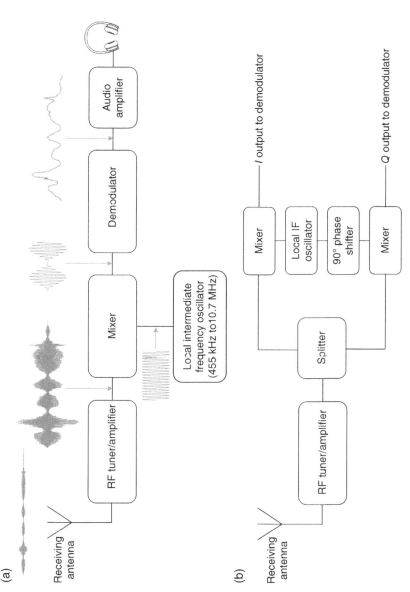

Figure 4.13 (a) Block diagram showing the superheterodyne method to extract the information (baseband signal) from the received radio wave. (b) Extension of the superheterodyne method to separately extract the *I* and *Q* modulations from the input wave.

waveform is passed into a splitter that passes equal amplitudes to two mixers connected to the IF oscillator but for one wave the oscillator phase is shifted by 90°. The outputs from the two mixers can be demodulated to obtain the separate I and Q components of the initial waveform.

4.3.4 Amplitude Modulated (AM) Transmission

In an AM transmission, the baseband signal is used to control the amplitude of the carrier and the basic modulation method was first demonstrated in 1900. Most non-aviation VHF transmissions have moved onto FM (see Section 4.3.6) due to its lower sensitivity to interference but UHF, VHF, and HF voice communications between aircraft and the ground still use AM. Partly this is historical but AM does have some advantages in aviation, for example it requires less power to transmit the same information. In addition, two stations transmitting at the same frequency with different signal strengths will both be demodulated by the receiver and although the message will be garbled, controllers will at least be aware that more than one station is transmitting. This is not the case with FM where due to an effect known as FM capture [3] the weaker signal will be completely suppressed.

With a QAM type modulator, AM is achieved by feeding the baseband signal into I and a multiple, k, of the same baseband signal into Q as illustrated in Figure 4.14a. The resulting modulated waveform will be AM and phase-shifted by an amount that depends on k. For example, if $k = 1$, the components are equal and the phase shift is 45° or if Q is held at zero ($k = 0$), the phase shift is zero. The phasor diagram for AM is shown in Figure 4.14b and consists of a vector varying in magnitude along a fixed angle whose value depends on k.

To analyze AM in more detail, consider a general modulation $m(t)$ applied to a carrier wave, with a frequency ω_c whose electric field is described by

$$e_c = E_c \sin(\omega_c t) \tag{4.14}$$

The resulting electric field can be written as

$$e_{am} = [E_c + m(t)] \sin(\omega_c t) \tag{4.15}$$

Usually the carrier wave amplitude is normalized to 1 so that Equation (4.15) can be written as

$$e_{am} = [1 + m(t)] \sin(\omega_c t) \tag{4.16}$$

Consider initially a modulation consisting of a cosine wave at a single frequency ω_m, that is:

$$m(t) = M \cos(\omega_m t + \phi) \tag{4.17}$$

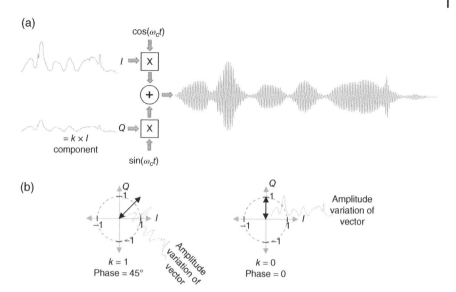

Figure 4.14 (a) Amplitude modulation using the QAM method. The same baseband signal is fed into both the *I* and *Q* inputs with the *Q* component multiplied by a factor *k*. (b) Phasor diagrams for amplitude modulation. The vector amplitude varies with the baseband amplitude at an angle that depends on *k*.

where a phase shift, ϕ, between the modulation and the carrier wave has been added for generality. In this simple case M is the modulation index and $M = 1$ corresponds to the modulation amplitude being equal to the carrier wave amplitude. This gives the maximum signal after demodulating while $M > 1$ (overmodulation) results in loss of information on demodulation. Figure 4.15b shows a carrier AM with different values of M. Combining Equations (4.16) and (4.17) gives the total electric field of the AM wave to be:

$$e_{am} = [1 + M \cos(\omega_m t + \phi)] \sin(\omega_c t) \tag{4.18}$$

and using the identity:

$$\cos(a)\sin(b) = \frac{\sin(a+b) - \sin(a-b)}{2} \tag{4.19}$$

Equation (4.18) becomes

$$e_{am} = \sin(\omega_c t) + \frac{M}{2}[\sin(\omega_c t + \omega_m t + \phi) - \sin(\omega_c t - \omega_m t - \phi)] \tag{4.20}$$

Expanding the total electric field in this way makes it clear that in the frequency domain the output of the transmitter consists of the unmodulated carrier at frequency ω_c and two AM signals. One has a frequency ω_m above the carrier

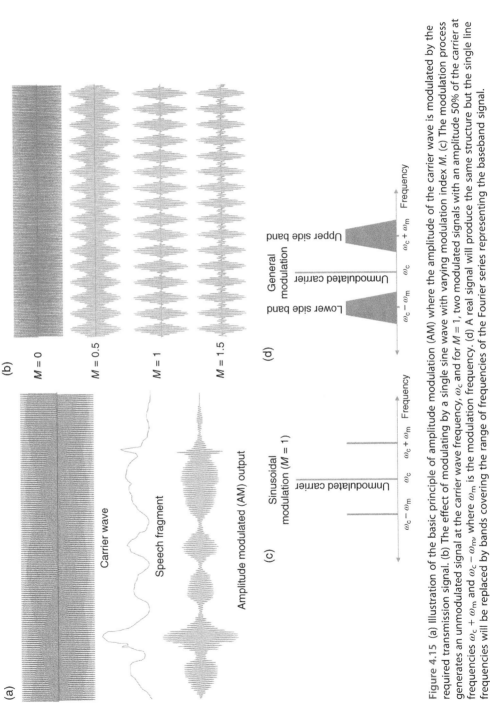

Figure 4.15 (a) Illustration of the basic principle of amplitude modulation (AM) where the amplitude of the carrier wave is modulated by the required transmission signal. (b) The effect of modulating by a single sine wave with varying modulation index M. (c) The modulation process generates an unmodulated signal at the carrier wave frequency, ω_c and for $M = 1$, two modulated signals with an amplitude 50% of the carrier at frequencies $\omega_c + \omega_m$ and $\omega_c - \omega_m$ where ω_m is the modulation frequency. (d) A real signal will produce the same structure but the single line frequencies will be replaced by bands covering the range of frequencies of the Fourier series representing the baseband signal.

frequency and another at a frequency ω_m below as illustrated in Figure 4.15c for $M = 1$. As shown by Equation (4.20) the phase shifts are also of opposite sign for the two modulated signals.

Since a general modulation signal can be expressed as a Fourier series, the above analysis can be carried out for each Fourier component so that the specific modulated frequencies above and below the carrier frequency are broadened into bands, as illustrated in Figure 4.15d and are referred to as upper and lower sidebands. Since a typical airband VHF communication frequency is 120 MHz and the frequency range required to accurately represent speech is ~300 Hz to 3.4 kHz, the sidebands are very narrow relative to the carrier frequency and close to it.

In aviation, VHF and UHF voice signals are transmitted with the full carrier and both sidebands, that is, the whole frequency spectrum from $\omega_c - \omega_m$ to $\omega_c + \omega_m$ (Figure 4.15d) is amplified and passed to the antenna and this is referred to as double sideband (DSB) transmission. HF voice communications on the other hand use suppressed carrier single sideband (SC-SSB) transmission, that is, the carrier is removed and just one sideband is transmitted, which is more efficient. This type of transmission is discussed in more detail in Section 4.6.1.

4.3.5 Channel Spacing in the VHF Band for AM Voice Transmission

Since the Second World War, short-range communications (up to ~200 km) from aircraft to ground and aircraft to aircraft have utilized VHF transmissions due to their low background noise and also to the fact that an efficient resonant quarter wave antenna is only about 30 cm long (see Section 4.4). The band of frequencies reserved for civil aviation is 108–136.975 with the range 108–117.95 reserved for radio navigation (VOR, ILS, etc.) and 118–136.975 for AM voice transmissions (Figure 4.16a). In aviation, voice communications normally use AM modulation and DSB processing, thus the whole frequency spectrum from the bottom of the lower sideband to the top of the upper sideband is used for each channel. To ensure a reasonable quality of voice reproduction the modulation frequency range 300 Hz to 3.4 kHz is used so that the whole bandwidth for a single channel is about 7 kHz as illustrated in Figure 4.16b.

The radio navigation band (108–119.75 kHz) has channels allocated 50 kHz (0.05 MHz) apart, giving 200 channels. This allows up to 200 radio navigation facilities (VOR, ILS, etc.) within line of sight range (~200 km), which is a high density of stations and the number of channels is not restrictive. Pre-1999 the band for voice transmissions (118–136.975 MHz) was allocated channels 25 kHz (0.025 MHz) apart giving 760 channels, which given the increasing density of traffic in some terminal areas was found to be restrictive. From 1999 the voice band channel spacing was reduced by a factor of three to 8.33 kHz giving 2280 channels. This channel spacing is only just wider than the 7 kHz required

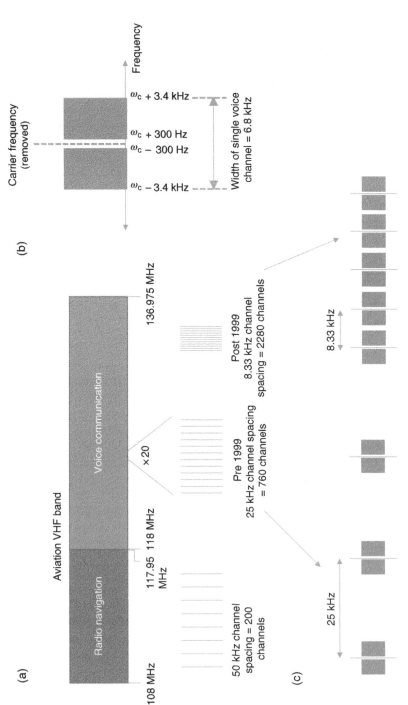

Figure 4.16 (a) Aviation VHF band for radio navigation and voice communication showing the channel separation. (b) Frequency range required for a single voice channel using DSB transmission. (c) Voice communication channel spacing pre-1999 and post-1999 relative to the single channel width.

for a single-channel DSB transmission as illustrated in Figure 4.13c, thus no further decrease in channel spacing is possible without fundamentally changing the transmission method used.

4.3.6 Frequency Modulation

Another form of modulation that imposes information onto a carrier wave uses the baseband signal to vary the instantaneous frequency of the carrier around the central frequency by an amount that is proportional to the amplitude of the analog signal. This type of modulation is used in the VHF omnidirectional range (VOR) for radio navigation (see Section 7.2). To see how FM can be achieved using the QAM method, first consider Figure 4.17a, which demonstrates that a change in frequency of a wave can be produced by a constantly changing phase angle. On an *IQ* phasor diagram (Figure 4.17b) this is represented by a rotating vector of constant length where the rate of rotation determines the magnitude of the change in frequency and the rotation direction determines whether the frequency is increased or decreased. That is, a clockwise (increasing phase angle) produces a decrease in frequency while anticlockwise (decreasing phase angle) produces an increase in frequency.

A rotating phasor can be generated by using a sinusoidal amplitude modulation of the *I* and *Q* inputs at the same amplitude and frequency as illustrated in Figure 4.17c. Following a single cycle of the *I* and *Q* inputs and plotting the phasor at every point shows that it rotates at a steady rate maintaining the same amplitude as required. Changing the frequency of both *I* and *Q* inputs produces a different frequency shift of the carrier wave. The full implementation of FM using the QAM method is illustrated schematically in Figure 4.17d. The baseband signal is used to drive a voltage-controlled oscillator whose frequency output depends on the voltage input. The oscillator output generates the quadrature *IQ* signals that modulate the carrier wave whose frequency shift is thus determined by the baseband signal input.

To look at FM in more detail, in a similar fashion to the analysis of AM, we consider a general modulation $m(t)$ applied to a carrier wave, with frequency ω_c whose electric field is described by

$$e_c = E_c \sin(\omega_c t) \tag{4.21}$$

The modulator combines the carrier and the baseband so that the transmitted signal is given by

$$e_c(t) = E_c \sin\left(\int_0^t \omega(\tau)d\tau\right) \tag{4.22}$$

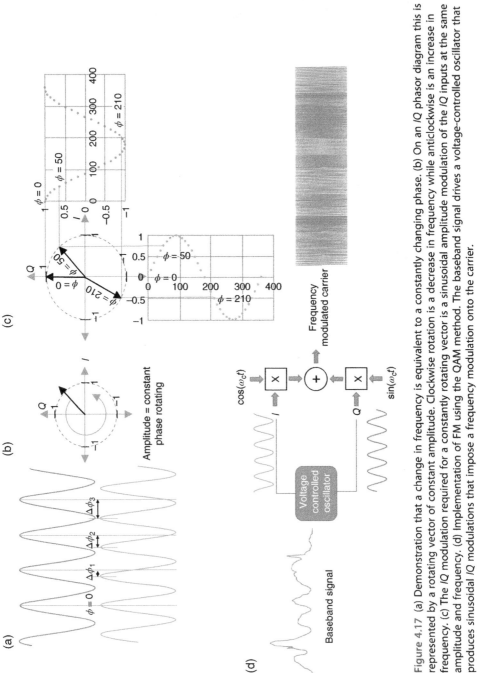

Figure 4.17 (a) Demonstration that a change in frequency is equivalent to a constantly changing phase. (b) On an *IQ* phasor diagram this is represented by a rotating vector of constant amplitude. Clockwise rotation is a decrease in frequency while anticlockwise is an increase in frequency. (c) The *IQ* modulation required for a constantly rotating vector is a sinusoidal amplitude modulation of the *IQ* inputs at the same amplitude and frequency. (d) Implementation of FM using the QAM method. The baseband signal drives a voltage-controlled oscillator that produces sinusoidal *IQ* modulations that impose a frequency modulation onto the carrier.

where $\omega(\tau)$ is the instantaneous frequency of the oscillator, τ is a time variable, and t is the instantaneous time under consideration. In terms of the carrier and modulation signals this can be written as

$$e_c(t) = E_c \sin\left(\int_0^t [\omega_c + \Delta\omega m(\tau)]d\tau\right) \tag{4.23}$$

$$e_c(t) = E_c \sin\left(\omega_c t + \Delta\omega \int_0^t [m(\tau)]d\tau\right) \tag{4.24}$$

where $\Delta\omega$ is the maximum shift in frequency produced by the modulation. Note that in Equation (4.24) the modulation is entirely within the argument of the trigonometric function and affects only the frequency and not the amplitude. To make the analysis more transparent we can consider the modulation as a single cosine wave given by

$$m(t) = E_m \cos(\omega_m t) \tag{4.25}$$

whose integral with respect to time is

$$\int_0^t m(\tau)d\tau = \frac{E_m}{\omega_m} \sin(\omega_m t) \tag{4.26}$$

and assuming, as before that the amplitude of $m(t)$ is normalized to 1, Equation (4.23) becomes

$$e_c(t) = E_c \sin\left(\omega_c t + \frac{\Delta\omega}{\omega_m} \sin(\omega_m t)\right) \tag{4.27}$$

In the case of FM, the modulation index, M, for a sinusoidal modulation of frequency, ω_m, is given by

$$M = \left(\frac{\Delta\omega}{\omega_m}\right) \tag{4.28}$$

and the effect of varying M on the total signal when $\omega_m = 0.05\omega_c$ is shown in Figure 4.18. It is seen that the modulation imposes a pattern on the carrier that repeats every 20 cycles as expected and a value of $M = 1$ corresponds to the frequency varying from 0 to $2\omega_c$. In practice, modulations of a few kHz are imposed on carriers ~100 MHz so that $\omega_m/\omega_c \sim 10^{-4}$ and values of $M \ll 1$ are used. When the baseband modulation is not a sine wave (the usual case) then the modulation index is given by:

$$M = \left(\frac{\Delta\omega}{\omega_{max}}\right) \tag{4.29}$$

where ω_{max} is the highest frequency component in the baseband signal.

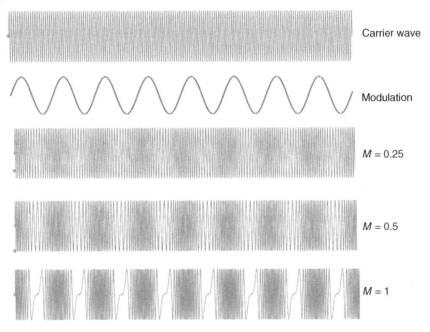

Carrier wave

Modulation

$M = 0.25$

$M = 0.5$

$M = 1$

Figure 4.18 Effect of frequency modulation of a carrier wave by a single-frequency sine wave for different values of modulation index.

The bandwidth requirement for a single channel FM broadcast is given by Carsons Rule [4], which states that the bandwidth containing 98% of the power of the signal, B_{FM}, is given by

$$B_{FM} \approx 2(\Delta f + f_{max}) \tag{4.30}$$

where $\Delta f = \Delta \omega / 2\pi$ and $f_{max} = \omega_{max} / 2\pi$. This is a reasonable approximation as long as there are no discontinuities in the baseband signal.

If the carrier is AM at a single frequency as well as frequency modulated, it will generate additional carrier frequencies shifted by the AM frequency above and below the main carrier frequency as shown in Section 4.3.4. These are known as subcarriers and can be used to transmit additional information. For example, in color television broadcasts the main carrier is frequency modulated to produce the black and white brightness signal and the subcarriers contain the color information. Thus, a black and white TV set will just use the main carrier while a color set will demodulate the subcarriers as well. Subcarriers are used to carry a 30 Hz FM in the VOR used for radio navigation as described in Section 7.2.

4.3.7 Modulation for Digital Data Transmission

The two types of modulation presented above are both analog in that the input signal is continuously varying as is appropriate for the transmission of speech. Increasingly, transmissions to and from aircraft are required to pass information in digital format. Examples include secondary surveillance radar (see Section 5.4), ACARS (see Section 4.8), TCAS (see Section 5.5), data from navigation satellites (see Chapter 8), and the internet. There are two basic types of digital modulation used in aviation, that is pulsed modulation where the carrier is emitted as a set of precisely timed pulses, or phase shift modulation where the carrier amplitude is continuous but its phase or frequency is shifted to indicate "1" or "0." Sometimes both types of modulation are used in the same transmission when a set of pulses is used as a preamble followed by a data block transmitted by phase-shift modulation, for example, mode S transponders (see Section 5.4).

4.3.7.1 Pulsed Modulation

In pulsed modulation, the carrier amplitude is switched between full amplitude ("1") and zero amplitude "0" as illustrated in Figure 4.19a. In its most general form, this type of modulation can be used to send timing signals in a preamble message prior to sending a digital data block and it is also used in radar to send radio pulses whose reflection is timed to get the range of a target. If it is used to send a block of digital information, the normal method is to encode the data by whether a pulse occurs in the first half of a specific time window (interpreted as a "1") or the second half of the window (interpreted as a "0") as illustrated in Figure 4.19b. This is known as pulse position modulation (PPM) and is used in Mode S transponder replies from aircraft (see Section 5.4.8).

To achieve pulsed modulation of a carrier using QAM, the Q amplitude is set permanently to "0" while the I amplitude switches between "0" and "1" as illustrated in Figure 4.19c. Thus, the end of the phasor switches between two points, that is, the center and "1" along the I axis. After the carrier is demodulated at the receiver, the output will reflect the pattern of 0s and 1s at the I input. The two discrete positions in the phasor diagram are referred to as constellation points and as shown below all digital modulation can be achieved by using specific constellation points.

4.3.7.2 Binary Phase Shift Keying (BPSK)

Binary phase shift keying (BPSK) encodes the data by the phase of the carrier whose amplitude is constant as illustrated in Figure 4.20a. A specific phase represents "1" and the opposite phase, that is, the carrier phase shifted by 180° represents "0." To implement BPSK using QAM, Q is maintained at zero and I is switched between −1 and +1 as illustrated in Figure 4.20b. Thus, unlike pulsed modulation, there is always a carrier transmitted but its phase is switched

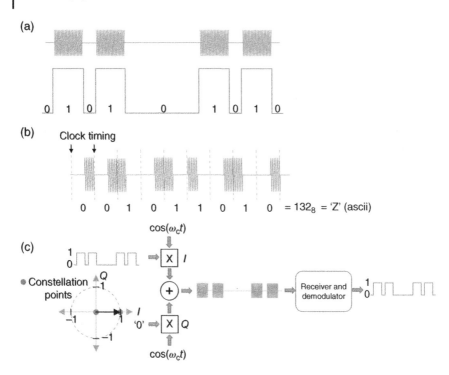

Figure 4.19 (a) Pulsed modulation where the carrier amplitude is switched between zero and full. (b) Pulse position modulation (PPM) where data is encoded by whether the pulse appears in the first or second half of a specific time window denoting "1" and "0", respectively. The example shown shows the transmission of the octal number 132_8 or the character "Z" in ascii. (c) *IQ* values required for QAM to produce pulsed modulation, that is Q is set permanently to "0" while *I* switches between "0" and "1." Thus, the end of the phasor switches between two "constellation points" in the phasor diagram.

by 180° at each transition. There are still two constellation points in the phasor diagram but the one at the center occurring with pulsed modulation is shifted to $I = -1$. The $-1/+1$ pattern is extracted at the demodulator and converted to a 0/1 pattern. BPSK is the modulation method used when a ground station makes a Mode S transponder selective call to an aircraft (see Section 5.4.7).

For both pulsed modulation and BPSK there are just two constellation points but by including switching of the Q channel as well it is possible to introduce more. For example, if Q was switched between -1 and $+1$ as well, then four combinations of I/Q are possible, that is, $-1 -1$, $-1 +1$, $+1 -1$, $+1$, $+1$ and these four input states would be represented by the four constellation points shown in Figure 4.21b, which also shows the four phases in the carrier. Thus, the transmission can transmit digits with base 4 (0 – 3) rather than binary. Allowing the

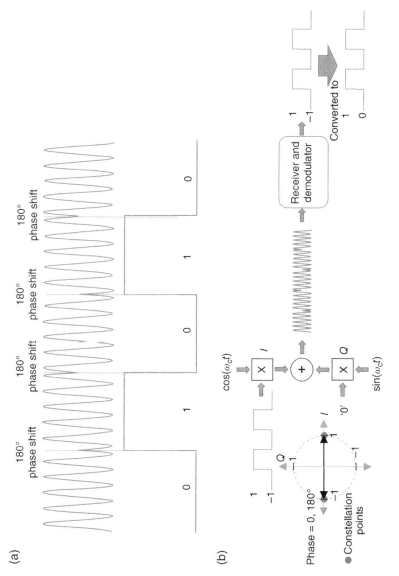

Figure 4.20 (a) Illustration of binary phase shift keying (BPSK) modulation where two phases of the carrier 180° apart are used to represent "0" and "1" in a data block. (b) Implementation of BPSK modulations using QAM. Q is maintained at "0" while I is switched between "1" and "−1" thus there are two constellation points.

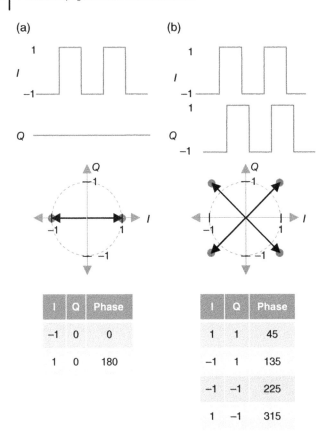

Figure 4.21 (a) BPSK modulation showing the constellation points and phases of the modulated carrier. (b) Switching *Q* as well as *I* between −1 and 1 allows four possible input states shown in the table along with the corresponding phase of the modulated carrier. The states are represented by the four constellation points on the phasor diagram. Thus, each piece of data transmitted is a 2-digit binary number.

magnitude of *I* and *Q* to vary as well as the sign introduces more constellation points and currently terrestrial digital broadcasts use transmissions with up to 256 constellation points so that an 8-bit number is transmitted in each digit.

4.3.7.3 Binary Continuous Phase Frequency Shift Keying (BCPFSK)

The abrupt phases change used in BPSK have some disadvantages and another method is to code the binary data by two frequencies around the carrier frequency with a continuous change in phase as the frequency is switched between the two values as illustrated in Figure 4.22a. The frequency is sampled at the

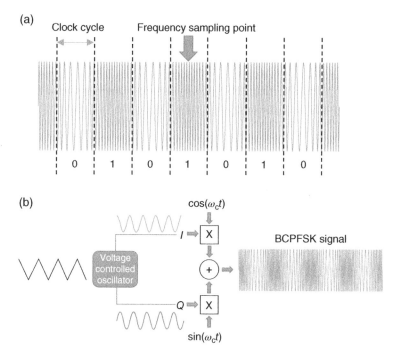

Figure 4.22 (a) BCPFSK modulation in which the two binary states are encoded by two different frequencies with a continuous change in phase in going from one frequency to the other. The frequency is sampled at the middle of each clock cycle. (b) The bit pattern shown in (a) can be produced by feeding a sawtooth wave into a voltage-controlled oscillator that generates the sinusoidal *I* and *Q* inputs in QAM.

middle of each clock cycle to determine the binary digit. This is basically FM with two values of frequency and as shown in Section 4.3.6, a frequency shift up is achieved by a continuously increasing phase angle of the carrier while a downward frequency shift is produced by a continuously decreasing phase angle. It was shown in Section 4.3.6 that FM can be produced by QAM with a rotating phasor of constant amplitude and this in turn can be generated by feeding sinusoidal signals into the *I* and *Q* inputs (see Figure 4.17). A clockwise rotating phasor produces a decrease in the carrier frequency while an anticlockwise rotating phasor produces an increase in frequency. The bit pattern shown in Figure 4.22a can be reproduced by feeding a sawtooth pattern into a voltage-controlled oscillator, which generates the *I* and *Q* inputs. BCPFSK modulation has been specified as the standard method for the transmission of digital data by universal access transceivers (see Section 5.4.11).

4.3.8 ITU Codes for Radio Emissions

In the last few sections, various types of modulation specialized to their respective transmission of information have been described, but this is just the small sample of the many types of signals and modulations used that is specific to aviation. In order to provide a shorthand description of radio transmission the ITU has developed an internationally agreed code shown in Figure 4.23. It consists of three blocks of letters and digits with the first block describing the bandwidth, the second providing details of the modulation used and the information carried, and the last providing more details of the transmitted information and the type of multiplexing used. An example of the code used for VHF voice communications is also shown. In the bandwidth box the symbol used to denote the frequency range (H, K, M, or G) is also used as a decimal point. So, for example, 125K denotes 125 kHz and 12M5 denotes 12.5 MHz, etc. The codes used for modulation types and information carried in the second block are shown in the tables in Figure 4.23 and the last block is never used for aviation signals. In practice, the bandwidth box is almost always absent as well so most aviation signals just use the three-character code in the middle box. Thus, for example, VHF and HF voice communications are labeled A3E and J3E while an NDB transmitting a Morse identification code would be labeled N3A.

4.4 Antennas

Aircraft are fitted with a large number of antennas designed to transmit and receive at frequencies ranging from the LF to the SHF band to provide communication, radio navigation, and radar. Figure 4.24 illustrates the range of antennas found on a commercial airliner and their placement and this section will describe the various antenna types in detail starting, in the following section, with some basic antenna theory.

4.4.1 Basic Antenna Theory

The radiation emitted by an oscillating current in a wire and how it propagates are described by Maxwell's equations, which in the case of antennas are normally solved in spherical polar coordinates, as described in Figure 4.25. The simplest type of radio wave source, illustrated in Figure 4.25, is an infinitesimal current element with a length $\delta\lambda \ll \lambda$, where λ is the wavelength, referred to as a Hertzian dipole. This is an artificial construct not achievable in practice but it is useful because the linearity of Maxwell's equations ensures that the fields generated by a wire antenna of any shape are the sum of fields produced by Hertzian dipoles joined together to reproduce the antenna. As derived in Appendix A (see Equations (A1.56) and (A1.57)), at distances, $r \gg \lambda$, described

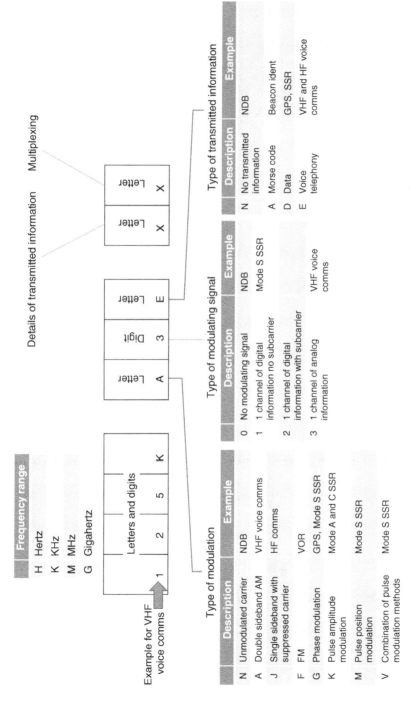

Figure 4.23 ITU codes used to describe types cf radio transmissions.

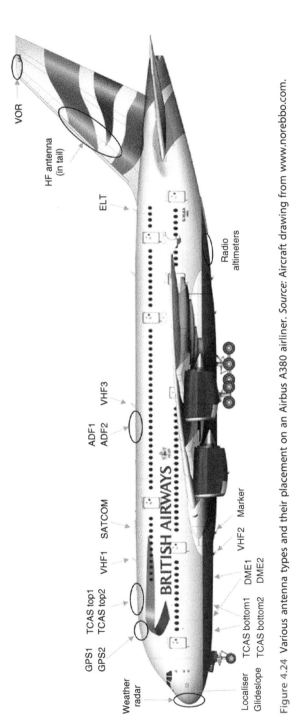

Figure 4.24 Various antenna types and their placement on an Airbus A380 airliner. *Source:* Aircraft drawing from www.norebbo.com.

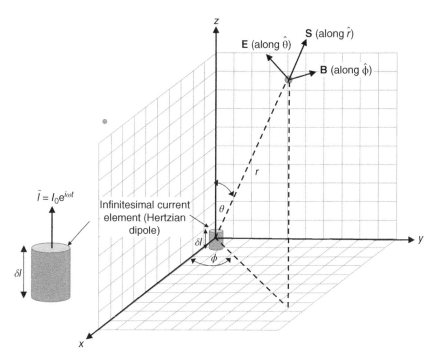

Figure 4.25 Spherical polar coordinates relative to a Cartesian coordinate frame and the **E** and **B** fields produced by a Hertzian dipole (Equations 4.31 and 4.32). The Poynting vector, **S** = **E** × **H**, which is orthogonal to E and H and is thus along r̂ is also shown.

as the *far field* of an antenna, the electric and magnetic fields generated by a Hertzian dipole are

$$\mathbf{B} = \frac{ik\mu_0\delta l\tilde{I}}{4\pi r}\sin\theta e^{-ikr}\hat{\boldsymbol{\phi}} \tag{4.31}$$

$$\mathbf{E} = \frac{ikZ_0\delta l\tilde{I}}{4\pi r}\sin\theta e^{-ikr}\hat{\boldsymbol{\theta}} \tag{4.32}$$

Here, the constant μ_0 (= $4\pi \times 10^{-7}$ NA^{-2}) is the magnetic permeability of free space, k is the wavevector, and $Z_0 = \sqrt{\mu_0/\varepsilon_0} = 377\Omega$ is referred to as the impedance of free space. The equations show a high degree of symmetry so that the magnitude of the electric and magnetic fields only differ by the factor Z_0, that is:

$$|\mathbf{E}| = Z_0|\mathbf{H}| \tag{4.33}$$

The current $\tilde{I}$ passing through the element oscillates at a frequency ω and is given by

$$\tilde{I} = I_0 e^{i\omega t} \tag{4.34}$$

Note that the fields predicted by Equations (4.31) and (4.32) are entirely imaginary but this simply specifies that they oscillate 90° out of phase with the real part of the current in the dipole. In order to evaluate measurable quantities, the magnitudes of the fields are used and mostly the quantity of interest is the radiated intensity, that is, the power per unit area, a distance r from the dipole, which is given by [5]:

$$\langle \mathbf{S} \rangle = \frac{1}{\mu_0} \mathrm{Re}(\mathbf{E} \times \mathbf{B}^*) \tag{4.35}$$

where $\mathbf{S}$ is the Poynting vector, which is aligned along the direction of propagation (see Figure 4.25). The $*$ on $\mathbf{B}$ denotes the complex conjugate and all quantities are time averaged over one cycle, thus from Equations (4.31) and (4.32) the radiated intensity, $R_I(\theta)$, is

$$R_I(\theta) = \frac{Z_0 k^2 |I|^2 \delta l^2}{(4\pi r)^2} \sin^2 \theta \tag{4.36}$$

where $|I|$ is the time-averaged current. The intensity thus follows the inverse square law with distance as expected and varies with elevation angle, θ, as $\sin^2 \theta$. This is illustrated on a polar plot in Figure 4.26a, which also shows the $\sin \theta$ dependence of the fields. The intensity is isotropic with azimuthal angle, ϕ, so the three-dimensional intensity pattern appears as a donut (Figure 4.26b).

From Equation (4.36), we can calculate the total radiated power, P_{tot}, from the dipole by integrating R_I over the surface of a sphere centered on the antenna,

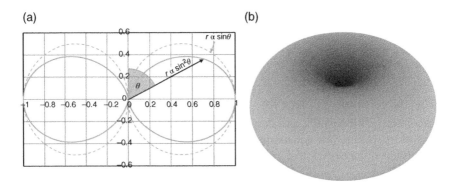

Figure 4.26 (a) Polar plot of the angular variation of intensity of radiation from a Hertzian dipole as a function of elevation (blue line) compared to the angular variation of the electric field (red dashed line). (b) Three-dimensional plot of radiation intensity from a Hertzian dipole. *Source:* Calculated by cocoaNEC2 [6].

which clearly should be independent of the radius of the sphere. When integrating over a spherical surface, S, as shown in Figure 4.27, the area element is given by

$$dS = r^2 \sin\theta d\theta d\phi \tag{4.37}$$

and to cover the whole surface requires that ϕ is swept from 0 to 2π and θ is swept from 0 to π. Thus, the integral of a function $F(r, \theta, \phi)$ over the surface is given by

$$\int_S F(r, \theta, \phi) dS = \int_0^{2\pi} \int_0^{\pi} Fr^2 \sin\theta d\theta d\phi \tag{4.38}$$

In the trivial case where $F = 1$, the integral simply returns the area of the surface ($4\pi r^2$).

Applying this to Equation (4.36) to obtain P_{tot} gives

$$P_{\text{tot}} = Z_0 k^2 |I|^2 \delta l^2 \int_0^{2\pi} \int_0^{\pi} \frac{\sin^2\theta}{(4\pi r)^2} r^2 \sin\theta d\theta d\phi \tag{4.39}$$

$$P_{\text{tot}} = \frac{Z_0}{16\pi^2} k^2 |I|^2 \delta l^2 \int_0^{2\pi} \int_0^{\pi} \sin^3\theta d\theta d\phi \tag{4.40}$$

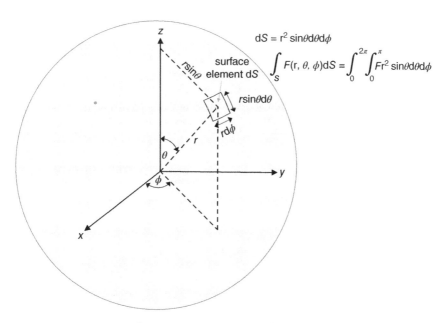

Figure 4.27 Integrating a function over a spherical surface in polar coordinates.

$$P_{\text{tot}} = \frac{Z_0}{16\pi^2} k^2 |I|^2 \delta l^2 2\pi \int_0^\pi \sin^3\theta d\theta \qquad (4.41)$$

The integral over θ gives 4/3 and replacing k with $2\pi/\lambda$, where λ is the wavelength gives

$$P_{\text{tot}} = \frac{2\pi Z_0}{3} |I|^2 \left(\frac{\delta l}{\lambda}\right)^2 \qquad (4.42)$$

Equation (4.42) shows that the total radiated power is proportional to I^2 and generally in an electrical circuit, power $= I^2 R$, so we can define a *radiation resistance*, R_{rad}, as the quantity:

$$R_{\text{rad}} = \frac{2\pi Z_0}{3} \left(\frac{\delta l}{\lambda}\right)^2 = 789 \left(\frac{\delta l}{\lambda}\right)^2 \Omega \qquad (4.43)$$

The role of radiation resistance can be understood by considering the equivalent circuit for an antenna shown in Figure 4.28. The load on the transmitter power supply can be considered as three loads in series, that is, the DC resistance and reactance of the antenna wire and the radiation resistance. In real antennas, the DC resistance of the wire is negligible and when the antenna is resonant (see below) the reactance goes to zero so the major component is the radiation resistance, which can be considered as the load on the transmitter required to generate the radio energy.

Beermat Calculation 4.2

A current of 1 A at a frequency of 100 MHz is passed through a Hertzian dipole of length 3 cm. Calculate the total radio energy emitted.

A frequency of 100 MHz corresponds to a wavelength, λ of 3 m (Equation 4.3). If the wire length is $\delta\lambda = 0.03$ m, then $\delta l/\lambda = 0.01$ and so from Equation (4.43) the radiation resistance is $R_{\text{rad}} = 0.079\,\Omega$. Thus, the power radiated $= I^2 R_{\text{rad}} = 0.079$ W.

DC resistance of
aerial wire (R_{dc})

Figure 4.28 Equivalent circuit for antenna.

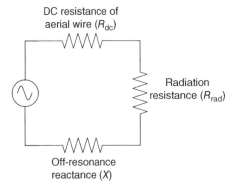

Radiation
resistance (R_{rad})

Off-resonance
reactance (X)

As shown in Beermat Calculation 4.2, a piece of wire satisfying the requirement to be a Hertzian dipole ($\delta l/\lambda \ll 1$) has a very low radiation resistance and emits a negligible amount of radio energy for normal currents. The power produced will increase as the square of the length but beyond a certain size the dipole can no longer be considered to be Hertzian and a new expression for the power generated and radiation resistance must be derived. In general, since the power radiated is $I^2 R_{rad}$, it is desirable for a transmitting antenna to have as high a radiation resistance as possible.

The efficiency of the antenna, which is often denoted by the symbol η, is defined by

$$\eta = \frac{P_{rad}}{P_{in}} \tag{4.44}$$

where P_{rad} is the power radiated by the antennas as radio waves and P_{in} is the power input to the antenna by the transmitter power supply. With reference to Figure 4.28, Equation (4.44) can be written as

$$\eta = \frac{I^2 R_{rad}}{I^2 R_{rad} + I^2 R_{dc}} = \frac{R_{rad}}{R_{rad} + R_{dc}} \tag{4.45}$$

where I is the current flowing in the antenna. The value of R_{dc} needs to be as small as possible relative to R_{rad}, otherwise a significant amount of transmitter power is dissipated as heat in the antenna wire. The efficiency is a much more important quantity in transmitting antennas since in a receiving antenna the currents flowing are tiny. Generally, the efficiency considers only the losses in the antenna itself and does not include Ohmic losses and reflected power in the feeder (see Section 4.4.4).

The final property of the Hertzian dipole we will consider in this section is the directive gain or directivity. Figure 4.26 shows that the radiated intensity is isotropic in azimuth (ϕ) but in elevation the intensity varies between a maximum at $\theta = 90°$ to zero at $\theta = 0$. The directive gain $G(\theta)$ is the intensity along a specific direction of interest divided by the total intensity averaged over a sphere, P_{av}. The average intensity is

$$P_{av} = \frac{P_{tot}}{4\pi r^2} \tag{4.46}$$

Thus, from Equation (4.42),

$$P_{av} = \frac{Z_0}{12r^2}|I|^2 \left(\frac{\delta l}{\lambda}\right)^2 \tag{4.47}$$

The intensity along a specific elevation, θ, is $R_1(\theta)$, given by Equation (4.36), so the directive gain is, from Equations (4.36) and (4.47):

$$G(\theta) = \frac{R_1(\theta)}{P_{av}} = \frac{3}{2}\sin^2\theta \qquad (4.48)$$

So, along $\theta = 0$, the gain is 1.5, that is, the intensity at a specific distance in the horizontal direction is 50% greater than the average intensity over a sphere at the same distance. This is normally expressed in decibels (dB), which give the ratio of two powers, P and P_0, according to the equation:

$$dB = 10Log_{10}\left(\frac{P}{P_0}\right) \qquad (4.49)$$

So, a directive gain of 1.5 has a dB value of $10Log_{10}(1.5) = 1.76$ dB. This is a weak directivity and as we will see below directional antennas can be designed to produce a high directivity $G(\theta, \phi)$ in both elevation and azimuth. In the case of a directional antenna, a single gain G is specified, which is the intensity along the center of the directional beam divided by the total intensity of the radio emission averaged over a sphere. In other words, it is the intensity at the center of the beam compared to the intensity that would be given by a hypothetical isotropic antenna emitting the same power. The gain applies in both transmission and reception. For example, a directional antenna with a gain of 1000 would produce an intensity at the center of the directional beam that was 1000× the intensity emitted by a hypothetical isotropic antenna in the same direction. The same antenna pointed at a directed beam would generate 1000× the signal that would be produced by an isotropic antenna. The definition of the parameters for a Hertzian dipole such as radiation resistance and directive gain have the same definition for all antennas and in the sections below we consider different practical antennas starting with "resonant dipoles."

4.4.2 Resonant Half-Wave Dipole and Quarter-Wave Monopole Antennas for VHF and UHF

The simplest practical antenna consists of a straight wire with a total length $\lambda/2$, where λ is the wavelength of the radio waves, broken in the middle and fed with an alternating current as illustrated in Figure 4.29. The resonant length of the antenna means that a standing wave in current is set up with Figure 4.29b showing the current distribution at one maximum of the cycle and Figure 4.29c showing the current distribution half a cycle later. Normally the antenna has to transmit over a range of frequencies so the length is chosen to be $\lambda/2$ for the wavelength corresponding to the center of the band. For example, as discussed in Section 4.3.5, the VHF voice communication band extends from 118 to 136.975 MHz, which corresponds to wavelengths in the range

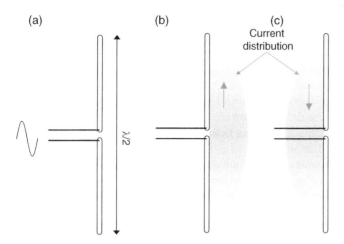

Figure 4.29 (a) Center-feed half wavelength dipole antenna. (b) Current distribution at the maximum of a cycle. (c) Current distribution half a cycle later. The arrows represent the direction of current flow.

2.19–2.54 m so a suitable dipole would have a length 1.183 m. In practice, the length will be slightly shorter to take account of the capacitance between the antenna and the ground, which changes the permittivity in the local environment.

Maxwell's equations are linear, so having worked out the electric field from a Hertzian dipole (Equation 4.32), the field generated by an arbitrary current distribution can be found by summing Hertzian dipoles that reproduce the current distribution of the real antenna. Consider a straight dipole of length l aligned in the z-direction, with a general current distribution, $\tilde{I}(z)$ (note that the harmonic time dependence is implied by the tilde over the I as in Equation (4.34)). One complication is that r and θ are both functions of z but if we assume that $r \gg l$, which is the same as the far field approximation, then all z positions can be considered to have the same r and θ so from Equation (4.32) the electric field of the dipole is

$$\mathbf{E} = \frac{ikZ_0}{4\pi r}\sin\theta e^{-ikr}\int_{-l/2}^{l/2}\tilde{I}(z)dz\hat{\boldsymbol{\theta}} \tag{4.50}$$

In the case of the half wave dipole:

$$\tilde{I}(z) = I_0 e^{i\omega t}\cos kz \tag{4.51}$$

and in order to transform the integral to one over θ we write $z = r\cos\theta$ and $dz = r\cos\theta \sin\theta \, d\theta$. The result is:

$$\mathbf{E} = \frac{ikZ_0\delta l\tilde{I}}{2\pi r}\frac{\cos\left(\frac{\pi}{2}\cos\theta\right)}{\sin\theta}e^{-ikr}\hat{\boldsymbol{\theta}} \tag{4.52}$$

So, apart from the factor of 2 in the constant the main modification to the field of the Hertzian dipole is that the elevation dependence has changed from $\sin\theta$ to $\cos\left(\frac{\pi}{2}\cos\theta\right)/\sin\theta$. As for the Hertzian dipole the intensity depends on the square of the magnitude of the electric field and the elevation angle dependence of the half-wave dipole compared to the Hertzian dipole is shown in Figure 4.30. The distribution is isotropic in azimuth (ϕ).

It is observed that the radiation pattern is slightly more directional and one can go through the same process as before to derive the directivity, or gain to be 2.16 dB. The radiation resistance is fixed at 73.1 Ω.

It is desirable for antennas mounted on aircraft to be as short as possible and at VHF and UHF frequencies the most common type is a quarter-wave monopole antenna, with a length $\lambda/4$, also known as a Marconi antenna after its inventor. It is fed with the RF signal at one end and mounted via an insulator on a conducting surface, which is assumed to be a perfect reflector. As seen in Figure 4.31, this gives it the same radiation pattern above the reflector as a half-wave dipole in free space and it can be represented as a real antenna above an image antenna of the same dimensions making it effectively a half-wave dipole. However, the current required to drive it is reduced by a factor of two and the impedance of the antenna is different. In addition, the directivity increases simply because the half of the beam below the $\theta = 0$ elevation has been removed.

The ground plane can either be the metal skin of the aircraft, or in the case of airframes made of composite materials, a metal liner that is placed on the inside

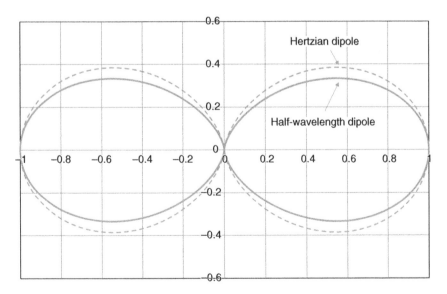

Figure 4.30 Polar plot of the angular variation of intensity of radiation from a half-wavelength dipole as a function of elevation (blue line) compared with that from a Hertzian dipole (dashed line).

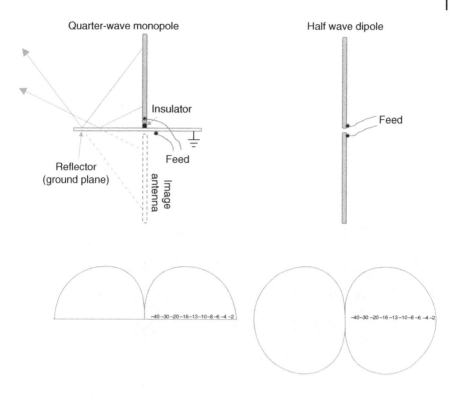

Figure 4.31 Quarter-wave monopole and radiation pattern in elevation compared to a half-wave dipole. Elevation plots were obtained using cocoaNEC2 [6].

under the antenna. It is also possible to mount quarter-wave antennas in the air by fitting them with a ground plane consisting of a number of grounded conductors placed around the bottom of the monopole projecting either horizontally or in a cone as shown in Figure 4.32d. In both cases the ground plane increases the gain in the horizontal direction but changing the angle of the wires alters the antenna impedance and sometimes non-horizontal wires are used to match the impedance to the coaxial feed cable (see Section 4.4.4).

Figure 4.32a shows a VHF quarter-wave monopole mounted at the top of the fuselage of a light twin for VHF voice communications, which is resonant at the wavelength corresponding to the center of the band, that is, 2.36 m with a quarter wavelength value of 0.59 m. The fuselage in this case is made of a composite nonconducting material and so a ground plane is installed under the antenna inside the fuselage. In the case of composite airframes, antennas can also be installed inside the fuselage and Figure 4.32b shows an internal UHF

(a) (b)

(c) (d) (e)

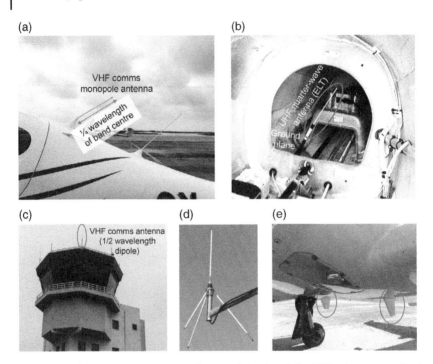

Figure 4.32 (a) Quarter-wave monopoles on a light aircraft for VHF voice communication. The length is resonant with the center of the band. (b) Quarter-wave monopole inside the composite fuselage of a light aircraft to transmit an emergency locator UHF signal at 406 MHz. The metal ground plane is also indicated. (c) Half-wave dipole for VHF voice communications on a control tower. (d) An elevated VHF quarter-wave monopole. *Source:* Reproduced from https://en.wikipedia.org/wiki/Monopole_antenna under Creative Commons license 3.0 [7]. (e) UHF DME and transponder blade antennas on the belly of a light twin.

quarter-wave monopole for transmitting an emergency locator signal at 406 MHz. Note the shorter length corresponding to the higher frequency of the UHF band. The ground plane under the antenna is also indicated. Figure 4.32c shows a half-wavelength dipole for VHF communications on a control tower and a VHF quarter-wave monopole with a radiating wire ground plane is shown in Figure 4.32d.

In the UHF band, an alternative to the thin wire monopole illustrated in Figure 4.32b is a blade antenna such as the distance measuring equipment (DME) and transponder antennas shown on the belly of a light twin in Figure 4.32e. These are also fed at the base against a ground plane and have a similar radiation pattern to the thin wire. They are, however, more robust and resistant to environmental phenomena such as hail and rain erosion.

4.4.3   Effect of Ground and Airframe on Radiation Pattern

So far, most of the radiation patterns plotted have been for propagation in free space but from ground-based antennas the radiation emitted at elevation angles, $\theta < 0$ reflects from the ground and interferes with the upward-going emission. This interference produces peaks and troughs in the emitted intensity along specific elevation angles as shown in Figure 4.33, which plots the radiation patterns at 120 MHz from a half-wave dipole placed at different heights above the ground. Constructive interference occurs whenever the path length difference between the direct wave and the reflected wave is an integral number of wavelength though the reflection from non-perfect ground will modify the effective path length. Midway between there will be destructive interference, so superimposed on the radiation pattern will be a series of peaks and troughs with the peaks occurring at angles that increment the path length difference by one. The nominal situation for the lowest dipole (Figure 4.33b), which is one wavelength above the ground, is shown in the inset in Figure 4.33b, which indicates that there should be three peaks superimposed on the radiation pattern. This is evident in the calculated pattern though the vertical one is partly obscured as it is superimposed on the elevation where the intensity goes to zero ($\theta = 0$). As the height of the antenna

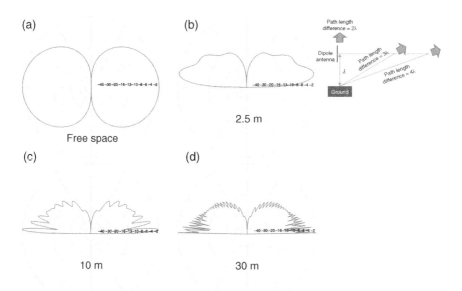

Figure 4.33 Radiation patterns of a half-wave dipole at different heights above the ground. (a) Radiation pattern in free space for comparison. (b)–(d) Radiation patterns for the center of the dipole at the indicated distances above the ground. The inset in (b) illustrates the elevation angle positions of constructive interference in the case of the antenna at 2.5 m (one wavelength off the ground). Elevation plots were obtained using cocoaNEC2 [6].

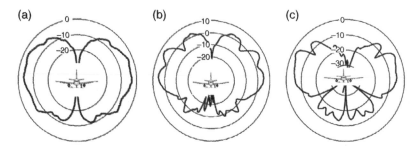

Figure 4.34 Radiation patterns in the roll plane of a monopole antenna at 120 MHz placed at different positions at the top of the fuselage measured using a scale model. (a) Forward of wings, (b) between the wings, and (c) between the wings and engines. *Source:* Reproduced with permission of John Wiley & Sons from Ref. [8].

gets higher, the wavelength increments in path length difference correspond to smaller changes in angle as seen by the increasing numbers of maxima and minima superimposed on the radiation pattern (Figure 4.32c and d). If the ground is flat, the azimuthal pattern will remain isotropic.

The reflective components in an aircraft will also modify the radiation pattern from antennas mounted on the airframe. The airframe or metal sheet immediately at the base of a quarter-wavelength antenna is used as an active element to create the ground plane but the overall airframe produces complex modifications to the radiation pattern in both elevation and azimuthal angles. Figure 4.34 shows measured radiation patterns from a scale model of an aircraft for a monopole VHF antenna placed at different positions at the top of the fuselage [8]. When measuring from a scale model, the wavelength of the radiation also has to be scaled by the same factor but the frequency shown, that is, 120 MHz, which is in the VHF communications band, is the value appropriate to the full-sized aircraft. It is evident that there is a fairly weak ground plane effect from the aircraft fuselage away from the wings (Figure 4.34a) compared to the radiation patterns shown in Figure 4.33, where the ground plane is infinite. Thus, the doughnut radiation pattern is largely preserved though the conducting fuselage is required to produce the image monopole. As shown in Figure 4.33b and c reflections from the wings and engines produce much more complex radiation patterns. Finding optimum positions to site antennas is a complex part of the aircraft design [8].

4.4.4 Feeders, Transmission Lines, Impedance Matching, and Standing Wave Ratio

The power from the output stage of the transmitter needs to be "fed" to the antenna, which may be some distance away. The feeder is generally described as a transmission line and can be just a pair of wires (Figure 4.35a) but due to its characteristic of screening the signal from electrical noise, the most common type of transmission line used in aircraft is a coaxial cable (Figure 4.35b). This

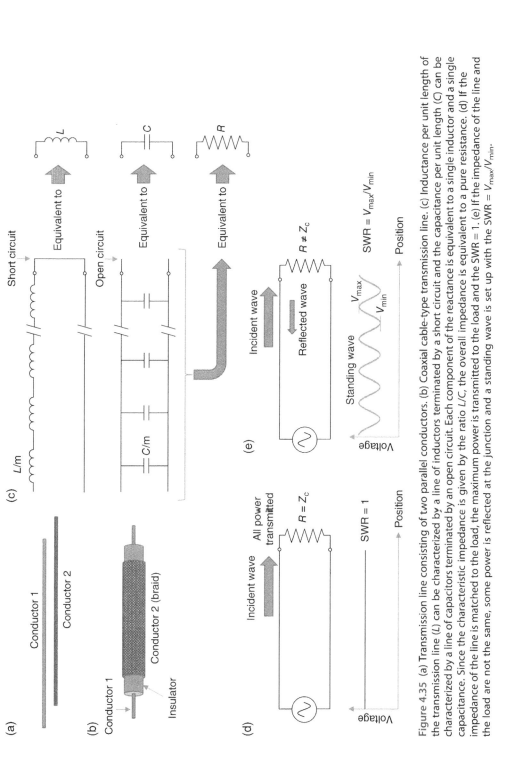

Figure 4.35 (a) Transmission line consisting of two parallel conductors. (b) Coaxial cable-type transmission line. (c) Inductance per unit length of the transmission line (L) can be characterized by a line of inductors terminated by a short circuit and the capacitance per unit length (C) can be characterized by a line of capacitors terminated by an open circuit. Each component of the reactance is equivalent to a single inductor and a single capacitance. Since the characteristic impedance is given by the ratio L/C, the overall impedance is equivalent to a pure resistance. (d) If the impedance of the line is matched to the load, the maximum power is transmitted to the load and the SWR = 1. (e) If the impedance of the line and the load are not the same, some power is reflected at the junction and a standing wave is set up with the SWR = V_{max}/V_{min}.

has a characteristic impedance, that is, the impedance sensed by a source connected to an infinite length of the line, which has an inductance per unit length (L) and a capacitance per unit length (C) as illustrated in Figure 4.35c. The characteristic impedance, Z_c, is given by (see Equation A2.7):

$$Z_c = \sqrt{\frac{L}{C}} \ \Omega \qquad (4.53)$$

The inductance per unit length can be represented by a line of inductors connected in series terminated by a closed circuit and the capacitance per unit length by a set of capacitors connected in parallel terminated by an open circuit. Each of these components of the impedance is equivalent to a single inductor and a capacitor connected to the output stage of the transmitter. The reactance of the inductor (X_L) and capacitor (X_C) vary with frequency (ω) in the opposite sense, that is, $X_L = \omega L$ and $X_C = 1/\omega C$. Since it is the ratio L/C that determines the characteristic impedance in Equation (4.53), Z_c is the same for any length of transmission line and the frequency dependence disappears, so the overall equivalent load of the transmission line is a pure resistance as illustrated in Figure 4.35c. This analysis is simplified as the pure resistance of the line itself has been ignored but this is usually very small compared to the inductive and capacitive reactance. Typical characteristic impedances of transmission lines are 50–100 Ω.

An oscillating signal fed into a transmission line can be considered to be a traveling wave. If the line is connected to a load that is the same as the characteristic impedance, the load will appear as an infinite length of transmission line, so the traveling wave will be transmitted without loss (Figure 4.35d). Thus, the maximum power is transmitted to the load when its impedance is matched to that of the line. Similarly, the maximum power is transferred into the transmission line when the output impedance of the transmitter matches that of the line. In a perfectly matched system, the voltage (and current) do not vary with position along the line. If there is an impedance mismatch at either connection, a part of the incident wave is reflected at the junction and a standing wave is set up along the line as illustrated in Figure 4.35e. In this situation, not all the power is transmitted into the load as some is reflected and the voltage (and current) vary sinusoidally with position along the line. The ratio of maximum voltage (V_{max}) to the minimum voltage (V_{min}) (or maximum to minimum current) is known as the standing wave ratio (SWR) and is a measure of the efficiency of the power transmitted into the load by the line. In the case of an impedance-matched load the SWR = 1. The load represented by the antenna is the radiation resistance, which can be calculated for each type of antenna as has been done in the previous sections for the case of the Hertzian dipole and the half-wave resonant dipole.

4.4.5 HF Antennas for Skywave Communications

As discussed in Section 4.2.3, long-range skywave communications utilize the HF part of the spectrum and the specific bands allocated to aviation extend from 2.85 to 23.35 MHz. This corresponds to wavelengths in the range 12.8–105 m and poses a problem in installing a resonant antenna on an aircraft as at the low frequency end even a quarter-wave Marconi antenna would need to be 26.3 m long. Resonant long wire antennas were used on vintage aircraft and strung between the top of the fuselage and the top of the vertical stabilizer (Figure 4.36a). This is also a suitable geometry to transmit horizontally polarized radio waves, which have maximum reflectivity from the ground and the ionosphere. As discussed in Section 1.5, some vintage aircraft could also deploy a long trailing antenna that could be unwound from a reel.

Such long wire installations are, however, unsuitable for high-speed aircraft and by the 1940s experiments explored the idea of using "shunt antennas" in which one end of the antenna is grounded to the airframe and the other end is fed by a high-current low-voltage signal. The installation is usually on the leading edge of the tail fin (Figure 4.36b) and the entire tail fin becomes the radiator. One problem with this type of system is the sharp variation of SWR with frequency that makes the antenna efficient at just one wavelength. This is overcome by including a matching unit close to the antenna within the fin that can change the impedance and allows the transmitter to efficiently drive the antenna throughout the HF band.

4.4.6 Low-Frequency Small Loop Antenna

The automatic direction finder (ADF) used for radio navigation (see Section 7.1) relies on the detection of radio signals from NDBs on the ground transmitting in

(a)   (b)

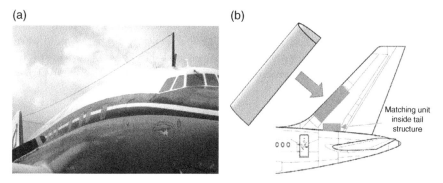

Matching unit inside tail structure

Figure 4.36 (a) Long-wire HF antenna on Super-Constellation airliner. *Source:* Reproduced from https://commons.wikimedia.org/wiki/File:Super_Constellation_HB-RSC_04.jpg under Creative Commons license 3.0 [7]. (b) HF shunt antenna installed on the leading edge of the tail.

the LF (30–300 kHz) and MF (300–3000 kHz) bands. Typical wavelengths are of the order of 1000 m and the antennas used to receive these signals are small loops typically around 1 cm. In this regime where the size of the antenna is very much smaller than the wavelength, the electrical signal in the loop can be derived using Faradays law (see Appendix A, equation A1.1) by considering the loop immersed in the time-varying magnetic field of the radio wave. Thus, the antenna couples to the magnetic component of the electromagnetic wave and small loop antennas are often referred to as magnetic loops. They can be considered to be the magnetic equivalent of a Hertzian dipole and also have a very small radiation resistance that depends on their size. The radiation resistance can be increased by using multiple loops in a coil and also by using a ferrite core, which effectively increases the surface area of each loop. Note that the geometry of the loops is not important and they can be circular or square, the critical parameter being the surface area, S. For a coil of N turns, the radiation resistance is given by

$$R_{rad} = \frac{8\pi^3 Z_0}{3} \left(\frac{NS}{\lambda^2} \right)^2 \tag{4.54}$$

Since $Z_0 = 377\ \Omega$, all the constants in Equation (4.54) can be gathered to produce the simplified expression:

$$R_{rad} = \left(\frac{177NS}{\lambda^2} \right)^2 \tag{4.55}$$

Even with multiple loops and ferrite cores, for typical LF wavelengths the radiation resistance is very small and the small loop makes a very inefficient transmitting antenna (see Equation 4.45), but it makes a reasonable receiving antenna. The free space radiation pattern (relevant for both transmitting and receiving) is identical to that for the Hertzian dipole though the polarization is orthogonal as illustrated in Figure 4.37. For a vertically polarized radio wave as transmitted by NDBs used for ADF (see Section 7.1), aligning the plane of the loop with the transmitting antenna maximizes the signal (see also Figure 1.14).

4.4.7 Directional Antennas in the VHF and UHF Bands

So far, all the antennas presented have had doughnut-shaped radiation patterns such as those shown in Figure 4.37 and in the azimuthal plane the radiated power is isotropic. Many transmissions in aviation require directed beams, which cannot be generated by a simple dipole-type antenna. This section describes directional antennas suitable for use in the VHF/UHF bands (30 MHz to 3 GHz).

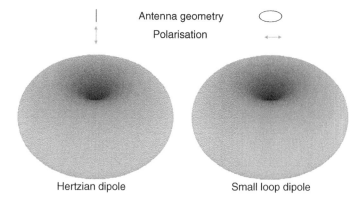

Figure 4.37 Comparison of free space radiation patterns (in terms of power) for a Hertzian dipole and a small loop antenna showing the antenna geometries and electric field polarization in comparison. *Source:* Calculated by cocoaNEC2 [6].

4.4.7.1 Yagi–Uda Antenna

The Yagi–Uda antenna, invented in 1926, consists of an array of antenna elements illustrated in Figure 4.38a in which there is a driven resonant half-wave dipole but the rest of the elements are passive, receiving and reradiating the emission from the active element. On one side, there is a single passive element called a reflector, which is slightly longer than the driven element and on the other side are one or more passive elements known as directors that are slightly shorter than the driven element. If the passive elements were resonant they would emit a wave 180° out of phase with the incident wave but making them nonresonant introduces an additional phase shift. The reflector being longer than resonant has an inductive reactance and so a phase shift that is opposite to the directors, which are shorter than resonant and have a capacitive reactance. The spacing of the elements, which is typically in the range $\lambda/10$ to $\lambda/4$ introduces an additional controllable phase shift. Thus, the antenna can be designed so that the waves from the driven and the passive elements all enhance on one side of the antenna and cancel on the other as illustrated for a driven element and a single director in Figure 4.38c. Since all the elements are resonant or close to resonant, there is a voltage node at the center of each so that the passive elements can be directly attached to a central supporting structure as illustrated in Figure 4.38b enabling a simple and cheap construction. Note that the length of the driven element will be slightly less than $\lambda/2$ to take into account the permittivity shift introduced by the neighboring elements.

Increasing the number of directors increases the directive gain of the antenna as shown in Figure 4.39, in which is plotted the angular power distribution from a basic three-element array (Figure 4.39a) and arrays with five (Figure 4.39b and

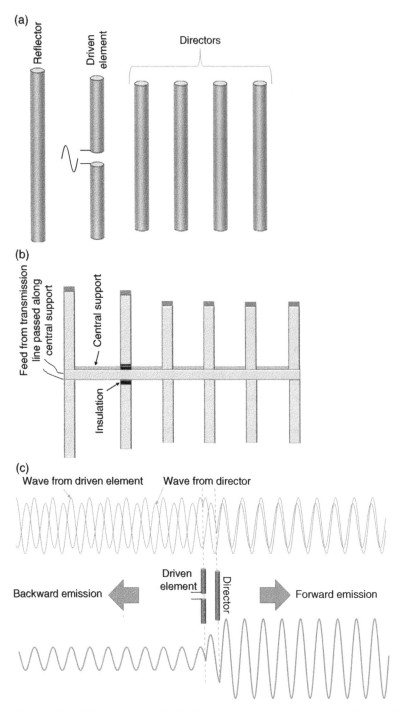

Figure 4.38 (a) Basic structure of a Yagi–Uda antenna with a resonant half-wave dipole-driven element, a passive reflector element on one side and one or more passive director elements on the other. (b) Practical construction of an antenna. Since all the elements have voltage nodes at the center, all the passive elements can be directly connected to a metal support. (c) Schematic of waves emitted by the driven element and a single director element showing the reinforcement of waves on one side and the cancelation on the other.

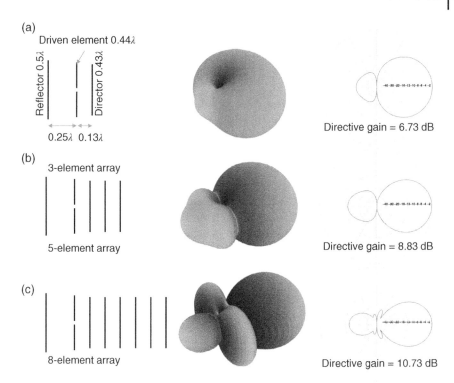

(a)
Driven element 0.44λ
Reflector 0.5λ
Director 0.43λ
0.25λ 0.13λ
Directive gain = 6.73 dB

(b)
3-element array
5-element array
Directive gain = 8.83 dB

(c)
8-element array
Directive gain = 10.73 dB

Figure 4.39 (a) Antenna configuration, three-dimensional plot of intensity, and elevation plot of intensity for a basic 3-element Yagi–Uda array. (b) As (a) for a 5-element Yagi–Uda array. (c) As (a) for an 8-element Yagi–Uda array. There is an increase in directive gain with the number of director elements. Note the appearance of side lobes for the higher gain antennas. Elevation and 3D plots were obtained using cocoaNEC2 [6].

8 (Figure 4.39c) elements. The increase in directive gain with the number of elements is evident and also for the higher gain antenna it is possible to observe the appearance of side lobes. These are weaker intensity maxima in directions other than the main lobe and are a common feature of directional antennas. It is found that increasing the number of reflectors does not have a significant effect on the directive gain.

4.4.7.2 Log-Periodic Antenna

Like the Yagi–Uda antenna the log-periodic antenna consists of a line of dipoles, but in this case every element is driven. The length of the dipoles in the array is set so that there is a constant ratio between the length of an element and its neighbor. Thus, if l is the length of element n, then:

$$\frac{l_n}{l_{n+1}} = k \tag{4.56}$$

where k is a constant and is a design parameter of the antenna. The same rule also applies to the distance, d, between the elements, that is:

$$\frac{d_n}{d_{n+1}} = k \qquad (4.57)$$

Figure 4.40a shows an example of a log-periodic antenna with $k = 1.25$. The driving waveform is reversed in going from element to element so that the emission is set to be 180° out of phase between neighboring elements, with nonresonant elements giving an additional phase shift. This is a similar situation to the Yagi–Uda antenna though in that case the 180° phase-shift in the emission between elements occurs naturally.

As shown in Figure 4.40b, each three-element section can be considered as a Yagi–Uda antenna at the driving or receiving frequency at which the central element is resonant. The directive gain is achieved in the same way, that is, an enhancement of the emission in the forward direction and a cancelation in the opposite direction. The log-normal antenna has a wide bandwidth as the length of the resonant element varies from the second longest in the array to the second shortest in contrast to the Yagi–Uda antenna in which there is a single resonant element. A common use for this type of antenna is in the localizer of an instrument landing system (ILS – see Section 7.4).

4.4.8 Directional Antennas in the SHF Band

4.4.8.1 Waveguides as Feeders
Section 4.4.4 discussed transmission lines as antenna feeders. As the frequency increases the losses in transmission lines also increase and in the SHF band (3–30 GHz) it is better to dispense with the central conductor in the transmission line and pass the signal through a hollow metal tube known as a waveguide. The geometry of a typical rectangular waveguide and the transverse polarization of the electromagnetic wave propagating down the tube is shown in Figure 4.41a with the field distribution in the x- and z-directions plotted in Figure 4.41b.

As shown in Appendix B this type of waveguide has a cutoff frequency ω_c, given by (see Equation A2.15):

$$\omega_c = \frac{\pi c}{a} \text{ or } f_c = \frac{c}{2a} \qquad (4.58)$$

and frequencies above ω_c (f_c) are transmitted efficiently while below the cutoff frequency the wave is attenuated and does not propagate. Generally, waveguides are used in single-frequency applications such as radar where the dimensions used allow only the fundamental mode to propagate and the rectangular cross-section maintains the polarization of the transmitted wave (see Appendix B). The electromagnetic wave inside the waveguide can be excited by driving a

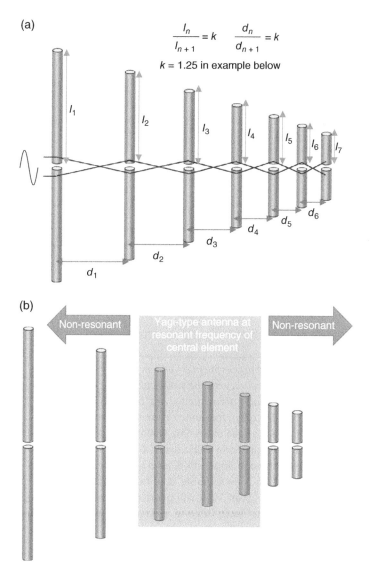

(a)

$$\frac{l_n}{l_{n+1}} = k \qquad \frac{d_n}{d_{n+1}} = k$$

$k = 1.25$ in example below

l_1 l_2 l_3 l_4 l_5 l_6 l_7

d_1 d_2 d_3 d_4 d_5 d_6

(b)

Non-resonant

Yagi-type antenna at resonant frequency of central element

Non-resonant

Figure 4.40 (a) Log-periodic antenna with $k = 1.25$. (b) Each three element section can be considered as a Yagi–Uda antenna for the frequency at which the central element is resonant.

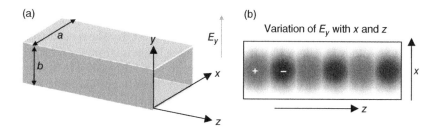

(a)

a

b

y

E_y

x

z

(b)

Variation of E_y with x and z

+ −

x

z

Figure 4.41 (a) Geometry of a rectangular waveguide and the polarization of the propagating electromagnetic field. (b) Schematic of the variation of E_y with x and z along the tube.

(a)

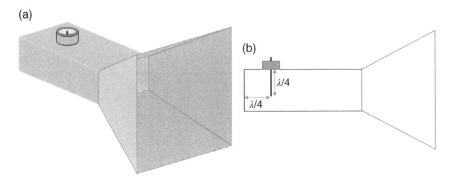

(b)

$\lambda/4$

$\lambda/4$

Figure 4.42 (a) Pyramidal horn attached to a short length of waveguide used as a microwave antenna. (b) Excitation source of the antenna consisting of a quarter-wave Marconi dipole.

quarter-wave Marconi antenna at one end (see Figure 4.42) or using a slot antenna described below.

Beermat Calculation 4.3

An airborne weather radar designed to operate at a frequency of 6 GHz uses a waveguide as a feeder to a parabolic antenna with the wave polarized in the y-direction shown in Figure 4.41. Estimate suitable dimensions a and b for the waveguide.

Equation (4.58) shows that for a given frequency there is a minimum value for a for the wave to propagate. In the case of a 6 GHz wave this is given by $a_{min} = c/2f = \lambda/2$ where λ is the wavelength. For a 6 GHz wave, $\lambda = 5$ cm so clearly a has to be greater than 2.5 cm. As shown in Appendix B, for $a > \lambda$ it is possible to get a second mode propagating, which is undesirable, so a needs to be less than 5 cm. On the other hand, if b is less than $\lambda/2$ it will prevent a wave propagating with orthogonal polarization thus ensuring the polarization is maintained. Therefore, suitable dimensions for the waveguide would be, for example, $a = 4$ cm, $b = 2$ cm.

4.4.8.2 Horn Antenna

Although it would appear that a waveguide provides a directed beam, it makes a poor antenna since the abrupt change of resistance at the end between the radiation resistance of the waveguide and free space ($Z_0 = 377\ \Omega$) produces significant reflection and a high SWR. In addition, due to the fact that the aperture has similar dimensions to the wavelength, the emerging beam is strongly diffracted so the directive gain is reduced. The transition can be made gradual by flaring out the ends to produce a horn, which reduces the reflection giving a more efficient antenna with a higher directive gain. Typical gains of horn antennas are in the range 10–20 dB and since there are no resonant elements they have a high bandwidth, that is, they can operate over a wide frequency range. Figure 4.42a shows a pyramidal-type horn attached to a short length of waveguide, which is a combination typically used as a microwave antenna. Figure 4.42b illustrates the excitation source consisting of a Marconi quarter-wave dipole in the optimum position, a quarter wavelength from the waveguide end with the waveguide wall acting as the ground plane. As well as stand-alone antennas, horns are often used as drivers for larger parabolic dish antennas described in the next section.

Interesting Diversion 4.1: Horn Antennas and Microwaves from Space

Horn antennas detecting microwaves from space have been at the heart of two of the most important discoveries in astrophysics. The first was the detection of microwave radiation emitted by neutral hydrogen at a frequency of 1420 MHz, $\lambda = 21$ cm in the UHF part of the spectrum by Ewen and Purcell in 1951 using a Horn antenna (Figure ID4.1.1). The emission arises when the spin of an electron

Figure ID4.1.1 Horn antenna used by Ewen and Purcell at Harvard University in 1951 for the first detection of radio emission from atomic hydrogen gas in the Milky Way at a wavelength of 21 cm. The antenna is now on display at the National Radio Astronomy Observatory, Green Bank, West Virgina, USA. *Source:* Reprinted under Creative Commons license 3.0 [7].

(Continued)

Interesting Diversion 4.1: (Continued)

in a hydrogen atom changes alignment from parallel to anti-parallel with that of the single proton nucleus and the tiny energy difference is emitted as a RF photon. The excited state is extremely stable and has a lifetime of around 10 million years, so it is very difficult to observe this emission on Earth but the huge clouds of neutral hydrogen gas distributed through our galaxy provide a detectable 21 cm emission whose intensity depends on the density of hydrogen gas.

The importance of this discovery is that the 21 cm radiation is not absorbed by dust, so for the first time the distribution of neutral hydrogen throughout the galaxy became detectable and could be mapped revealing the spiral structure that we are familiar with. The first such maps were made in 1952. In addition, measuring Doppler shifts of the 21 cm line reveal how the hydrogen is moving relative to Earth.

The second discovery was the cosmic microwave background (CMB) by Penzias and Wilson in 1964 using the huge Horn antenna shown in Figure ID4.1.2.

Figure ID4.1.2 The 15 m Horn antenna at Bell telephone laboratories in Holmdel with which Penzias and Wilson first detected the cosmic microwave background radiation in 1964. *Source:* Reprinted under Creative Commons license 3.0 [7].

In this application, the receiving end of the horn was fitted with a radiometer that measured the total power being received by the antenna. In radiometer applications, it is normal to describe the total power received as an equivalent temperature and the signal is the thermal noise that would be generated by a resistor at that temperature (the Nyquist–Johnson noise – see Appendix C). When all other sources had been eliminated, Penzias and Wilson found that there was a residual excess equivalent temperature of about 4 K, which was attributed to the predicted CMB radiation resulting from the creation of the Universe. This can be regarded as the temperature of space and more recent satellite-based radiometry measurement have fixed the precise value at 2.725 K.

At the point of creation, the Universe consisted of a tiny volume at an enormous temperature and has since been expanding and cooling. It is important to realize that it is space itself that has been expanding and the Universe can be considered as an expanding closed cavity containing radiation (a black body) in which the internal temperature has been dropping from the enormous value at the start till around 14 billion years later it has reached the 2.7 K that has been measured today. The spectrum of a black body can be precisely calculated and the curve for 2.725 K is shown in Figure ID4.1.3 revealing that the emission is in the microwave region with the peak at 160 GHz ($\lambda = 1.875$ mm). The SHF band is shown for comparison.

The discovery by Penzias and Wilson was regarded as the smoking gun evidence for the so-called "big bang" theory, which is currently the generally accepted model for the origin of the Universe. It earned them the 1978 Nobel Prize in Physics.

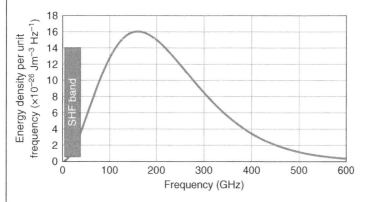

Figure ID4.1.3 Black body spectrum at 2.725 K showing emission in the microwave region. The SHF band is shown for comparison.

4.4.8.3 Parabolic Dish Antenna

A well-established way to obtain a directed beam is to use a parabolic reflecting surface, which is smooth on the scale of the wavelength. As illustrated in Figure 4.43a, a parabolic surface has the property that beams traveling parallel to the axis will all be specularly reflected to a single focal point, which in the case of a radio antenna is where the pick-up antenna is situated. Conversely, a point source at the focus will produce a beam parallel to the axis, so this is where the feeder antenna is placed. An additional useful property of the parabolic reflector is that the distance traveled by any beam from the focus and back to the focal plane is constant so that transmitted beams are all in phase.

Feeders for parabolic dishes are directional antennas, with horns being the most common type and the ideal radiation pattern is one that exactly fills the dish with the intensity dropping abruptly to zero at the edges, which is impossible to achieve in practice. If the feeder beam is wider than the parabolic reflector (Figure 4.43b), the dish is said to be overfilled and intensity is wasted. If it is narrower (Figure 4.43c) then some of the reflecting surface is not used. In practice, for a feed horn antenna, the optimum filling is given by a radiation pattern that drops to 10 dB below the peak value at the dish edge.

As shown in Equation (A3.13), the gain of a parabolic dish antenna is given by

$$G = \frac{4\pi A}{\lambda^2}\eta \qquad (4.59)$$

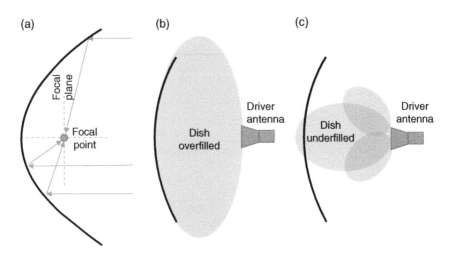

Figure 4.43 (a) A parabolic surface will bring all beams parallel to the axis to a single focal point by specular reflection. (b) Overfilling of the dish by the feed antenna. (c) Underfilling of the dish by the feed antenna.

where A is the geometric area of the dish, λ is the wavelength of the radio waves, and η is the antenna efficiency. In the case of a dish with a circular aperture of diameter d, this becomes

$$G = \left(\frac{\pi d}{\lambda}\right)^2 \eta \tag{4.60}$$

making it clear that it is the size relative to the wavelength that is the important parameter. Typical values for η are in the range 0.55–0.70 so, for example, a 1 m diameter dish operating at 3 cm (10 GHz), which is typical for an airborne weather radar would have a gain of around 2000 or 33 dB.

The radiation pattern from a parabolic dish is concentrated into the main lobe with a high gain but side lobes are still produced as a result of diffraction. Any aperture illuminated by or emitting a coherent beam will produce multiple weaker maxima around the main beam due to interference by diffracted beams from the edges of the aperture as illustrated in Figure 4.44a for an optical aperture. Whenever there is a path length difference of an integer number of wavelengths at a measurement plane between beams from opposite sides of the aperture, there will be coherent interference and an intensity maximum, though this will be much weaker than the central beam. If the measurement plane is a large distance away compared with the aperture diameter d, the bright beams occur at angles, θ, given by (for small angles):

$$\sin\theta = \frac{n\lambda}{d} \tag{4.61}$$

where λ is the wavelength and n is an integer with positive and negative values. In the case of visible light optics, the bright beams are referred to as interference fringes, while for radio transmission from a parabolic dish (Figure 4.44b) they are referred to as side lobes. Parabolic dishes also have back lobes of emission due to overfilling of the dish by the driver antenna.

The angular width of radar beams is characterized by the half power beam width (HPBW), that is, the total angular spread between the angle where the central beam has dropped to half the maximum power (-3 dB) to the equivalent angle on the other side as illustrated in Figure 4.44c. For a parabolic antenna, the HPBW, θ, is given by

$$\theta = \frac{k\lambda}{d} \tag{4.62}$$

where k depends on the filling efficiency and the shape of the dish but in the optimum case is $180/\pi$. A typical practical value is ~70 and using the weather radar example above, where $d = 1$ m and $\lambda = 3$ cm, gives $\theta = 2.1°$.

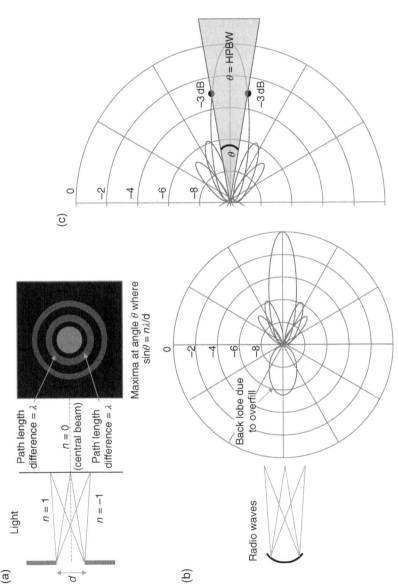

Figure 4.44 (a) Illustration of aperture diffraction. An intensity maximum occurs whenever the path length difference of beams from either side of the aperture is equal to an integer number of wavelengths. In optics the maxima are referred to as interference fringes. (b) The same process from an aperture antenna such as a parabolic dish produces side lobes. There is also a back lobe due to overfilling of the dish by the driver antenna. (c) Illustration of the half-power beam width characterization of a radar beam.

Beermat Calculation 4.4

Calculate the gain, both as a simple factor and in dB and the HPBW of the Lovell radio telescope at Jodrell Bank, whose parabolic dish diameter is 76.2 m, assuming an efficiency of 0.65 and a k factor of 70 when it is being used to detect the 21 cm wavelength hydrogen line (see Interesting Diversion 4.1).

The gain, G, is given by Equation (4.60) and is $(\pi^*76.2/0.21)^2 \times 0.65 = 844\,884$.

This is expressed in dB using Equation (4.49), that is $dB = 10Log_{10}(G) = 59.2$ dB.

The HPBW, θ, is given by Equation (4.62) and is $\theta = 70^*0.21/76.2 = 0.2°$.

Thus, as expected, this huge dish has a very high gain and a very narrow beam width.

4.4.8.4 Slotted Array

An alternative method to a parabolic dish for producing a pencil-type beam in the SHF band is to use an array of slot antennas. A single rectangular slot in a conducting plate can act as an antenna if an oscillating current is made to flow around it by driving it as indicated in Figure 4.45a. The radiation pattern is the same as a half-wave dipole oriented in the same direction but the polarization of the emitted wave is orthogonal (Figure 4.45b). The length of the slot determines the resonant wavelength in the same way as the length of a dipole and the width of the slot effects the bandwidth. The analysis is often given in terms of Babinet's principle, which states that the diffraction pattern from an opaque shape surrounded by a transparent medium is the same as a hole of the same shape and size in an opaque medium, though the intensity of the transmitted beam will be different. It is more instructive, however, to describe the operation in terms of the voltages and current around the slot.

Assume initially that the voltage is applied half way along the length of the slot. It must go to zero at the two ends so it will have a peak somewhere along

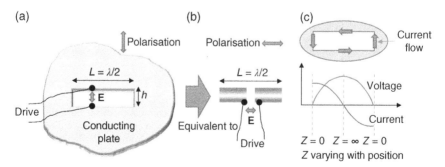

Figure 4.45 (a) Slot antenna in a conducting plate and driver. (b) The slot is equivalent to a dipole and gives a similar radiation pattern though the polarization of the emission is orthogonal. (c) The impedance of the slot antenna is controlled by the position of the driver.

the length of the slot as indicated by the red curve in Figure 4.45c. The voltage will drive a current around the edges as shown and in the middle of the slot the two currents in opposite directions cancel giving zero, while at the two edges the net current is maximum in opposite directions as indicated by the blue curve. Since the current is zero at the central position where the voltage is maximum, the impedance is infinite while at the edges where the current is maximum at zero voltage the impedance goes to zero. Thus, the impedance can be controlled by driving the slot at a suitable position along its length and can be matched to the feeding transmission line.

The normal way to drive a slot antenna is to use a waveguide, which simply means cutting a slot in the tube carrying the radiation. An analysis of the electric field in the waveguide and the resulting currents around the slot show that the position of the slot determines the emitted power. For example, a slot aligned with the length of the waveguide and placed at the center will not radiate, so slots need to be offset as indicated in Figure 4.46a. If a series of offset slots is

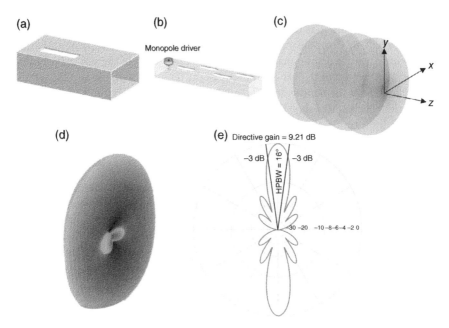

Figure 4.46 (a) Slot antenna excited by a waveguide, which must be offset from the centerline to radiate. (b) An array of slots cut into a waveguide and excited by a monopole source. (c) The radiation pattern from the entire array will result from the overlapping donut patterns from the individual slots. (d) Constructive and destructive interference will concentrate the pattern into a disc around the waveguide. (e) Azimuthal plot of the emission from an array of six slots of length 2 cm excited by a resonant 7.5 GHz wave ($\lambda = 4$ cm) in the waveguide revealing a directive gain of 9.21 dB and a HPBW of 16°. Azimuthal and 3D plots were calculated using CocoaNEC2 [6].

cut into the waveguide as shown in Figure 4.46b, the radiation pattern from the entire array will result from the overlapping donut patterns from the individual slots (Figure 4.46c). Constructive interference in the x–y plane and destructive interference moving away from this plane focuses the radiation pattern into a disc as shown in Figure 4.46d. The emitted intensity as a function of elevation angle from an array of six offset slots of length 2 cm excited by a resonant 7.5 GHz (λ = 4 cm) wave in the waveguide is plotted in Figure 4.46e and displays a directive gain of 9.21 dB with a HPBW of 16°. The side lobes are produced by constructive interference at angles where the path length difference of the emission from adjacent slots equals one wavelength, two wavelengths, etc. As the number of slots is increased the directive gain increases and the beam width decreases. Note that the radiation pattern shown assumes isolated slot arrays in space and in a practical system the emission will be just into the half space away from the waveguide.

To make a pencil beam that can be used for radar, linear slot arrays can be assembled as in Figure 4.47a to produce a two-dimensional array. If the square array is in the x–z plane, then constructive and destructive interference between the disc emissions from each line will produce a pencil beam along the y-direction as shown. As usual there will also be side lobes at angles where the path length difference in the emission from different waveguides equals an integer number of wavelengths. The array of slotted waveguides can either be fed by a set of monopoles connected to a common SHF driver as shown in Figure 4.47a or by coupling slots cut into an orthogonal waveguide illustrated in Figure 4.47b. A slotted array antenna used in an airborne weather radar (see Section 5.3) is shown in Figure 4.47c.

4.4.8.5 Patch or Microstrip Antenna

A simple SHF antenna that has some directivity is the patch or microstrip antenna, which consists of a flat conducting plate on a dielectric with a backing ground plane as illustrated in Figure 4.48a. This type of antenna is normally used to transmit the signal for radio altimeters (see Section 5.6). Its simplicity and flat design make it particularly suitable for mounting on the outside of the fuselage, which can also act as the ground plane. Figure 4.48b shows the instantaneous voltage distribution across the antenna in the direction of the feed line along with the currents, which are in opposite directions through the antenna and ground plane and thus cancel. The fringe electric fields that bulge from the edges of the antenna in the y-direction and are in phase on both sides produce the emission with a polarization along y. Each edge has a radiation pattern similar to a slot antenna and so the combined pattern from the two edges is the addition of two slots in phase, which gives it a directivity around 5–7 dB.

The center frequency, f_c, of the antenna is given by

$$f_c = \frac{c}{2L\sqrt{\varepsilon_r}} \tag{4.63}$$

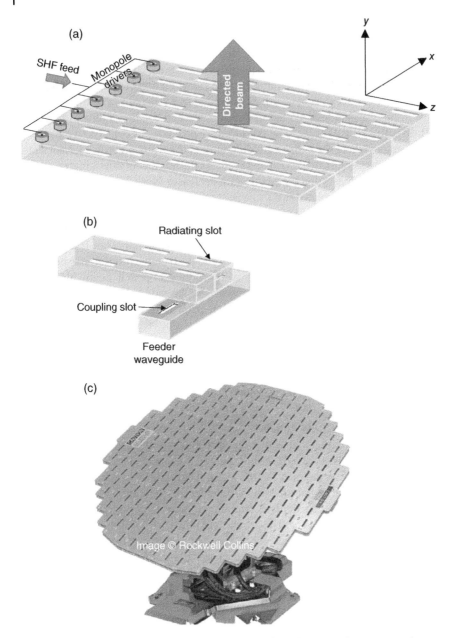

Figure 4.47 (a) Slotted waveguides assembled into a two-dimensional array to produce a pencil beam suitable for radar in the direction shown. There will also be associated side lobes. The feed for the array is a set of monopoles fed by a common SHF driver. (b) The array can also be fed by an orthogonal slotted waveguide. (c) A Rockwell Collins LT 005-WRAU airborne weather radar antenna. *Source:* Reproduced with permission of © Rockwell Collins.

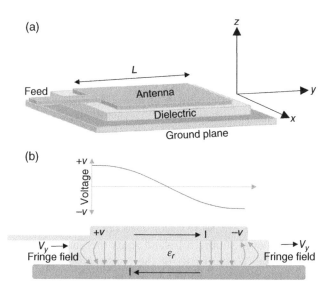

Figure 4.48 (a) Basic structure of a microstrip antenna consisting of the antenna element on a dielectric medium and backed by a ground plane. (b) Electric fields and currents in the antenna showing the fringe fields that produce the radiation polarized in the *y*-direction.

where c is the speed of light, L is the length of the antenna, and ε_r is the relative permittivity of the dielectric. In general, it is best to use a dielectric with a low relative permittivity close to 1, so that the fringe fields penetrate further in the *y*-direction giving higher emission. Thus, for a frequency of 4.3 GHz, which is typical for a radar altimeter, the length needs to be 7 cm. The bandwidth is determined to some extent by the width of the patch but is very narrow – typically around 3% of the center frequency so the antenna is restricted to applications such as the radar altimeter, which utilize a narrow frequency band.

4.5 VHF Communications System

As discussed in Section 4.3.5, the VHF voice communication band is in the frequency range 118–136.975 MHz and is divided into 2280 channels each 8.33 kHz wide. Propagation of VHF waves is by space wave with no penetration of terrain and is limited to line of sight only giving it a maximum usable range of about 200 nautical miles (see Section 4.2.2.2). For flights over densely populated areas, such as Europe, there are sufficient Air Traffic Control (ATC) centers so that aircraft are always within line of sight range of a control center and VHF voice communication is all that is required. As an aircraft travels, it is passed from one controller to another en route to stay within range. For the purposes of ATC, airspace is divided into "Flight Information Regions" and these are further subdivided into an Upper Information Region (UIR), for traffic above FL195, though this may vary and a Lower Information Region (LIR) for traffic below. Every portion of the navigable atmosphere belongs to a specific FIR with smaller countries allocated a single FIR within their borders and larger countries encompassing several FIRs. The map in Figure 4.49 shows the division of airspace in Europe into FIRs, each providing an ATC service. Within each FIR there is at least one control center that provides, at minimum, a permanent flight information service (FIS) and an alerting service that can process emergency calls and alert the appropriate search and rescue services.

Some of the FIRs are over oceans across which it is not possible to maintain VHF radio contact and satellite or HF skywave communications must be used (see below). These are designated as oceanic control areas (OCAs) and there is an oceanic control center with HF capability within each one. For example, in Figure 4.49, the Shannon FIR over Southern Ireland abuts the Shanwick FIR, which is an OCA covering the Western part of the North Atlantic Ocean. An interactive map of all FIRs globally is published by the ICAO [9].

As an example, consider the communications used by an IFR flight in an unpressurized twin-engine piston aircraft with a maximum flight level of FL100 between two regional airports, that is, Coventry in the United Kingdom to Quimper in France (see Figure 4.49). This involves flight in just two FIRs, that is London control and Brest control with the changeover half way across the English Channel. Table 4.1 shows a list of the stations to be contacted during the flight and it is evident that a large number of frequencies need to be used. Modern VHF radios have a standby frequency as well as the frequency in use so that a single button press will change to the next required frequency. In addition, there is a second communication channel with two frequencies, one in use and one standby, that can be selected by a single button press. Thus, four frequencies can quickly be selected by at most, two button presses. These are shown on the Primary Flight Display (PFD) of an aircraft with a glass cockpit in Figure 4.50. When tuning the radio, it is the standby frequency that is altered while the active frequency remains unchanged. Then, when it is time to change from one station to another the swap frequency button is pressed to make the standby frequency become the active one, which is displayed in green on the G1000 PFD shown in Figure 4.50.

Before start-up the VHF radio would be tuned to the Automatic Terminal Information Service (ATIS), which is a regularly updated recorded message providing airport information such as the runway in use, pressure settings, weather, specific traffic alerts, etc. Next, the tower frequency would be made active to request start-up clearance from ATC and when ready the pilot calls for taxi instructions. The tower will issue all clearances for departure, taxi, and takeoff and stay with the aircraft till it is climbing away and then issue an instruction for the pilot to change to the approach frequency for joining the en route phase. At

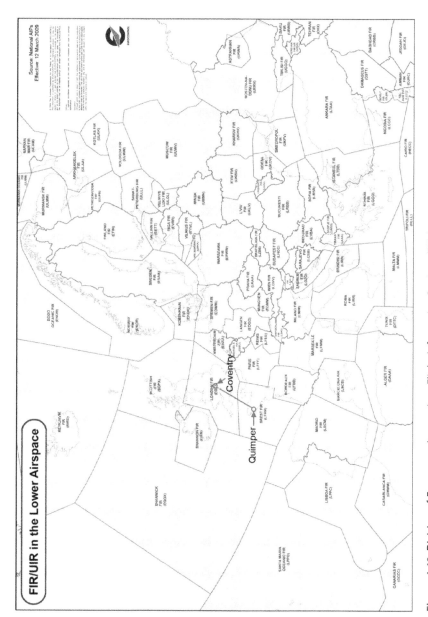

Figure 4.49 Division of European airspace into Flight Information Regions (FIRs).

Table 4.1 Example of VHF ATC control in a flight between two regional airports.

Station and call sign	Frequency (MHz)	Role	Phase of flight
Coventry ATIS (Automatic terminal information service)	126.050	Recorded message updated regularly providing airport information such as runway in use and weather	Prior to engine start
Coventry tower	118.175	Controls traffic during taxi, takeoff, and landing	Taxi and takeoff
Coventry approach	123.825	Controls traffic in the vicinity of the airport	Post takeoff prior to switching to en-route service
London control	Frequency allocated in flight	En route control	En route
London control	Frequency allocated in flight	En route control	En route
...etc. (several frequency changes)			
Brest control	Frequency allocated in flight	En route control after crossing from London to Brest FIR	Crossing from London to Brest FIR
Iroise approach	135.825	Controls traffic in the vicinity of Quimper airport among others	Approach to land
Quimper tower	118.625	Controls traffic during taxi, takeoff, and landing	Landing and taxi

some point the approach controller at Coventry will issue an instruction for the pilot to change frequency to a controller in the FIR control center, in this case, London control, who will issue the clearance to climb to the planned altitude and take over control of the flight. As the flight progresses the pilot will be passed from controller to controller, each assigned a separate frequency, who will issue instructions for the route to follow, sticking as closely as possible to the filed flight plan. Then, midway across the channel the pilot will be passed from London control to Brest control who will take over the management of the flight. When approaching the destination, Brest control will pass the flight over to Iroise approach. Quimper airport does not have a radar so control of flights is handled by a radar service located elsewhere. Finally, on approach to land, the Quimper tower frequency will be selected to handle the landing and taxi instructions.

Figure 4.50 Communication and radio navigation frequency displays and controls in a G1000 Primary Flight Display (PFD).

Thus, this short flight encompassing just two FIRs has to contact at least eight different ATC frequencies and given the huge volume of traffic in some parts of the World it is evident that the band is crowded. The move to an 8.33 kHz channel spacing in 1999 helped but the growth of traffic has meant that crowding on the band is again becoming a problem and, as discussed in Section 4.3.5, it is not possible to increase the channel spacing further without fundamentally changing the transmission method used. A solution is to replace some of the voice messages with data transmissions in a system known as controller–pilot data link communications (CPDLC) as discussed in Section 4.8.

4.6 Long-Range HF Communications System

4.6.1 Coverage and Frequency Bands

Away from populated land masses, the limited range of VHF line-of-sight transmissions provides sparse coverage as illustrated in Figure 4.51 below showing the VHF coverage in the North Atlantic area [10]. To provide communication links over oceans and the polar regions, the long ranges provided by skywave radio propagation via the ionosphere must be utilized. The range limitation of VHF can also be overcome by satellite communications (see following section) but as discussed below satellite and skywave communications are complementary and both systems are maintained.

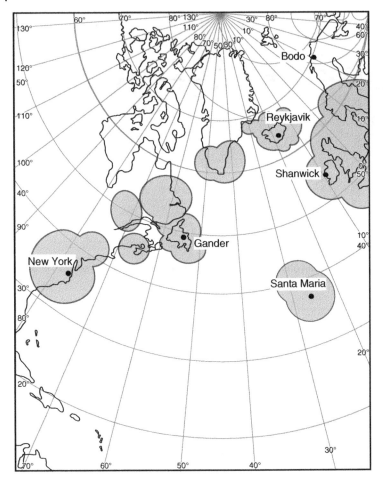

Figure 4.51 VHF coverage in the North Atlantic area. *Source:* Reproduced with permission of Elsevier from Ref. [10].

As discussed in Section 4.2.3.2 the maximum usable frequency (MUF) for an HF skywave transmission is given by Equation (4.12), that is:

$$f_{muf} = \frac{f_{crit}}{\sin\theta}$$

where f_{crit} is the critical frequency above which normally incident radio waves will be totally absorbed rather than reflected by the ionosphere and is given by the remarkable simple formula $f_{crit} = 9\sqrt{N}$ (Equation 4.11), where N is the electron density in the layer of the ionosphere being used in m^{-3}. The angle, θ, is the reflection angle off the ionosphere required for the wave to travel from the transmitting station to the aircraft (see Figure 4.10). Since the electron density varies between night and day and with other environmental factors such as solar activity, the MUF shows significant diurnal variation as well as random fluctuations. A typical variation of the MUF over a 24-hour period for a transmission from Madrid to New York (~3100 nautical miles) covered by the Santa Maria OCA is shown in Figure 4.52. It is evident that there is at least a factor of two difference between day and night MUFs and in addition a 10–20% random fluctuation during each period. For this reason, an OWF is chosen that is 85% of the seasonal average of the MUF. Normally, skywave transmissions do not need to be over such a long range as an aircraft will switch from one OCA to another. For example, an aircraft flying from Madrid to New York would switch from the Santa Maria to the New York Oceanic control mid-Atlantic. Thus, typical frequencies used across the Atlantic might be 12 MHz during the day and 6 MHz at night.

Unlike the VHF airband, which is completely utilized between the upper and lower limits containing 2280 contiguous channels, there are discrete bands within the HF band allocated to aircraft communication shown in Figure 4.53, with frequencies in between utilized by other consumers including maritime communications, short wave broadcasting, and radio amateurs.

Some of the aeronautical bands are very narrow (100 kHz) not allowing frequency space for many channels within them. As shown in Section 4.3.5 an AM signal with both sidebands requires ~7 kHz bandwidth to transmit speech and it is desirable to reduce this for HF transmissions. All the required modulation information is contained in a single sideband so HF transmitters remove the upper or lower sideband. In addition, since the carrier contains no modulation information this is removed also as transmitting it is waste of power. Thus, HF transmissions are referred to as single sideband suppressed carrier (SSB-SC) or using the ITU code (see Figure 4.23) as a J3E transmission. This makes processing of the received signal more complicated as the carrier has to be added back in so that the signal can be demodulated but this is a standard technique in radio [10].

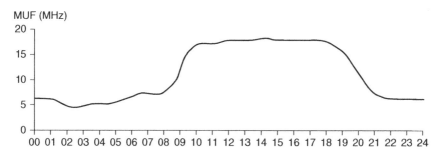

Figure 4.52 Typical variation in MUF over a 24-hour period for a transmission from Madrid to New York covered by the Santa Maria oceanic control area. *Source:* Reproduced with permission of Elsevier from Ref. [10].

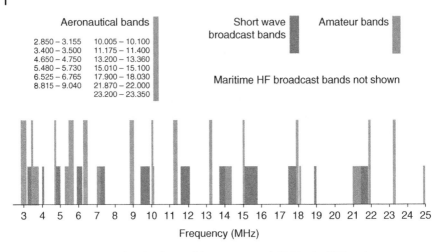

Figure 4.53 Frequency bands allocated for aeronautical ATC in the HF band.

4.6.2 Selective Calling (SELCAL)

Another issue with the lack of a carrier is that it makes squelch systems more difficult to set up. The squelch filter is set automatically or manually by the pilot so that any signal after the demodulator below a certain threshold is not passed to the headphones and static can be suppressed giving silence in between voice messages. Due to the problems of getting a reliable squelch setting there is a high background noise level on HF radio frequencies and listening for sparse transmissions for long periods of time is tiring. Aircrews usually turn down the audio level of their HF receiver and they are alerted that a transmission is about to be received by a selective calling (SELCAL) system. When the ground station operator wishes to communicate with an aircraft, he enters into the SELCAL encoder the 4-letter code of that aircraft, which is included in its flight plan, and transmits that code over the assigned radio channel. The letters are encoded by an audio frequency tone assigned to each letter. All aircraft monitoring that channel receive the SELCAL broadcast, but only the one that has been programmed with that 4-letter code will respond by sounding a chime to alert the crew who will set their volume control to listen to the voice traffic.

4.6.3 HF Ground Station Network

The main provider of long-range oceanic HF communication was ARINC (Aeronautical Radio Inc.), which was established in 1929 and started to develop a network of ground stations. The system was taken over by Rockwell Collins in 2013 who currently maintains 15 ground stations whose location is shown in Figure 4.54. These provide global coverage by HF radio links for voice and data communication (see following section) as well as VHF links.

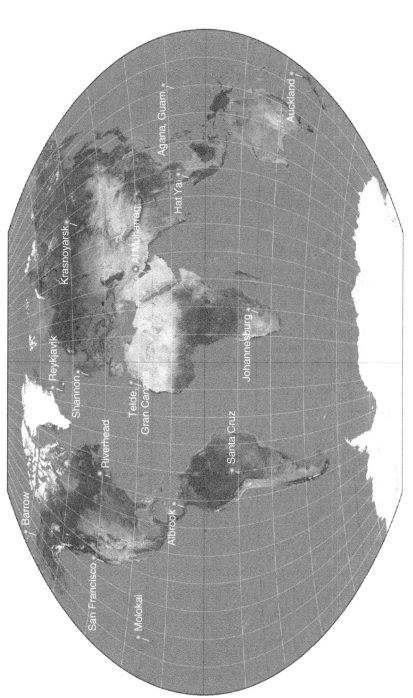

Figure 4.54 Network of ground stations providing global HF radio links. *Source:* Reproduced with permission of https://en.wikipedia.org/wiki/ Winkel_tripel_projection#/media/File:Winkel_tripel_projection_SW.jpg. Licensed under cc by sa 3.0.

4.6.4   HF Data Link (HFDL)

Radio communications in the HF band began solely as voice transmissions but for decades there has been a growing need to transmit digital data. In 1998, ARINC set up a digital data link service on its HF network (HFDL) to satisfy this requirement with a global radio coverage. The digital data is encoded by BPSK (see Section 4.3.7.2) and is transmitted at 300, 600, 1200, or 1800 bits per second (bps) depending on the channel frequency in use. A similar system known as VDL has been available at VHF frequencies since the 1980s but digital transmissions are much more prevalent in the HF band. Satellite communication links (Satcom) can send digital data at much higher rates but there are some weaknesses in the satellite system (see below).

To enable several aircraft to communicate with the same ground station at the same time, a combination of frequency division multiplexing (FDM) and time division multiplexing (TDM) is used. A highly simplified version of the process for individual bit sampling is illustrated in Figure 4.55. Each aircraft that logs onto the station is assigned a sub-frequency channel, indicated by blue (Channel 1), red (Channel 2), and green (Channel 3) for the three aircraft in Figure 4.55. The data is streaming over the airborne link at a rate in the range 300–1800 bps and there is thus a time slot, τ, given by 1/(data rate) associated with each bit. The multiplexer samples each channel for a time slot $\tau/3$ and loads the measured bit pattern onto a local link that runs at three times the speed of the HF radio link. The demultiplexer then samples the incoming data at the higher speed and feeds each bit in turn onto a separate channel thus recreating the bit pattern broadcast by the aircraft. Given the relatively low rate of the HF link and that local links can be extremely fast, it is clear that many aircraft channels can be separated using this method. Data is passed from the station to the aircraft in the same way. In practice, there has to be additional synchronization signals passed back and forth and it is possible to set up an asynchronous system such that data are passed in larger blocks after the exchange of alerting signals (handshaking).

The fundamental limit to the data transfer rate is set by the relatively low HF frequency, which is required to achieve the long ranges available with skywave communication. For normal airline operational control the bit rate of HFDL is sufficient and it has many advantages including reliable global coverage and relatively low cost. The original avionics is available on all new aircraft and there is a growing demand for the service. The main drawback is that it is unable to provide high-end digital services such as the internet.

4.7   Satellite Communications

As aviation has moved into the digital age the main problem with HF communication is its inability to transmit data at the high rates that people are used to in terrestrial communications. For transmission of digital data, the carrier wave

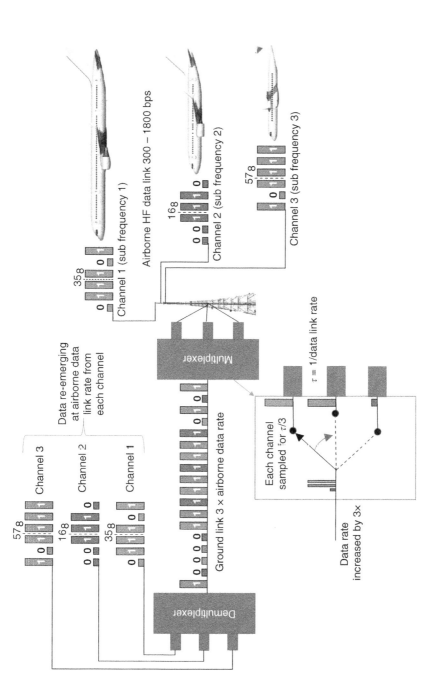

Figure 4.55 Schematic illustration of frequency division multiplexing (FDM) and time division multiplexing (TDM) allowing several aircraft to pass digital data to a ground station at the same time.

needs to be at a much higher frequency than the modulation frequency and a high data rate demands a very high carrier wave frequency. This rules out sky-wave propagation and for terrestrial broadcasts at higher frequency the range is restricted to line of sight only. From the early 1970s, this range limitation of high-frequency transmissions began to be removed by using satellites. A satellite orbiting at a height of 22 236 miles (35 786 km) above the Earth's equator is *geostationary*, that is, it has the same angular velocity as the spinning Earth below and so it stays fixed at one point in the sky as seen from the surface. It has line of sight with everything below and a constellation of a few geostationary satellites can cover the globe.

A schematic of the system used to communicate between aircraft and ground via satellites is shown in Figure 4.56. An operator who can be an air traffic controller or airline operations staff will send a message to the network run by the satellite operator (for example, Inmarsat), which will provide not only voice communication but digital services such as the Aircraft Communications Addressing and Reporting System (ACARS: see Section 4.8) or the internet. The message (along with all the other services) is sent to a ground station

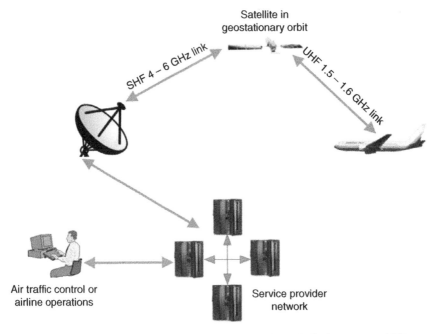

Figure 4.56 Typical system for satellite communication (SATCOM). The operator (ATC or airline staff) sends a message to the service provider network that also transmits digital data (e.g. ACARS or internet). The message is sent to a ground station whose dish is aligned with the satellite in geostationary orbit. The ground station transmits the voice message and data via an SHF link and the satellite relays everything to the aircraft in flight via a UHF link.

satellite dish, which is pointing permanently at the geostationary satellite. Data and voice are transmitted to the satellite via an SHF band (3–30 GHz) signal and then relayed to the aircraft in flight via a UHF (300 MHz to 3 GHz) link. Current systems are able to handle digital data at a rate of up to ~11 Megabits/second (11 Mbps), which is 6000 times faster than HFDL. The need for ever higher data transmission rates is pushing the RF higher and in the latest system under development this will reach 40 GHz, corresponding to 1 cm waves normally used in radar.

Despite the advantages of Satcom there are problems with the system, for example, the inaccessibility of satellites for repairs if they fail. In addition, the poles still have unreliable satellite coverage making it necessary to use HF radio communication in those regions. Thus, for global digital transmissions, HFDL will remain as a complementary system for the foreseeable future. The need for this has been especially demonstrated in the case of emergencies and natural disasters. There are numerous instances including the 2004 Indian Ocean tsunami, in which damaged or overloaded satellite and telephone systems have meant that humanitarian and emergency agencies have had to make extensive use of HF radio to coordinate their missions.

4.8 Aircraft Communications Addressing and Reporting System (ACARS)

ACARS was first implemented in 1976 essentially as an automated reporting system for times spent in different phases of flight. The system accepts inputs from sensors on the wheels, doors, etc. to determine whether the flight is Out of the gate, Off the ground, On the ground, or In the gate (OOOI) and reports the data back to the airline operations. The communication is digital and uses whatever radio link is in use, that is VDL, HFDL, or Satcom. In addition to removing the burden for flight crews to record all these information, the system provides for smooth operation, for example, by alerting replacement crews the optimum time to report for flight. The digital capabilities were soon expanded and the system can now be used to obtain additional information from airline operations such as weather reports at destination aerodromes, flight plans, amendments to flight plans, etc. It can also communicate with other ground facilities including the airline maintenance department and the aircraft or engine manufacturer. For example, ACARS can transmit a continuous report on the state of health of the engines to the manufacturer to organize maintenance schedules, troubleshoot problems, or provide data to implement design improvements. ACARS also communicates with ATC and is used to implement a new mode of communication between pilots and ATC known as CPDLC.

A major problem with voice communications, which were largely developed in an environment of one-on-one between pilot and controller, is that in a busy airspace a large number of pilots are on the same frequency. Any aircraft transmitting will block the frequency for all others and as the traffic increases, the probability that a pilot accidentally cuts off the transmission of another also increases. In addition, each exchange between the controller and a pilot takes a specific amount of time and a traffic volume can be reached where it is no longer possible to pass the necessary messages to all aircraft within the time required. A solution is to increase the number of controllers and have each one on a separate frequency but that also introduces its own problems including the increased time required to handover flights from one controller to another. The CPDLC system alleviates these problems by using a data link with ATC to pass on clearances and other instructions to pilots and for pilots to request level changes, etc. This also relieves some of the stress on flight crews as they do not have to concentrate on picking out their call sign from almost continuous ATC talk and the text information that appears on the screen cannot be misheard. ACARS controllers are also interfaced to a printer so that a hard copy of important information can be obtained. Figure 4.57 shows an ACARS clearance displayed on the Flight Management Display on a commercial airliner.

As an addendum to this section it is worth discussing the role that satellite communications and ACARS played in the attempts to find the missing Malaysian Airlines flight MH370 that disappeared on 7 March 2014. The system is a commercial service provided by Inmarsat that has to be paid for and Malaysian Airlines had opted only for the engine monitoring service that reports the state of engine health to Rolls Royce. The transponder was switched off and the ACARS system was disabled by switching off the SATCOM and VHF channels, which stopped all ACARS transmissions to the ground station. In this

Figure 4.57 Clearance issued by ACARS on a flight from Sydney to Darwin. *Source:* Reproduced with permission of Jeroen Hoppenbrouwers.

circumstance, the satellite continues to send a simple handshaking signal to the aircraft known as a *ping* every hour to check whether the aircraft is still online and the aircraft continues to respond. From the time taken between the ping and the reception of the aircraft response it is possible to determine the distance from the satellite. Six pings were sent after the loss of contact and from these it was possible to map out an arc traveled away from the geostationary satellite for six hours in two directions, North and South. Distance measurements alone, however, could not distinguish which track was flown. Some innovative work by Inmarsat engineers analyzed the shift in frequency of the return signal due to the Doppler effect to determine the aircraft speed relative to the geostationary satellite. By comparing this with other Malaysian airline flights on the same route they were able to say with a reasonable degree of certainty that it was the Southern track that was taken, which was a significant help to the Search and Rescue operation. Tragically, the mystery of what happened to flight MH370 has never been solved but the incident has reignited the debate over whether full ACARS implementation should be compulsory to act as a kind of "continuous black box" in flight.

Problems

1 It is decided to install an LF NDB operating at 400 kHz at the center of an approximately circular island 20 nautical miles in diameter. If the beacon is to have a range of 50 nautical miles calculate the required power of the transmitter.

2 It is required to send a skywave signal via the F-layer from Madrid to New York, which is a distance of 3100 nautical miles. On the day in question the F-layer is at a height of 350 km with an electron density of 6×10^{11} m^{-3}. Calculate the OWF for the transmission.

3 An aircraft flying at 5000 ft above sea level and over the sea is unable to make contact via VHF transmissions with an ATC station on the coast 130 nautical miles away. It is in contact with another aircraft flying at 2000 ft and 40 miles closer to the station and asks the pilot of the second aircraft to relay a message to the ATC station. Is the second aircraft able to assist?

4 A small loop antenna with an air core is used to detect the emission from a NDB operating at 400 kHz. If the loop contains 100 turns and has a diameter of 2 cm, calculate the radiation resistance. If the wire in the coil has a total resistance of 0.5 Ω, determine the efficiency of the coil as a transmitting antenna.

5   A log-periodic antenna has its longest element length equal to 1 m and has eight other elements. If the k value for the antenna is 1.2, determine the operating bandwidth.

6   A 15 m diameter parabolic antenna is used to communicate with a geostationary satellite over a 5 GHz link. Assuming an efficiency of 0.6 and the maximum possible k value. Determine the gain in dB and the HPBW. The geostationary satellite is 36 000 km away from the dish. What is the diameter of the radio beam defined by the HPBW at the satellite?

7   A rectangular waveguide is being designed as a feeder for a radar system operating at 4 GHz. Calculate the limits for the height and width of the rectangle in order to carry a single-mode polarized wave.

References

1 Zolesi, B. and Cander, L.R. (2014). *Ionospheric Prediction and Forecasting*. Springer.

2 Stern, H.P.E., Mahmoud, S.A., and Stern, L.E. (2004). *Communication Systems: Analysis and Design*. Pearson Prentice Hall.

3 Leentvar, K. and Flint, J.H. (1976). The capture effect in FM receivers. *IEEE Transactions on Communications* **24**: 531–539.

4 Carson, J.R. (1922). Notes on the theory of modulation. *Proceedings of the IRE* **10**: 57–64.

5 Griffiths, D.J. (1999). *Introduction to Electrodynamics*, 3e. Prentice Hall chapter 11.

6 Chen, K. (2012). cocoaNEC2.0 software.

7 Creative commons (1953). http://creativecommons.org/licenses/by-sa/3.0/ legalcode (accessed 12 June 2018).

8 Macnamara, T.M. (2010). *Introduction to Antenna Placement and Installation*. Wiley.

9 ICAO. http://gis.icao.int/FIRWORLD (accessed 12 June 2018).

10 Tooley, M.H. and Wyatt, D. (2007). *Aircraft Communications and Navigation Systems: Principles, Operation and Maintenance*. Elsevier/Butterworth-Heinemann chapter 5.

5

Primary and Secondary Radar

5.1 Primary Radar

Primary radar range finding is based on measuring the time between a transmitted pulse in the UHF or SHF band and the reception of the reflected pulse from the target. After each pulse is transmitted, the antenna is switched to receive mode to detect the return signal. In the case of *primary* radar the return is a direct reflection of the transmitted pulse and so the carrier within the reflected pulse is at the same frequency as the transmitted one. Another mode known as *secondary radar* is also used where the transmitted pulse triggers a transponder at the target, which then transmits its own pulse or set of pulses in return that are of different frequency. The timing between the transmitted and return pulses gives the distance of a target and the transmitted beam is directional so that the vector to the target is also known. Additional information can be returned from the target in the case of secondary radar, which is described in Section 5.4. In this section, the focus is on primary radar.

The transmitter and receiver share the same directional antenna and as discussed in Section 4.4.8, directional antennas in the SHF band are either parabolic dishes, usually with waveguide feeders (Figure 5.1a), or slotted waveguide arrays (Figure 5.1b) with the latter type commonly installed in the nose of aircraft as the antenna for weather radar. Both types of antenna emit a directional beam whose beam width is defined as the angular range at which the power in the beam is at least half (−3 dB) of the maximum as illustrated in Figure 5.1c, that is, the half power beam width (HPBW). In the case of a parabolic dish antenna, it was shown in Section 4.4.8.3 (see Equation 4.62) that the HPBW, θ, is given by

$$\theta = \frac{k\lambda}{d} \tag{5.1}$$

Aircraft Systems: Instruments, Communications, Navigation, and Control,
First Edition. Chris Binns.
© 2019 John Wiley & Sons, Inc. Published 2019 by John Wiley & Sons, Inc.
Companion website: www.wiley.com/go/binns/aircraft_systems_instru_communi_Navi_control

(a)

(b)

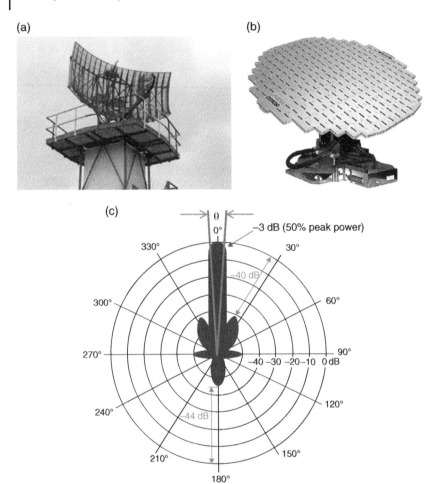

(c)

Figure 5.1 (a) Parabolic antenna commonly used in a ground radar installation. (b) LT 005-WRAU slotted planar array antenna used in airborne weather radar produced by Rockwell Collins. *Source:* Reproduced with permission of © Rockwell Collins. (c) Typical polar plot of radiated power from a directional antenna showing the definition of beam width.

where λ is the wavelength of the transmitted beam, d is the diameter of the dish, and k is a factor that depends on the filling efficiency of the feeder and the shape of the dish but in the optimum case is $180/\pi = 57.3°$. Thus, shorter wavelengths or larger antennas produce narrower beams.

The gain, G, of a directional antenna is defined as the ratio of the power in the directional beam to the power that would be transmitted in the same area if the antenna emitted an isotropic beam and an antenna has the same gain whether it is being used to transmit or receive. For a parabolic dish, it was shown in Section 4.4.8.3 (see Equation 4.59) that the gain is given by

$$G = \frac{4\pi A}{\lambda^2}\eta \tag{5.2}$$

where A is the geometric area of the dish, λ is the wavelength of the radio waves, and η is the antenna efficiency, which is typically in the range 0.55–0.7. Equation (5.2) makes it clear that it is the size of the dish relative to the wavelength that is the important parameter. Slotted array antennas tend to have the lowest k values, producing narrower beams and less intense side lobes.

The characteristic timing associated with a radar pulse train is shown in Figure 5.2. The carrier wave is emitted as a set of pulses with a given pulse width (PW) and pulse repetition time (PRT). Typical PWs are ~1 μs while typical carrier frequencies are in the SHF band (3–30 GHz), so a single pulse will contain thousands of oscillations of the carrier. We can also define the pulse recurrence frequency (PRF = 1/PRT) giving the number of pulses emitted per second. The interval between the end of one pulse and the beginning of the next pulse is known as the resting time, which is divided into two sections. The first period in the resting time is the dead time required for the antenna to switch from transmitting to receiving mode and the majority of the resting time is the listening time during which the antenna can detect returns from the target.

In order to assign a return to a specific transmitted pulse, the maximum range that an object can be identified is the distance that light travels in half the PRT. The speed of light is 162 000 nautical miles/second and a useful measure of range is the radar mile, which is the time that light takes to travel 1 nautical out and back (i.e. 2 nautical miles), which is 12.34 μs.

Thus, the maximum range, R_{max}, in nautical miles that a radar can identify targets is given by

$$R_{\text{max}}(\text{nautical miles}) = \frac{PRT\,(\mu s)}{12.34} \tag{5.3}$$

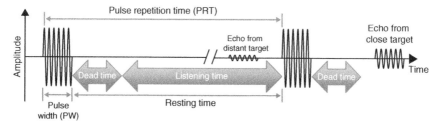

Figure 5.2 Characteristic times associated with a radar pulse train.

In practice, R_{max} will be less than the theoretical maximum due to the finite time required to switch the installation from receive to transmit mode (dead time) and other factors. In addition, as with all transmissions at VHF or higher frequency, there has to be line of sight between the transmitter and the target so there is a "horizon" limiting range given by Equation (4.6), that is,

$$\text{Distance to horizon} = 1.23 \times \left(\sqrt{H_{TX}} + \sqrt{H_{RX}} \right) \tag{5.4}$$

where H_{TX} and H_{RX} are the height of the transmitter and target in feet.

There is also a minimum range since the pulse has to be transmitted and the set switched to receive mode in the time taken for the target to return the signal. The theoretical minimum range, that is, assuming the dead time is zero (unachievable in practice) is the distance that light travels in half the PW. For example, a 3.3 µs long pulse would allow a theoretical minimum range of 500 m.

To obtain the bearing of a target from the station, the antenna is rotated and each reflection is plotted on a circular screen (plan position indicator or PPI) on a radial corresponding to the direction of the antenna when the signal was received. In addition to bearing and range information, the Doppler shift of the received signal gives the radial velocity of the object, which coupled with the azimuthal velocity given by the movement between two pulses yields the total velocity vector of the target. Air traffic control (ATC) radars also act as secondary radars and trigger aircraft transponders that supply identifying information and other data for the aircraft (see Section 5.4). If a target is not moving on a PPI and has no Doppler shift, then it is stationary and this return can be removed from the PPI thus removing clutter produced by local reflections from buildings, etc.

The radial location of the target can only be determined within the distance that light travels during the PW and the radial resolution is defined as half this distance (see Figure 5.3a). For example, a PW of 6 µs would give a radial resolution of 900 m. The azimuthal resolution is given by the length subtended by half the beam width at a given range and so is distance dependent. For a close approximation it is given by $r\sin\theta/2$ where r is the range of the target and θ is the HPBW of the antenna (see Figure 5.3b). So, for example, a typical long-range radar beam width of 6° would give an azimuthal resolution of 3 miles at a range of 60 miles. These resolutions may also be convoluted with the pixel width on the radar display (PPI).

Two additional important parameters, illustrated in Figure 5.4, are the dwell time (T_D) of the radar beam, which is the time that it takes the directional beam to sweep across the target, and the hits per scan, which is the number of pulses transmitted and received during the dwell time. If the antenna beam width is θ

(a)

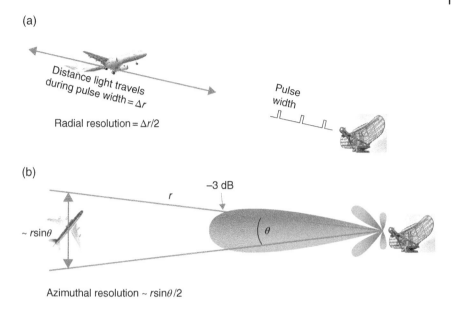

Distance light travels during pulse width = Δr

Radial resolution = $\Delta r/2$

Pulse width

(b)

−3 dB

r

~ $r\sin\theta$

θ

Azimuthal resolution ~ $r\sin\theta/2$

Figure 5.3 (a) Radial resolution is determined by the pulse width. (b) Azimuthal resolution is determined by the beam width.

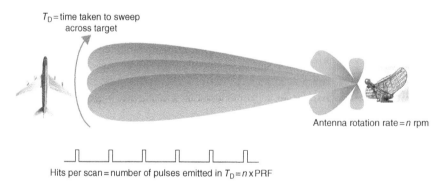

T_D = time taken to sweep across target

Antenna rotation rate = n rpm

Hits per scan = number of pulses emitted in $T_D = n \times PRF$

Figure 5.4 Illustration of the dwell time and hits per scan.

and the rate of rotation of the antenna is n revolutions per minute, the dwell time is given by

$$T_D = \frac{\theta \times 60}{360 \times n} \tag{5.5}$$

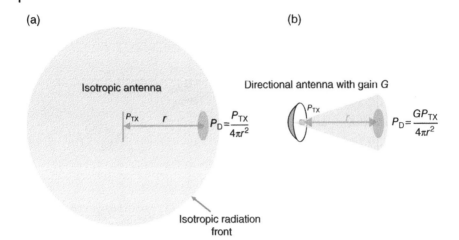

(a)   (b)

Figure 5.5 (a) Power density P_D a distance r away from a hypothetical lossless isotropic antenna transmitting power P_{TX}. (b) Power density P_D a distance r away from a directional antenna transmitting power P_{TX}.

The hits per scan, N, are given by

$$N = T_D \times PRF \tag{5.6}$$

Weaker returns given by more distant targets require a larger number of hits per scan to retrieve the signal from the background noise. In addition, acquiring distant targets requires a long PRT and thus a low PRF, so long-range radars have slower antenna rotation rates than their shorter range counterparts.

The strength of the return signal is given by the *radar equation*, which is derived as follows. The power density ($W\,m^{-2}$), P_D, in the radio wave a distance r from a hypothetical lossless isotropic transmitter with a power P_{TX} (see Figure 5.5a) is given by

$$P_D = \frac{P_{TX}}{4\pi r^2} \tag{5.7}$$

A radar emits a directional beam and the ratio between the power density in the direction of interest and the power density from an isotropic antenna is defined as the antenna gain, G. Thus, the power density a distance r away from a radar transmitter with an antenna gain, G (see Figure 5.5b), is

$$P_D = \frac{GP_{TX}}{4\pi r^2} \tag{5.8}$$

The power density, P_D, is incident on the target, which then acts as a secondary source to transmit radio waves back to the receiver and generally the emission can be considered to be isotropic. Each target has a cross-section

Table 5.1 Typical radar cross-sections of objects.

Object	Cross-section, σ (m^2)
Light aircraft	1
Helicopter	3
Medium weight jet airliner	40
Boeing 747	100
Large insect (or stealth fighter)	10^{-4}

σ, which is the area of a spherical perfect reflector that emits the same power as the real target. Some typical cross-sections of objects are given in Table 5.1.

Thus, the target can be considered to be an isotropic emitter transmitting a power $P_D \times \sigma$, so from Equation (5.8) the power emitted by the target, P_t, is

$$P_t = \frac{\sigma G P_{TX}}{4\pi r^2} \tag{5.9}$$

Since the target is an isotropic emitter, the power density received back at the radar station a distance r away is

$$P_D = \frac{P_t}{4\pi r^2} \tag{5.10}$$

So, from Equation (5.9),

$$P_D = \frac{\sigma G P_{TX}}{16\pi^2 r^4} \tag{5.11}$$

The actual received signal power, P_{RX}, is

$$P_{RX} = P_D \times A_{eff} \tag{5.12}$$

where A_{eff} is the effective aperture of the receiver antenna. It can be shown that the effective aperture can be written in terms of G and the wavelength of the radio waves, λ, as (see Equation (A3.12):

$$A_{eff} = \frac{\lambda^2 G}{4\pi} \tag{5.13}$$

So from Equations (5.11)–(5.13) the received power from the target at the antenna is

$$P_{RX} = \frac{\lambda^2 \sigma G^2 P_{TX}}{64\pi^3 r^4} \tag{5.14}$$

The derivation of Equation (5.14) assumes attenuation only by the inverse square law and does not consider any additional attenuation by absorption

---

Beermat Calculation 5.1

An ATC radar with an average transmitting power of 1 kW working at 3 GHz and an antenna gain of 4000 is tracking a medium weight airliner 10 nautical miles away. What is the received power at the antenna from the radar return?

The answer is given by Equation (5.14). The wavelength, λ, is 10 cm, the distance, r, is 18 500 m, and from Table 5.1 the target cross-section is 40 m². The transmitter power, P_{TX}, and antenna gain, G, are both given, so Equation (5.14) yields that the received power, $P_{RX} = 2.75 \times 10^{-11}$ W.

---

and scattering in the atmosphere. In dry air containing no dust, absorption and scattering do not contribute significantly to attenuation but at the high-frequency end of the SHF band, there is very strong attenuation of radar beams by precipitation. Consider a radar beam of intensity I_0 incident on a slab of precipitation with a thickness x as illustrated in Figure 5.6a. The transmitted intensity, I, is given by

$$I = I_0 e^{-\mu x} \tag{5.15}$$

where μ is the attenuation coefficient, which has units (length)$^{-1}$. Normally the units chosen for μ are km^{-1} and because of the large variation in its magnitude it is convenient to express its value in dB km^{-1}. Thus, for example, if at a specific frequency and density of rain $\mu = 1$ dB km^{-1}, then $\mu = 1.26$ km^{-1} in linear units and a radar beam traveling through 1 km of that precipitation would have an intensity $I/I_0 = e^{-1.26} = 0.28$, that is, 72% of the original intensity will be lost.

Figure 5.6b plots the attenuation coefficient in dB km^{-1} as a function of wavelength for different levels of precipitation determined by theoretical calculations using Mie theory [1]. It is observed that there is a strong onset of attenuation starting above 3 GHz, thus long-range radars have to use frequencies of 3 GHz or less. In the case of short-range radars, for example, those used to monitor airport surface movement, higher frequencies can be used. Airborne weather radars (AWRs), which detect backscatter from precipitation use frequencies that cover the onset of the strong attenuation, that is, in the range 4–12 GHz.

The calculations in Figure 5.6b were carried out assuming the distribution of raindrop sizes reported by Marshall–Palmer [2]. They carried out measurements of the sizes for different rainfall rates and discovered that, with the exception of small drops the distribution closely follows the power law:

$$N_D(D) = N_0 e^{-\Lambda D} \tag{5.16}$$

where D is the diameter, $N_D \delta D$ is the number of raindrops per unit volume with sizes between D and $D + \delta D$, N_0 is the value of N_D at the small drop limit, and Λ is a fit parameter. The units of N_D and N_0 are [length]$^{-4}$ since they are a number per unit volume within a given size range. The measured distributions depend on the

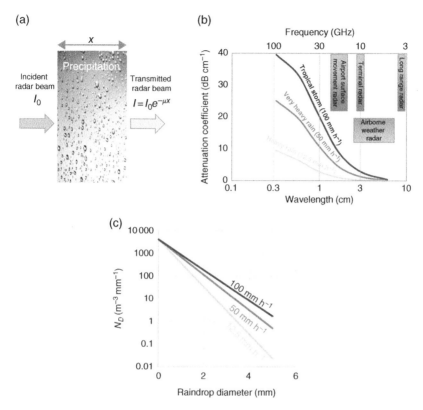

Figure 5.6 (a) Attenuation of radar beam by precipitation described by the attenuation coefficient, μ. (b) Attenuation coefficient in dB km^{-1} as a function of wavelength for different levels of precipitation. (c) Marshall–Palmer distribution of raindrop sizes for the precipitation rates shown in (b) [2].

rainfall rate and fit the power law when $N_0 = 4096$ m^{-4} irrespective of rate and Λ is given by

$$\Lambda = 4100R^{-0.21}\,\text{m}^{-1} \tag{5.17}$$

where R is the rainfall rate in mm h^{-1}, which is the standard unit. The distributions of raindrop sizes for the three calculated attenuation curves displayed in Figure 5.6b are plotted in Figure 5.6c. We will return to this distribution in Section 5.3 as it is relevant to AWR.

5.2 Ground Radar

Ground radar installations have widely varying ranges (1–300 nm) depending on their purpose and the design and operating parameters, such as frequency, pulse timing sequence, and rate of rotation, are determined by the range required.

Table 5.2 Design parameters for different types of ground radar.

Function	Range (nautical miles)	Frequency (GHz)	Wavelength (cm)	Pulse repetition time (µs)	Pulse width (µs)	Rotation rate (rpm)
Area surveillance (Figure 5.7a)	Up to 300	0.6–3	10–50	Up to 4000	2–4	5–6
Terminal area surveillance (Figure 5.7b)	Up to 80	0.6–3	10–50	Up to 1200	1–3	12–15
Approach surveillance (Figure 5.7c)	Up to 25	3–10	3–10	Up to 350	0.5–1.2	20
Airport surface movement (Figure 5.7d)	Up to 6 (light precipitation)	15–17	1.76–2	Up to 100	0.03	60

As shown in the previous section, long-range radars have to use lower frequencies (≤3 GHz) to avoid heavy attenuation by rain. The pulse repletion time (PRT) needs to be sufficiently long for the return from a distant target to be detected and in addition, the return signal is very weak, so long pulse lengths are used to facilitate extraction of the return signal from background noise. The long pulse lengths exclude the radar from detecting close targets. A long-range radar also needs a high antenna gain, which requires a large antenna (see Equation 4.59). As discussed in the previous section, the rate of rotation of a long-range radar antenna needs to slow in order to increase the dwell time and the number of hits per scan. Conversely, short-range radars can use higher frequencies, shorter PRTs, shorter pulses, and smaller antennas with higher rotation rates. The basic design parameters for radars of different ranges are listed in Table 5.2 and examples of various radars are shown in Figure 5.7.

5.3 Airborne Weather Radar

AWR fitted in the nose has become standard equipment on commercial transport aircraft and has led to a significant improvement in safety and passenger comfort as it allows pilots to go around or over storms. The nose cone in front of the antenna is plastic or glass fiber, which is transparent to SHF band radio waves. Older systems use a parabolic antenna (Figure 5.8a) and frequencies in the range 4–8 GHz (λ = 3.8–7.5 cm) but newer systems use a slotted array

(a) (b)

(c) (d)

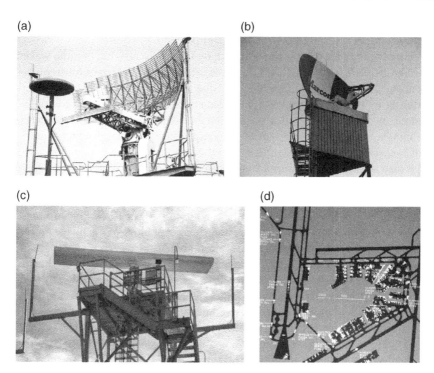

Figure 5.7 (a) Area surveillance radar. (b) Approach surveillance radar. *Source:* Reproduced under License cc by sa 2.5. (c) Airport surface movement radar. *Source:* Reproduced with permission of SAAB Group. (d) Display of the ground radar system, Schiphol Airport by Mark Brouwer. *Source:* Reproduced under License cc by sa 2.5.

(a) (b)

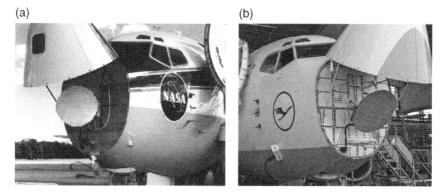

Figure 5.8 (a) Airborne weather radar using parabolic antenna (4–8 GHz, $\lambda = 3.8$–7.5 cm). (b) Airborne weather radar using slotted array antenna (8–12.5 GHz, $\lambda = 2.4$–3.8 cm). *Source:* wikimedia.org.

antenna (Figure 5.8b), which has a higher azimuthal resolution and works in the frequency range 8–12.5 GHz (λ = 2.4–3.8 cm).

Primarily the radar detects precipitation by the backscattering of radio waves by water droplets or ice crystals. The rise in attenuation coefficient for frequencies above 3 GHz shown in Figure 5.6b is due to the increase in absorption and scattering by the water droplets. The intensity removed by scattering is redistributed in all directions, so a proportion is scattered back to the antenna. The choice of frequency is essentially a compromise between producing sufficient backscattered intensity and keeping the attenuation to an acceptable level and is also optimized for returns from large raindrops. The hierarchy of reflectivity by various types of precipitation is illustrated in Figure 5.9 revealing that wet hail produces the strongest return and dry snow the weakest. Essentially, the precipitation with the largest droplet sizes containing water gives the strongest reflection. A trained operator can interpret the morphology of the image on the screen to deduce a good deal of information about storms.

AWR is a primary radar using the same pulsed emission and timing method as ground radar except the beam scans backward and forward in a fan ahead of the aircraft at a tilt angle that is set by the pilot (Figure 5.10a). The radar mounting is normally attitude stabilized to the horizontal plane while the aircraft pitches and rolls so that the same tilt angle is maintained throughout maneuvers. As illustrated in Figure 5.10b, the tilt function enables the radar to be used to scan the clouds ahead at different heights or by tilting downward to show the surface

Figure 5.9 Relative radar reflectivity at AWR frequencies of different types of precipitation.

Wet hail

Increasing reflectivity

Rain

Ice crystals

Wet snow

Dry hail

Dry snow

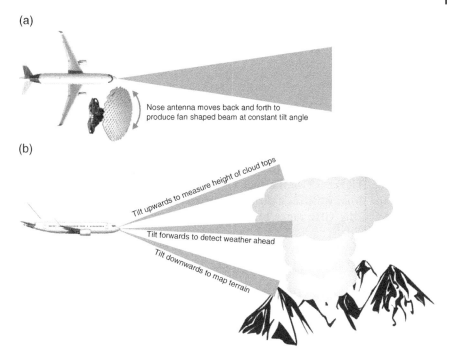

(a)

Nose antenna moves back and forth to produce fan shaped beam at constant tilt angle

(b)

Tilt upwards to measure height of cloud tops

Tilt forwards to detect weather ahead

Tilt downwards to map terrain

Figure 5.10 (a) Fan-shaped beam produced by a weather radar antenna scanned back and forth. (b) The antenna scans back and forth at an adjustable tilt angle (stabilized to the horizontal plane) to facilitate terrain or weather mapping.

topography ahead. In addition, by tilting upwards till the cloud disappears it is possible to measure the height of the cloud tops of clouds.

Normally AWRs incorporate an automatic gain control that reduces the gain for closer objects so that there is the same intensity of returns from similar objects irrespective of distance. The cockpit display may be a stand-alone screen or may be included in the navigation display in a glass cockpit. On the controller (see Figure 5.13), the pilot can select the range in nautical miles, which changes the PRT, the gain, the tilt or the mode, that is, optimized for weather (WX) or ground mapping (MAP). The mode can also be set to standby where the beam is not transmitted, which is important when the aircraft is on the ground as the intensity of the beam can harm personnel and damage electronics in other aircraft and vehicles.

A careful consideration of the intensity of the backscattered (reflected) radiation from precipitation enables quantitative information such as the rain rate to be determined. The reflected intensity can be calculated by modifying the radar equation (5.14), which was derived for a single target with a cross-section, σ. The reflection from precipitation is due to a large number of raindrops distributed through the scanned volume, V, with a range of cross-sections. The single cross-section of the basic radar equation is thus replaced by an average cross-section over all raindrops given by

$$\bar{\sigma} = V \sum_i \sigma_i \qquad (5.18)$$

where σ_i is the cross-section of an individual drop. As illustrated in Figure 5.11a, the wavelength of an SHF band radar beam is much longer than the size of an individual raindrop, so spatially the instantaneous field is approximately constant across any drop. In this limit, the interaction of the electromagnetic field with the raindrops is referred to as Rayleigh scattering. Considering an individual raindrop (Figure 5.11b, the spatially constant electric field at any instant polarizes the drop, which is a dielectric material and the temporal oscillation of the field in the GHz range turns it into an oscillating dipole (a Hertzian

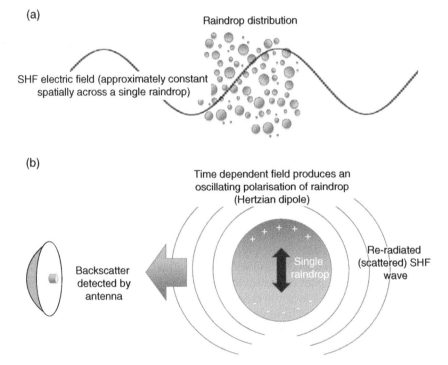

(a)

Raindrop distribution

SHF electric field (approximately constant spatially across a single raindrop)

(b)

Time dependent field produces an oscillating polarisation of raindrop (Hertzian dipole)

Backscatter detected by antenna

Single raindrop

Re-radiated (scattered) SHF wave

Figure 5.11 (a) The wavelength of an SHF radar beam is much longer than the size of a raindrop so that the electric field can be considered to be spatially constant across an individual drop. (b) The time oscillation of the incident electric field produces an oscillating polarization of a raindrop tuning it into a source of radiation at the same frequency (Hertzian dipole). The secondary radiation is the scattered beam and some is backscattered to the antenna.

dipole – see Section 4.4.1). The oscillating dipole acts as a source of electro-magnetic radiation, which has the same frequency as the incident radar beam and is distributed isotropically. This radiation is the scattered wave, some of which is backscattered to the antenna.

The backscattering cross-section for a dielectric sphere in the Rayleigh scattering regime with no absorption (i.e. all attenuation is due to scattering) is a standard calculation described, for example, in Ref. [3] and is given by

$$\sigma_i = \frac{\pi^5}{\lambda^4}|K|^2 D_i^6 \tag{5.19}$$

where D_i is the raindrop diameter and $|K|^2$ is related to the dielectric constant, ε, of the water (or ice) in the drop by

$$|K|^2 = \left|\frac{\varepsilon - 1}{\varepsilon + 2}\right|^2 \tag{5.20}$$

For example, $|K|^2 = 0.93$ for water and $|K|^2 = 0.2$ for ice. Substituting Equation (5.19) into Equation (5.18) gives

$$\bar{\sigma} = V\frac{\pi^5}{\lambda^4}|K|^2 \sum_{i=1}^{n} D_i^6 \tag{5.21}$$

where n is the number of raindrops in the scanned volume. The sum over individual raindrops is not in a form that can be calculated and it is replaced by the integral:

$$\sum_{i=1}^{n} D_i^6 = \int_0^{\infty} D^6 N_D(D)\mathrm{d}D \tag{5.22}$$

where $N_D(D)$ is the size distribution of the raindrops. For example, this could be the Marshall–Palmer size distribution given in Equation (5.16). In the literature, the sum over raindrop diameters or the integral in Equation (5.22) is often called the *radar reflectivity* and is given the symbol Z while the cross-section per unit volume, $\bar{\sigma}/V$, is denoted by the symbol η. Thus, in this terminology, Equation (5.21) can be written as

$$\eta = \frac{\pi^5}{\lambda^4}|K|^2 Z \tag{5.23}$$

The volume of rain probed by the radar beam at one position in the scan depends on the distance, r, from the antenna as illustrated in Figure 5.12 and is given by the radial resolution multiplied by the area of the beam at r. As shown in Figure 5.3a, the radial resolution is the distance that light travels during half a PW, τ, that is, $c\tau/2$, where c is the speed of light. The radar beam cross-section is an ellipse with angular width θ and ϕ in orthogonal directions (Figure 5.12), so

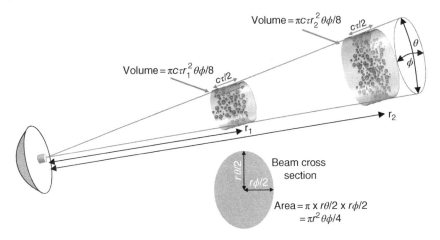

Figure 5.12 Illustration of volume sampled at a single point in the scan by the radar beam as a function of distance from the antenna.

at a distance r, the cross-sectional area is $\pi r^2 \theta \phi/4$. Thus, the volume, V, probed a distance r away is

$$V = \frac{\pi c \tau r^2 \theta \phi}{8} \qquad (5.24)$$

Thus, substituting Equations (5.24) and (5.22) into Equation (5.21) gives the average cross-section for returns from the raindrops in the scanned volume to be

$$\bar{\sigma} = \left(\frac{\pi c \tau r^2 \theta \phi}{8} \right) \frac{\pi^5}{\lambda^4} |K|^2 \int_0^\infty D^6 N_D(D) \mathrm{d}D \qquad (5.25)$$

This average cross-section replaces the single target cross-section in the basic radar Equation (5.14) to give the power of the return signal for a given transmitted power, that is,

$$P_{\mathrm{RX}} = \frac{\lambda^2 G^2 P_{\mathrm{TX}}}{64\pi^3 r^4} \left[\left(\frac{\pi c \tau r^2 \theta \phi}{8} \right) \frac{\pi^5}{\lambda^4} |K|^2 \int_0^\infty D^6 N_D(D) \mathrm{d}D \right] \qquad (5.26)$$

$$\text{or } P_{\mathrm{RX}} = \frac{G^2 P_{\mathrm{TX}} c \tau \theta \phi \pi^3 |K|^2}{512 \lambda^2} \frac{Z}{r^2} \qquad (5.27)$$

Note that in the case of reflection from raindrops the received power decays as $1/r^2$ as opposed to $1/r^4$ for a single target since the scanned volume (and thus the raindrop cross-section) increases as r^2 (Equation 5.24). Equation (5.27) is known as the weather radar equation and can be written as

$$P_{RX} = C\frac{|K|^2}{r^2}Z \tag{5.28}$$

where C is a constant for a specific weather radar transmitter and antenna.

The return signal can be used to determine the rain rate, R (in mm h^{-1}). In the absence of wind and turbulence (i.e. no up or downdraughts) R is given by [4]:

$$R = 6\pi \times 10^{-4}\int_0^\infty D^3 v(D)N_D(D)\mathrm{d}D \tag{5.29}$$

where $v(D)$ is the terminal velocity distribution for drops of size D in m s^{-1}, which is normally considered to be a power law of the form:

$$v(D) = cD^\gamma \tag{5.30}$$

Since the radar reflectivity:

$$Z = \int_0^\infty D^6 N_D(D)\mathrm{d}D \tag{5.31}$$

uses the same raindrop size distribution, comparing Equation (5.31) with that for the rain rate, Equation (5.29) shows that we can write the following power law relationship connecting Z and R:

$$Z = aR^b$$

and the most commonly used form is [5]:

$$Z = 200R^{1.6} \tag{5.32}$$

Thus, knowing the distance to the region of rain it is possible to determine the rate of rainfall directly from the return signal using Equations (5.28) and (5.32).

In the WX mode, the returns are color coded on the cockpit display in terms of the intensity of the precipitate assuming it is raindrops. The colors and the level of precipitation they represent are shown in Table 5.3.

A schematic of a typical stand-alone AWR scan of a thunderstorm cloud is shown in Figure 5.13 and interpreting the displays takes some care. For example, the back of a red/magenta area is not necessarily the back of the intense precipitation as the front part attenuates the beam so there are weak returns from deeper in the cloud as illustrated by the radar shadow indicated in Figure 5.13. Also, green and yellow areas may be from very intense precipitation of dry hail, which has a lower reflectivity than water drops. A general indicator of turbulence is where color zones are thin, that is, the rainfall gradient is very steep. Pilots receive training how to interpret the general morphology of weather radar images to obtain information about

Table 5.3 Color coding of airborne weather radar returns.

Color	Intensity of return	Rainfall rate
Black (no display)	Very light/none	$<0.7 \text{ mm h}^{-1}$
Green	Light	$0.7-4 \text{ mm h}^{-1}$
Yellow	Medium	$4-12 \text{ mm h}^{-1}$
Red	Strong	$>12 \text{ mm h}^{-1}$
Magenta	Maximum	Shows area of most intense rainfall and also indicates turbulence

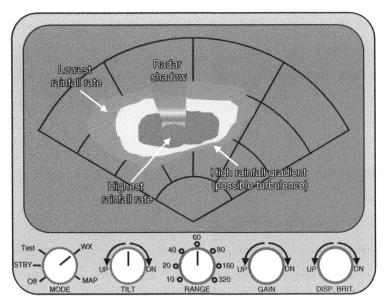

Figure 5.13 Schematic of a stand-alone weather radar display showing color coding of radar returns from a typical thunderstorm cloud. Intense rainfall produces a shadow masking returns from deeper in the cloud.

turbulence and possible hail so that the aircraft can be flown around storms to avoid damage from either.

The tilt function can be used to determine the height of cloud tops by adjusting the tilt angle of the antenna upwards till the cloud return disappears as illustrated in Figure 5.14. The tilt angle is zero when the center of the beam is horizontal so the angle of the top of the cloud relative to the aircraft, α, is given by

$$\alpha = \text{TILT} - \frac{\theta}{2} \tag{5.33}$$

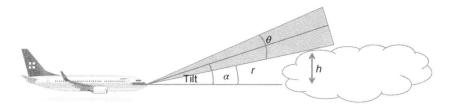

Figure 5.14 Using the tilt function to measure the height of cloud tops relative to aircraft altitude.

where TILT is the angle at which the cloud just disappears and θ is the total beam width in degrees. The range, r, of the cloud is known, so the height of the top of the cloud, h, above the aircraft altitude is given by:

$$h = r\sin\alpha \tag{5.34}$$

Since the angles are small it is a good approximation to assume $\sin\theta = \theta$, where θ is in radians (= degrees $\times\pi/180$). Since 1 nautical mile ≈ 6000 ft, the above equation reduces to the following simple formula in units that are convenient to the pilot:

$$h\,(\text{ft}) \approx r\,(\text{NM}) \times \theta\,(°) \times 100 \tag{5.35}$$

Thus, it is easy to estimate whether it is possible for the aircraft to climb over the storm.

Beermat Calculation 5.2

The AWR on an aircraft flying at FL290 (29 000 ft above ISA sea level) detects a storm cloud at a range of 40 nautical miles. The cloud just disappears from the display at an upward tilt angle of +5° and the AWR beam width is 6°. If the service ceiling of the aircraft is 39 000 ft is it able to fly above the storm?

$$\text{The cloud angle} = \text{TILT} - \frac{1}{2} \times \text{beam width} = 5 - 3 = 2$$

Using Equation (5.35) the top of the cloud top is $2 \times 40 \times 100 = 8000$ ft above the aircraft, that is, at 37 000 ft, so the aircraft is able to fly above the storm. Note that using the exact Equations (5.33) and (5.36) yields a height of the top of the cloud 8482 ft above the aircraft, so there is 6% error in using the simple equation.

In the latest generation systems, the weather radar antenna scans the tilt angle as well as the azimuth and uses additional sensors to determine the moisture

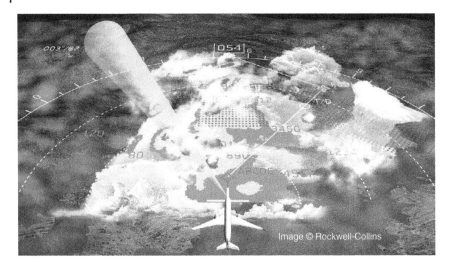

Figure 5.15 Latest generation three-dimensional airborne weather radar display on the navigation screen in a glass-cockpit. *Source:* Reproduced with permission of © Rockwell-Collins.

content and temperature profile in any storms ahead of the aircraft. The data is processed by a computer to provide a three-dimensional image of the storm clouds and interprets the images for the pilot giving positions of lightning and hail. The image is displayed on the navigation screen of a glass cockpit and is superimposed on the terrain and other navigational information. An example is shown in Figure 5.15, which shows a screen shot from the Rockwell-Collins WXR-2100 system.

Interesting Diversion 5.1: Mapping the Surface of Venus using Synthetic Aperture Radar

Prior to spacecraft arriving at the planet Venus, nothing was known about its surface topography. Telescope observations from the Earth show a featureless disk due to a dense atmosphere of mostly CO_2 with an estimated 20 km-thick cloud layer that obscures the surface. The atmosphere is believed to be the result of a runaway greenhouse effect and the surface temperature and pressure are 480 °C and 90 bars, respectively. In this harsh environment, the maximum time that a lander has survived is two hours, achieved by Venera 13 in 1982.

3.7 m diameter
parabolic antenna

Figure ID5.1.1 Artist's impression of the Magellan spacecraft in orbit around Venus. The 3.7 m diameter parabolic antenna was used to transmit radar pulses at a frequency of 2.385 GHz (12.6 cm) that could penetrate the thick Venusian atmosphere and map the topography of the surface. *Source: NASA.*

The atmosphere, however, is transparent to 10 cm waves at the top of the UHF band and in the early 1980s a proposal to send a spacecraft equipped with a radar transmitter to orbit the planet and map the topography was put forward. The project developed into the Magellan spacecraft (Figure ID5.1.1) that was launched in May 1989 from the space shuttle and was injected into an elliptical polar orbit around Venus in August 1990. The spacecraft was equipped with a 3.7 m diameter parabolic antenna and transmitter operating at 2.385 GHz (12.6 cm). During each orbit at the closest approach to the surface (<300 km) the radar would map the surface and as it moved away from the planet, the same antenna would transmit the data back to the Earth. On the following orbit at the closest approach the planet would have rotated slightly and the radar would map a neighboring strip to the previous one. The planet rotates once on its axis every 243 days, so in this period the entire surface could be surveyed by Magellan.

The science demanded a lateral resolution of around 100 m and to achieve this from a 300 km distance requires an angular beam width of 0.02°. According to Equation (5.1), a parabolic antenna operating at a wavelength of 12.6 cm would need to have a diameter of at least 360 m to produce such a narrow beam, which is clearly impractical for a space mission. At the same wavelength, the 3.7 m diameter antenna used on Magellan, assuming perfect antenna

(Continued)

Interesting Diversion 5.1: (Continued)

performance, would have a beam width of 1.95° producing an imaged area 10 km wide at the surface. In practice, the radar was pointed at the surface at an angle and the real performance resulted in an elliptical area around 20 km across illuminated by the radar beam (Figure ID5.1.2a). A single return pulse processed in real time would produce a single value for the average range within the probed area thus giving a 20 km resolution.

For a stationary target such as terrain, the resolution can be greatly improved using a technique known as synthetic aperture radar (SAR). The terrain produces a range of return times and Doppler shifts and this information is embedded in each pulse by its return time, width, and phase shift. Each pulse is stored and then large numbers of pulses are processed and their evolution as the spacecraft moves is fitted to a model of the terrain. As illustrated in Figure ID5.1.2a, this is done by dividing a target area into a number of cubic "voxels" (3D equivalent to a pixel in a 2D image). Starting with an empty grid, selected voxels are then filled with a reflective medium till the calculated evolution of return PWs and phase shifts has an optimum fit to the measured data.

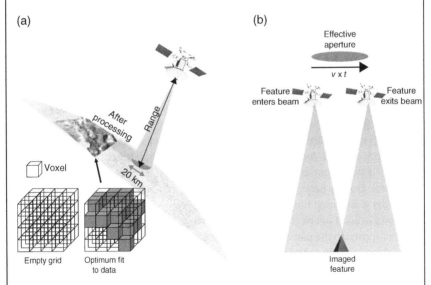

Figure ID5.1.2 (a) The SAR method stores pulses and processes their evolution in time. The landscape is modeled by filling "voxels" with a reflective medium to produce a topology whose calculated evolution of pulse widths and phase shifts best fits the data. (b) The method provides the same resolution as real-time imaging by a large antenna whose aperture has a diameter $v \times t$, where v is the spacecraft velocity and t is the time a feature spends illuminated by the beam.

As illustrated in Figure ID5.1.2b, this is equivalent to scanning the area with a huge antenna since each feature on the surface keeps producing information embedded in the return pulses from the time that it is first illuminated by the beam to the time it exits from the beam. Processing all the information it provides during this time is equivalent to real-time scanning by a single dish whose aperture diameter is $v \times t$, where v is the spacecraft velocity and t is the time the feature spends in the beam. The data obtained using SAR on Magellan were combined with radar altimetry to get a true 3D mapping of the surface.

An example of the superb images of the Venusian topography obtained by processing the radar data from Magellan is the 3D perspective of the volcano Maat Mons from a view point 634 km away shown in Figure ID5.1.3. Lava flows are seen extending for hundreds of km. Note that the vertical scale has been exaggerated by a factor of 10. The simulated color in the image is based on the color photographs taken at the surface by the Russian Venera 13 and Venera 14 landers.

Figure ID5.1.3 3D perspective image obtained by processing Magellan SAR data of the volcano Maat Mons with the vertical scale exaggerated by a factor of 10. The viewpoint is from 634 km away and the simulated color is based on color photographs taken by the Russian Venera 13 and Venera 14 landers. *Source:* NASA.

5.4 Secondary Surveillance Radar (SSR)

An issue with primary radar alone is that the only information from the target is the position, velocity, and strength of the return. During the Second World War, a system was developed to identify "friend from foe" (IFF) where a radio signal from the ground activated a system on the aircraft that transmitted a radio response containing information to identify it. This has developed into the secondary surveillance radar (SSR) system that is used to provide additional information about traffic that cannot be obtained by primary radar alone. The device on the aircraft that transmits a signal in response to receiving one from the ground station is called a transponder (merging of transmitter/responder). The system has evolved in sophistication and gone through three main evolutionary stages labeled, mode A, mode C, and mode S. Each advance in the technology has maintained downward compatibility so that the system can operate in an airspace containing aircraft that have mode A, C, or S transponders fitted. Communication between ground stations and aircraft occur in all three modes while mode S provides, in addition, the capability of aircraft to communicate with each other. A ground station SSR antenna is normally attached to a rotating primary surveillance radar (PSR) antenna as illustrated in Figure 5.16 and

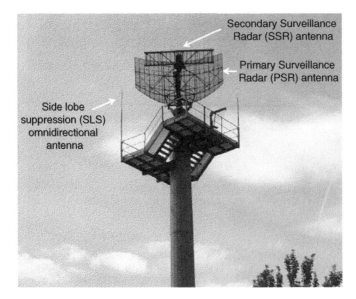

Figure 5.16 Rotating Primary surveillance radar (PSR) antenna with attached separate rotating secondary surveillance radar (SSR) antenna. Also visible is the fixed omnidirectional dipole antenna used for side lobe suppression (SLS). *Source:* Reproduced with permission of EASAT.

transmits directional *interrogation* pulses to the aircraft at a frequency of 1030 MHz (UHF band). The system also requires omnidirectional pulses transmitted at the same frequency to facilitate side lobe suppression (SLS) as described below and these are transmitted by dipole antennas indicated in Figure 5.16.

5.4.1 Mode A and Mode C Interrogation Pulses

In a single mode A or mode C interrogation sequence, two pulses with a width 0.8 μs wide are transmitted by the directional antenna and are labeled P1 and P3. Their spacing is 8 or 21 μs apart depending on whether the interrogation is mode A or mode C. A separate fixed omnidirectional antenna transmits a third pulse labeled P2 in all directions, again at a frequency of 1030 MHz, 2 μs after P1. The pulse sequences for the two types of interrogation are shown in Figure 5.17. The rate at which this sequence is transmitted is typically 1200 times per second and the repetition frequency of the reply from the aircraft is the same as that of the interrogation pulse sequence.

The purpose of the omnidirectional P2 pulse is for SLS, that is, to ensure that a response from the aircraft is not triggered by one of the side lobes of the

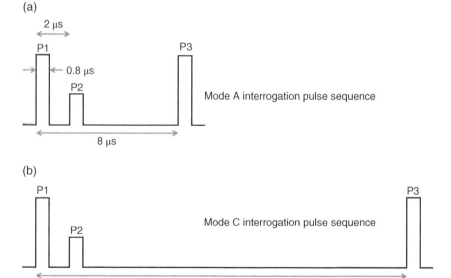

Figure 5.17 Mode A and mode C interrogation pulse sequences transmitted by the ground antennas. P1 and P3 are transmitted by the rotating directional antenna while P2 is transmitted by the fixed omnidirectional antenna.

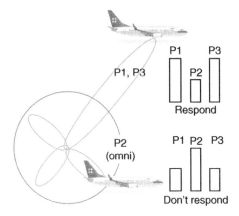

Figure 5.18 Method for detecting whether interrogation pulses are coming from a side lobe of the directional antenna. In this case the omnidirectional P2 pulse will be stronger than the directional P1 and P3 pulses and the aircraft does not respond.

directional beam transmitting the P1 and P3 pulses. The aircraft detects this by comparing the strength of the P1 and P3 with the P2 pulse as illustrated in Figure 5.18. If the directional pulses are coming from a side lobe, their strength will be lower than that of P2 and the aircraft does not respond. If P1 and P3 are stronger than P2, then a response is triggered.

Once an aircraft receives a valid interrogation pulse sequence, it responds in mode A or C depending on the time interval between the P1 and P3 pulses. The response is transmitted by an omnidirectional antenna at the slightly higher frequency of 1090 MHz so that there is no possibility of confusion from pulses reflected directly from the aircraft.

5.4.2 Mode A Reply from the Aircraft

If the aircraft receives a mode A interrogation pulse (Figure 5.17a), it sends a mode A reply. This is a 4-digit octal code allocated to the aircraft by the ATC unit and set up on the controller in the cockpit by the pilot with the format of the transmitted word shown in Figure 5.19a. There are two framing pulses (F1 and F2) 20.3 μs apart and within these are four groups of three pulses (labeled ABCD) plus an unused pulse (labeled X). The three pulses within each group (A–D) are binary code for an octal digit in the range 0–7_8 (000_2–111_2) and the order of the digits in the pulse train is shown in Figure 5.19a. For each octal digit, A, B, C or D, the bits are labeled 1, 2, or 4, that is, the labels show the value of the bit. The presence of a pulse at the labeled position represents "1" and an absence "0." For example, the octal code 4361 will be represented by the bits shown in the table in Figure 5.19b and the transmitted pulse train will be as in Figure 5.19c. The number allocated by the ATC is known as a "squawk" code, which comes from earliest military use of transponders where the secret

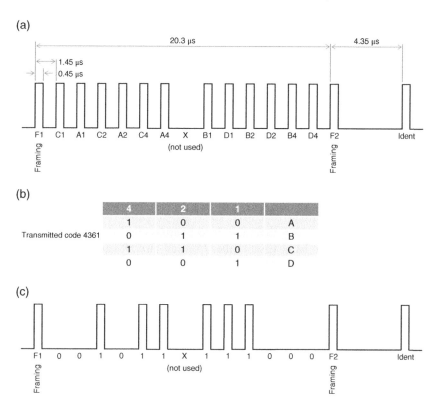

Figure 5.19 (a) Format of word transmitted by aircraft transponder in mode A when interrogated by a valid signal from the ground station. (b) Binary code for octal digits ABDC for code. (c) Pulse train transmitted for squawk code 4361.

project had the code name "parrot." The 4-digit code allocated by the ATC is displayed on the controller's PPI next to the primary return from the aircraft.

The extra pulse in the sequence labeled "Ident" occurring 4.35 μs after the F2 framing pulse is transmitted along with the rest for 15–20 seconds if the pilot presses an "ident" button on the transponder control panel. This is picked up by the receiver on the ground and if present causes the aircraft icon on the controller's screen to be highlighted.

5.4.3 Mode C Reply from the Aircraft

If the aircraft receives a mode C interrogation pulse (Figure 5.17b) and its transponder is mode C compliant, it sends a mode C reply. This includes the 4-digit squawk code allocated by ATC and in addition a second word that reports the altitude of the aircraft, which is then displayed on the controller's screen (PPI).

This altitude is either obtained from the altimeter or a separate aneroid capsule fitted with an encoder (see Section 2.8.4) and is independent of the altimeter subscale setting. It is always referenced to ISA sea level pressure, that is, 1013.25 mbar and reported to the nearest hundred feet (in other words the altitude is reported as a flight level). The format of the word reporting the altitude is also a 4-digit octal number and uses Gillham code to represent the flight level (see Figure 2.18).

5.4.4 Conflicts Between Mode A and Mode C Replies from Different Aircraft

Since all interrogation transmissions from the ground are at 1030 MHz and all returns from aircraft are at 1090 MHz there are potential conflicts if there is dense traffic or there is more than one airport within range sending SSR interrogation pulses. In the latter case, replies from an aircraft can be received by more than one ground radar station and for the stations that did not send the interrogation message the reply will be unsynchronized with the primary radar (Figure 5.20a). This fault is known as "false replies from unsynchronized interrogator transmissions" (FRUIT).

The other possible conflict occurs when two or more aircraft are within the ground antenna beam width, which, as shown in Section 5.1, can be two miles wide at the target. In this case the individual replies can overlap at the receiver, which is known as "Synchronized Garbling" (Figure 5.20b).

5.4.5 Mode S

By the 1960s, the rapid increase in air traffic was beginning to overwhelm the existing ATC systems, with problems including the conflicts between mode A and mode C replies discussed in the previous section. In the process of overhauling ATC, mode S (for "select") SSR was developed by the Lincoln Laboratory at MIT, starting in the early 1970s. The aim was to produce a system that can specifically address each aircraft and to facilitate this, every aircraft is now assigned a unique 24-bit code during its registration process, which allows 16 777 216 to be accommodated. Given that new technology takes some time to be adopted, the system was designed to be compatible with mode A- and mode C-equipped aircraft so that mode S transmissions were transparent to them and they would continue to respond with mode A and mode C replies. It was also decided to maintain the same frequencies for transmission, that is, 1030 MHz for interrogation and 1090 MHz for response. Thus, the digital data required for mode S has to share the same channel and be transmitted in the gaps between the pulses used for mode A and mode C transmissions. Additional functions fulfilled by mode S include the traffic collision avoidance system (TCAS) described in Section 5.5 and an improvement in height

(a)

(b)

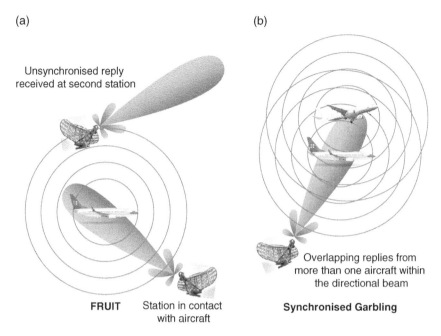

Unsynchronised reply
received at second station

FRUIT   Station in contact
with aircraft

Overlapping replies from
more than one aircraft within
the directional beam

Synchronised Garbling

Figure 5.20 (a) False replies from unsynchronized interrogator transmissions (FRUIT) conflict occurring when an aircraft is within the range of more than one radar station. (b) Synchronized garbling that occurs when the transmissions overlap from more than one aircraft within the radar beam.

resolution to 25 ft increments. Currently, all aircraft flying under instrument flight rules (IFR) must be mode S compliant.

Mode A and mode C interrogations are entirely "all call," that is, every aircraft within range is addressed with every transmission whereas mode S has three fundamental interrogation modes, that is:

1) All call (all aircraft are interrogated as in modes A and C)
2) Selective call to a unique aircraft address
3) Broadcast (send message without requiring a response)

These are discussed in more detail below.

5.4.6 Mode S All Call Interrogation

The all call interrogation format is shown in Figure 5.21 and includes the same P1, P2, and P3 pulses, transmitted by a 1030 MHz carrier, used in modes A and C (see Figure 5.17) with an additional P4 pulse that is either 0.8 or 1.6 μs wide. If P4 is 0.8 μs wide, aircraft with mode S transponders ignore the call while

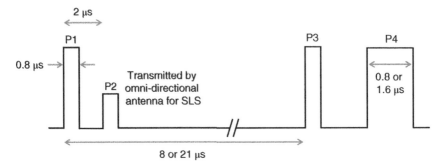

Figure 5.21 Format of the mode S all call signal from the ground station.

mode A and C transponders will respond with their ATC code (and altitude for mode C) as described in Sections 5.4.2 and 5.4.3. If P4 is 1.6 µs wide, mode S transponders that receive the all call interrogation reply with their unique address and these are all stored on the ATC system. The format of the reply from the mode S transponder is described below. If an all call signal is sent, it alternates between the two widths of P4 so that all aircraft are addressed.

5.4.7 Mode S Selective Call Interrogation

The interrogator pulse format for the selective broadcast (i.e. to a specific aircraft) mode is shown in Figure 5.22. This time the P1 and P2 pulses are sent by the main directional antenna and are of equal strength so that the SLS system on mode A and mode C transponders forces them to ignore the interrogator. A reference pulse of width 1.25 µs indicates the start of the data word, which

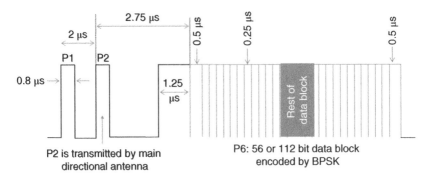

Figure 5.22 Format of the mode S selective call signal to a specific aircraft from the ground station. The P2 pulse is transmitted by the main directional antenna and has the same strength as the P1 pulse so the interrogation is ignored by mode A and mode C aircraft.

is a series of 56 or 112 binary data bits framed between two 0.5 µs pulses. Each data bit occupies 0.25 µs and the logic 1/0 levels within the data block are encoded by BPSK modulation (see Section 4.3.7.2), that is, by shifting the carrier phase by −90° for logic 1 and +90° for logic 0. The data block includes the 24-bit address of the specific aeroplane.

The ground interrogators also have a unique code specific to that ground station that is sent within the data block. The normal system for using mode S is that the ground station will send out an all call and collect all aircraft codes in its space. The individual aircraft transponders then lock themselves out of responding to all calls and wait for specific calls with the correct code. The ground station will also ignore responses with a different ground station code, thus avoiding FRUITing. The selective calls can then issue instructions to flight crews without the need for voice communications.

5.4.8 Mode S Reply from Aircraft

The mode S reply from aircraft transponders is transmitted with a 1090 MHz carrier wave as for mode A and mode C and the data format is shown in Figure 5.23. There is a 4-pulse preamble followed by a binary data block that is 56 or 112 µs long with each bit occupying 1 µs. The 1/0 coding is different to the interrogator and as illustrated in Figure 5.23, a "0" is represented by a pulse appearing in the second half of the 1 µs interval while a "1" is represented by a pulse in the first half of the interval. This type of coding is known as Pulse

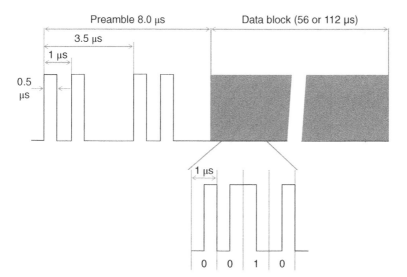

Figure 5.23 Format of the mode S reply.

Position Modulation (PPM: see Section 4.3.7.1). The data block contains the 24-bit address of the aircraft, formatting, parity, and control bits and in the case of the 112-bit block, there is an additional message field which can provide additional flight information as described in Section 5.4.10.

5.4.9 Traffic Surveillance by Mode S

Mode S enables the provision of surveillance of local traffic independent of primary radar by aircraft sending air to air interrogation signals broadcast on the ground station broadcast frequency of 1030 MHz to other aircraft. Transponders on aircraft in the vicinity send replies at 1090 MHz and the time difference between the interrogation and response is used to calculate the slant range while the bearing is calculated using a directional antenna, which can either be four discrete elements or a pair of blade antennas (Figure 5.24). The phase difference between the signal at each blade gives the receiving angle from which the bearing can be calculated given the heading of the receiving aircraft.

To track aircraft fitted with older mode A and mode C transponders, all call interrogations (Figure 5.21) are sent once per second and in addition to providing range and bearing information as described above, mode C replies will also report the height of interrogated aircraft. Any mode S responses to the all call

Figure 5.24 Pair of blade antennas on the belly of a light twin used to provide mode S traffic surveillance.

broadcast will contain the unique address of the aircraft as well as the height information (in 25 ft increments). Having received a mode S response each aircraft will, in addition to the all call broadcast, transmit mode S selective call signals once per second to each local mode S equipped aircraft, which reduces the probability of overlapping or garbled replies. The mode S surveillance function, which feeds traffic information into the traffic collision and avoidance system (TCAS) described in Section 5.5, is normally provided by an independent set of antennas, one at the top and one at the bottom of the fuselage, to the antennas providing the normal ground-based mode S service.

5.4.10 Squitters and Automatic Dependent Surveillance Broadcast (ADS-B)

The system developed for Mode S has enabled a new form of communication directly between aircraft, which is not based on interrogation and response but on unsolicited broadcasts from aircraft known as squitters. Since any aircraft flying under IFR will have a sophisticated navigation system, it can report its position more precisely than can be obtained by primary radar and it does this by broadcasting frequent squitters. These come in various forms with the simplest being an 8 µs long 4-pulse preamble followed by a 56-bit data block coded using PPM, which is an identical format to the mode S reply shown in Figure 5.23. This type of broadcast, known as a mode S acquisition squitter, is sent once per second and contains just the aircraft 24-bit address plus control and parity bits giving a simple "I am here" message (Figure 5.25a).

The capability of the system has been enhanced by extending squitters to contain 112 bits in the data block and the extra 56 bits is used to send additional positional and flight information employing the format illustrated in Figure 5.25b. The extended squitters (ES) illustrate a new method to provide traffic information referred to as the automatic dependent surveillance broadcast (ADS-B) system. It is automatic because the information is sent without any operator intervention, dependent because the information is derived from other onboard systems such as GNSS, surveillance because it provides radar-type information though with higher precision and broadcast as the information is continuously sent and can be received by any suitably equipped aircraft or ground station.

The information contained in the 56-bit ADS-B message block for the four types of ES is shown in Figure 5.25c along with the broadcast rate. The code is highly compacted to try and pack as much information as possible into 56 bits and requires further processing to extract usable data. For example, the latitude and longitude are returned as 17-bit numbers in the airborne position squitters but to report a position to a 5 m accuracy requires 7 decimal digits plus an alphanumeric character (N/S or E/W). It takes four bits to encode a decimal digit and

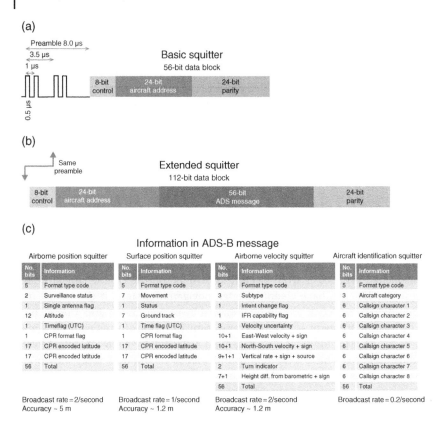

(a)

Preamble 8.0 μs
3.5 μs
1 μs

Basic squitter
56-bit data block

0.5 μs

| 8-bit control | 24-bit aircraft address | 24-bit parity |

(b)

Same preamble

Extended squitter
112-bit data block

| 8-bit control | 24-bit aircraft address | 56-bit ADS message | 24-bit parity |

(c)

Information in ADS-B message

Airborne position squitter

No. bits	Information
5	Format type code
2	Surveillance status
1	Single antenna flag
12	Altitude
1	Timeflag (UTC)
1	CPR format flag
17	CPR encoded latitude
17	CPR encoded latitude
56	Total

Broadcast rate = 2/second
Accuracy ~ 5 m

Surface position squitter

No. bits	Information
5	Format type code
7	Movement
1	Status
7	Ground track
1	Time flag (UTC)
1	CPR format flag
17	CPR encoded latitude
17	CPR encoded latitude
56	Total

Broadcast rate = 1/second
Accuracy ~ 1.2 m

Airborne velocity squitter

No. bits	Information
5	Format type code
3	Subtype
1	Intent change flag
1	IFR capability flag
3	Velocity uncertainty
10+1	East-West velocity + sign
10+1	North-South velocity + sign
9+1+1	Vertical rate + sign + source
2	Turn indicator
7+1	Height diff. from barometric + sign
56	Total

Broadcast rate = 2/second
Accuracy ~ 1.2 m

Aircraft identification squitter

No. bits	Information
5	Format type code
3	Aircraft category
6	Callsign character 1
6	Callsign character 2
6	Callsign character 3
6	Callsign character 4
6	Callsign character 5
6	Callsign character 6
6	Callsign character 7
6	Callsign character 8
56	Total

Broadcast rate = 0.2/second

Figure 5.25 (a) Format of a basic squitter containing 56 bits in the data block. (b) Format of extended squitters containing an additional 56 bits in the data block for the ADS-B message. (c) Information contained in the 56-bit ADS-B message.

the alphanumeric could be represented by a single bit (e.g. $N = 0$, $S = 1$), so the full coordinates would require 29 bits. The 17-bit number given represents the accurate local position and the higher-order bits describing the global position are removed with the resulting ambiguity in global position removed by other information. The 17-bit format is referred to as compact position reporting (CPR).

The ADS-B system enables comprehensive data about any flight to be obtained by a simple ADS-B receiver and it is expected that by 2020 it will replace primary radar as the main surveillance method for commercial air traffic. The main advantage is its much higher precision of position reporting, which is down to around 5 m for airborne aircraft. Primary radar will continue to play an important role, however, due to its ability to monitor all traffic whatever its capabilities.

5.4.11   Universal Access Transceivers (UAT) and ADS-B

Having established the power of ADS-B as a method for the surveillance and control of air traffic the development of a new datalink system that is independent of mode S began in 1995, that is, the universal access transceiver (UAT). The ICAO set up the standards and recommended practices (SARPS) for the new system in 2007 and it has been assigned the frequency 978 MHz and digital data modulation rate of 1.041667 Mbits s^{-1}. The type of modulation used is specified as binary continuous phase frequency shift Keying (BCPFSK), which was described in Section 4.3.7.3. With this type of modulation, the binary data is encoded by two different frequencies centered around 978 MHz (±312.5 kHz). The transition between a "1" and "0" or vice versa occurs with a continuous change of phase and the frequency is sampled at the midpoint of each clock cycle as illustrated in Figure 5.26.

The organization of UAT messages in the time domain is illustrated in Figure 5.27, which start with the fundamental time structure one-second long known as the *frame* (Figure 5.27a). A new frame begins at each UTC second (see Section 5.4.12) and is segmented into two sections, labeled the ground segment, which is reserved for ground to air transmissions and the ADS-B segment, which supports all transmissions. The frame is subdivided into 4000 "message start opportunities" (MSOs) each 250 μs long, which mark the points where the start of a message can be scheduled and can be considered as the fundamental "clock" of the frame.

The ground segment is 752 MSOs (188 ms) long including a guard segment of 48 MSOs (12 ms) to provide separation from the ADS-B segment. The remaining 176 ms is available to transmit data and this interval is divided into 32 ground message slots each 5.5 ms long. Different ground stations are assigned different slots with each one broadcasting once per second in its assigned slot to provide

Figure 5.26 Binary continuous phase frequency shift keying (BCPFSK) used to encode data used in the universal access transceiver (UAT). Two frequencies 312.5 kHz either side of the carrier frequency of 978 MHz are used to denote "0" and "1."

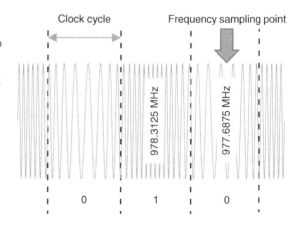

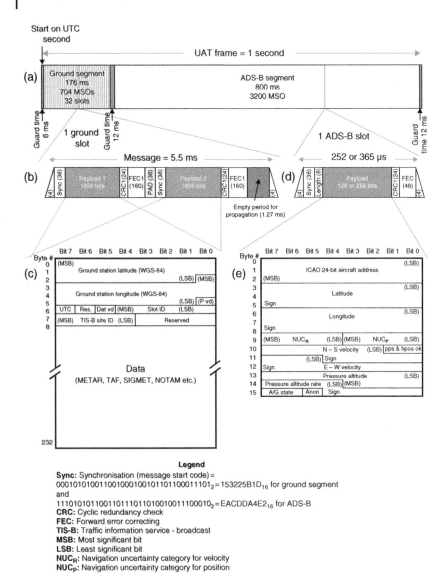

Start on UTC second

UAT frame = 1 second

(a)
Ground segment
176 ms
704 MSOs
32 slots

ADS-B segment
800 ms
3200 MSO

Guard time 6 ms

1 ground slot

Guard time 12 ms

1 ADS-B slot

Guard time 12 ms

Message = 5.5 ms

252 or 365 µs

(b)
Sync (36) | Payload 1 1856 bits | CRC1 (24) | FEC1 (160) | PAD (8) | Sync (36) | Payload 2 1856 bits | CRC1 (24) | FEC1 (160)
(4)

(d)
Sync (36) | Length (8) | Payload 128 or 256 bits | CRC1 (24) | FEC (48)
(4)

Empty period for propagation (1.27 ms)

(c)

Byte #	Bit 7	Bit 6	Bit 5	Bit 4	Bit 3	Bit 2	Bit 1	Bit 0
0	(MSB)							
1		Ground station latitude (WGS-84)						
2							(LSB)	(MSB)
3								
4		Ground station longitude (WGS-84)						
5							(LSB)	(P vd)
6	UTC	Res.	Dat vd	(MSB)		Slot ID		(LSB)
7	(MSB)	TIS-B site ID	(LSB)		Reserved			
8								

Data
(METAR, TAF, SIGMET, NOTAM etc.)

232

(e)

Byte #	Bit 7	Bit 6	Bit 5	Bit 4	Bit 3	Bit 2	Bit 1	Bit 0
0								(LSB)
1		ICAO 24-bit aircraft address						
2	(MSB)							
3								(LSB)
4		Latitude						
5	Sign							
6		Longitude						(LSB)
7								
8	Sign							
9	(MSB)	NUC_R		(LSB)	(MSB)	NUC_P		(LSB)
10		N – S velocity				(LSB)	pps & hpos ok	
11						(LSB)	Sign	
12	Sign	E – W velocity						
13		Pressure altitude						(LSB)
14	Pressure altitude rate			(LSB)	(MSB)			
15	A/G state		Anon	Sign				

Legend
Sync: Synchronisation (message start code) =
$00010101001100100010010110110001110_2 = 153225B1D_{16}$ for ground segment
and
$1110101011001101110110100100111100010_2 = EACDDA4E2_{16}$ for ADS-B
CRC: Cyclic redundancy check
FEC: Forward error correcting
TIS-B: Traffic information service - broadcast
MSB: Most significant bit
LSB: Least significant bit
NUC_R: Navigation uncertainty category for velocity
NUC_P: Navigation uncertainty category for position

Figure 5.27 (a) Basic time structure of a 1 second UAT frame that starts on each UTC second. (b) Time structure of a single ground station message slot. (c) Information broadcast within the data payload parts of the ground message slot. (d) Time structure of a single ADS-B message slot. (e) Information broadcast within the 128-bit data payload of an ADS-B message slot. The 256-bit version will contain additional information.

interference-free communication between several ground stations and an aircraft. The time structure within a single slot is shown in Figure 5.27b and the package consists of two blocks of data each starting after a 36-bit synchronization sequence of:

$$000101010011001000100101101100011101_2 = 153225B1D_{16}$$

This alerts the system to the start of a message segment 1856 bits long and then follows a cyclic redundancy check (CRC) of 24 bits and a forward error collection block of 160 bits. The error detection and correction scheme follows the method published by Irving Reed and Gustavo Solomon in 1960 [6]. A pad of 36 bits separates the first part of the message from the second and the whole cycle repeats again to complete the message. Including a 4-bit packer at each end, the entire sequence is 4196 bits long, which at a rate of 1.041667 Mbits s^{-1} corresponds to 4.028 ms leaving a 1.472 ms gap before the start of the next slot. This corresponds to 238 nautical miles at light speed, which is further than the maximum line of sight range expected so the gap ensures that transmissions from distant stations will arrive within their allocated slot. The data payloads (Figure 5.27c) contain the coordinates of the ground station, the slot (1–32) being used and a site identifier followed by a substantial block of data. The data blocks can be used to broadcast a flight information service (FIS-B) for weather and other important flight information.

The ADS-B segment within each one second frame has 800 ms available for data and thus 3200 MSOs for message slots. Unlike the ground segment in which a specific slot is assigned to a ground station, the ADS-B slots are chosen randomly by aircraft and ground vehicles to provide separation of transmissions. An entire message packet, shown in Figure 5.27d), is shorter than the ground one and contains a single data payload.

The start of an ADS-B message is also signaled by a 36-bit synchronization sequence, which is the bit inverse of the sequence at the start of ground messages, that is,

$$111010101100110111011010010011100010_2 = EACDDA4E2_{16}$$

This is followed by a length identifier as ADS-B messages have two possible lengths followed by the data payload, which is 128 or 256 bits long. There follows a CRC and an FEC sequence, which is shorter than the ground one due to the smaller amount of data. The entire package is 252 or 365 µs long.

The information within the basic 128-bit message, shown in Figure 5.27e, reports the ICAO 24-bit address of the aircraft along with its position, altitude, airspeed, and vertical speed. Also included is the navigational uncertainty category (NUC) for the position and speed, which is a measure of the estimated error. The entire 128-bit message is known as the state vector component and is part of every ADS-B message. The longer version containing 256 bits in the data block will broadcast additional information including the aircraft call sign and the next trajectory change point.

There is some overlap between the ADS-B information from mode-S ES and UATs, but the UAT also provides additional information such as FIS-B. Another important fundamental difference is that a UAT-based system is essentially bidirectional and ground stations collate aircraft information and rebroadcast it to provide a traffic information service broadcast (TIS-B) as described below. The full capability to send and receive messages is known as ADS-B in and ADS-B out.

5.4.12 Surveillance by ADS-B

It is envisaged that ADS-B will become the primary course of aircraft surveillance and traffic information in the next few years with either S-mode ES at 1090 MHz or UATs at 978 MHz broadcasting the information. The aerospace environment needs to accommodate aircraft with capabilities ranging from no ADS-B to full ADS-B in and out with data transmitted either by UAT or Mode S ES. Figure 5.28 illustrates how the various levels of information are combined

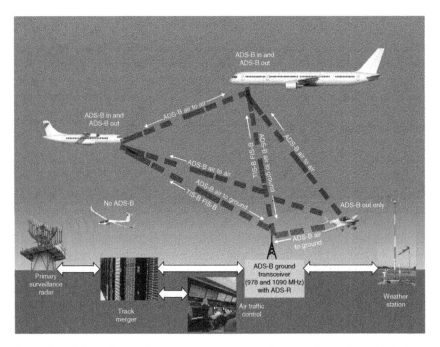

Figure 5.28 Information exchange in an aerospace environment in which ADS-B is the primary source of surveillance.

to give comprehensive traffic surveillance. Aircraft having both ADS-B in and ADS-B out capability transmit to each other and to ground stations with information on each other's positions and velocities. Aircraft having full capability can build up a partial database on other full capability aircraft in the vicinity that are transmitting on the same frequency. The transmissions at both frequencies are also received by ground stations that send the information to a track merger that combines the data with that from primary radar. The full list of all aircraft positions irrespective of their capability is then broadcast as ADS-B messages at both frequencies to all aircraft in the vicinity. Thus, the data from aircraft transmitting data at 978 MHz is rebroadcast to aircraft using 1090 MHz and vice versa. This is known as ADS-Rebroadcast (ADS-R). In addition, aircraft that are using a UAT and with ADS-in capability are able to receive the FIS-B service from ground facilities in range. The FAA has specified that by 1 January 2020, a minimum capability of ADS-B out will be required for flight in airspace that currently requires a transponder.

The traffic information received by ADS-B is used to drive a cockpit traffic information display either as a stand-alone screen (Figure 5.29) or

Figure 5.29 Traffic information from a TIS-B service from a ground station displayed on a stand-alone screen in a light aircraft. The arrow-shaped symbols show the position and direction of travel of ADS-B-equipped aircraft. The bullet-shaped symbols show the position and direction of travel of non-ADS-B-equipped aircraft. *Source:* Reproduced with permission of AOPA.

superimposed on a glass cockpit navigation display. Between 1999 and 2006, the use of the ADS-B surveillance system was trialed by the FAA in Alaska by fitting around 400 commercial and general aviation aircraft with ADS-B equipment [7]. More recently in 2017 in the United Kingdom, the CAA set up a trial of real-time air traffic displays based on ADS-B at a number of general aviation airfields in which it is not economically viable to install primary radar [8].

Another function that has been built in to UAT ground transmissions is the possibility of passively determining aircraft positions using the transmissions themselves [9]. If an aircraft could measure its distance from three different ground stations, this would provide a unique position fix. The ground segment of the signal broadcast by all ADS-B ground stations is synchronized to UTC seconds, which have atomic clock precision. Thus, measuring the time difference between transmission and reception from each station provides a "pseudo-range" to the aircraft, that is, a range with an associated large error due to the clock on the aircraft not being synchronized to UTC seconds. It is not possible to determine the position accurately by using the pseudo ranges from three ground stations but since the offset between the aircraft clock and UTC seconds is constant for all ground stations, the timing of signals from a fourth station will determine the time offset between the aircraft clock and UTC seconds and hence an accurate position fix. This is the same system as that used by the space-based global navigation satellite system (GNSS: see Chapter 8) and it could provide a local earth-based position determination within localized regions with good coverage by ground stations. The over-dependence on the relatively vulnerable space-based system has led to the search for alternative methods of position finding using terrestrial facilities.

5.5 Traffic Collision Avoidance System (TCAS)

The previous section described various methods whereby local traffic information can be provided on a cockpit display including mode S interrogation and ADS-B based on mode S ES and UATs. A collision avoidance system must use this surveillance data to determine whether any of the local aircraft represent a collision threat, a method for displaying the level of this threat to the pilot and if necessary determine what evasive action to take.

The local traffic information obtained by whatever surveillance method is in use, including the position and velocity of aircraft in the vicinity, is passed to algorithms that determine if a threat exists. For any identified threat, a second set of algorithms formulates an appropriate response. Each aircraft in the vicinity is characterized by one of four levels of threat [10] determined by linear extrapolation of the current horizontal and vertical velocities of all aircraft into the future. The position of local traffic and the level of threat is indicated on the

(a)

(b)

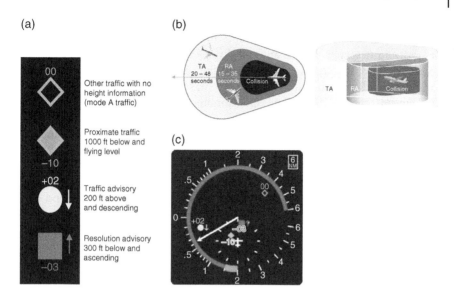

(c)

Other traffic with no height information (mode A traffic)

Proximate traffic 1000 ft below and flying level

Traffic advisory 200 ft above and descending

Resolution advisory 300 ft below and ascending

Figure 5.30 (a) Symbols used to depict the four levels of threat in a TCAS display along with the relative height in hundreds of feet and climb or descent. (b) The time to closest point of approach (CPA) and a schematic representation of the volume of space involved to issue a traffic advisory (TA) or resolution advisory (RA). (c) Target vertical speeds required to avoid a collision are indicated by the green range on the display.

cockpit display by one of the four standardized symbols displayed in Figure 5.30a. Also shown above or below the symbols is the relative height in units of hundreds of feet and an arrow at the side of the symbol indicates if the intruder is climbing or descending.

Traffic that is too distant to be a likely threat ("other") is shown as a hollow white or cyan diamond while closer ("proximate") traffic is indicated by a filled diamond. The collision threat is assessed in terms of the closest point of approach (CPA). The system does not know positions and trajectories with sufficient accuracy to predict that a collision will actually occur but if the CPA falls below a specified threshold then it is deemed that a collision threat is present. The level of threat is then determined by the time to CPA on current trajectories. If the time to CPA is within 20–48 seconds (depending on altitude), the symbol for that intruder is changed to a yellow filled circle and an audible alert ("traffic traffic") is activated in the cockpit. This is known as a traffic advisory (TA) and it is triggered by an intruder entering a teardrop-shaped volume of space shown in Figure 5.30b. The longest time used in the calculation is at the lowest altitudes where traffic is generally slower. The TA alerts the crew to search for the traffic and maintain visual separation. If the intercept course is maintained and the time to CPA decreases to 15–35 seconds, again depending on altitude, the system issues a resolution advisory (RA) and the traffic symbol changes to a red square. At the same time an audible alert such as "climb climb" or "descend descend" sounds and the display highlights in green the range of vertical speeds required to avoid a collision as shown in Figure 5.30c. For example, the display shown in Figure 5.30c has triggered an RA due to the aircraft ahead that is 300 ft below and climbing. The vertical speed indication shows that a descent of at least $1500 \, \text{ft min}^{-1}$ must be initiated to avoid a collision. The display shown is an older stand-alone version and in newer aircraft the TCAS information is incorporated into the primary flight display and navigation display.

The avoidance algorithm first decides the sense of the avoidance maneuver (climb or descend) assuming that the pilot takes five seconds to respond and that a vertical acceleration of $\pm 0.25 \, \text{g}$ can be achieved till a climb rate of $\pm 1500 \, \text{ft min}^{-1}$ is reached. Then, assuming the intruder does not change its own trajectory, the CPA is calculated for the case of a climb or descend and the option that gives the maximum CPA is selected as illustrated in Figure 5.31. If the intruder aircraft is also mode S equipped, the datalink is used to coordinate the maneuver so that an opposite sense RA is issued to the intruder.

Having decided the sense of the RA, TCAS algorithms then calculate the strength of the response by finding the minimum vertical rate that will achieve a target minimum vertical separation at the CPA, which is normally 400 ft. The situation is monitored during the maneuver and, if necessary, a new climb rate is indicated on the display along with a verbal alert such as "increase climb."

The current version of TCAS, designated TCAS II 7.1 only issues instructions to climb or descend as surveillance based on passive mode S detection does not have sufficiently accurate bearing information to issue reliable instructions for lateral avoidance maneuvers. Future versions of TCAS that include more accurate ADS-B surveillance will be able to issue lateral instructions also.

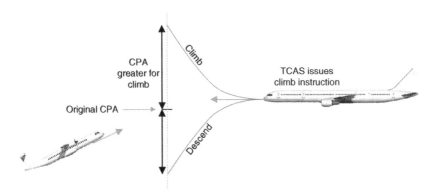

Figure 5.31 The sense of the RA is decided by the option that gives the greatest CPA.

5.6   Radio Altimeter

The Radio Altimeter provides an accurate measurement of the height of the aircraft above the ground up to a maximum of 5000 ft, though for most radio altimeters used on commercial aircraft the limit is 2500 ft. In addition to providing continuous information about clearance above terrain when the aircraft is close to the ground, the radio altimeter provides outputs for the autopilot during autoland and the flight management computer. Although some systems work similarly to radar, that is, pulses are emitted and the time between the emitted and return pulse measured to determine the distance above the ground, most radio altimeters used in commercial airliners emit a continuous wave at frequencies in the range 4200–4400 MHz ($\lambda \sim 7$ cm) in the SHF band. The continuous wave is emitted as a directional beam from a horn or microstrip antenna (see Section 4.4.8.5) on the underside of the aircraft and a similar antenna placed a short distance away acts as a receiver. The receiving and transmitting antennas are situated a sufficient distance apart to prevent direct pickup so that the signal received is entirely due to reflected radio waves from the ground. This means that the distance the beam travels from the transmitting to the receiving antenna is slightly greater than twice the height off the ground as illustrated in Figure 5.32a. Commercial airliners have two independent radio altimeters and thus four antennas.

The principle of operation of a continuous wave radio altimeter is illustrated in Figure 5.32b. The frequency of the transmitted beam is swept up and down in a sawtooth pattern typically over a frequency change of 100 MHz repeating about 300 times a second. Because of the time difference between the transmitted and returning beam, there will be a frequency shift between the beam transmitted and the one detected at the receiving antenna. Mixing the transmitted and reflected signals will produce an intermediate or "beat" frequency that is the difference between the two. The beat frequency difference is proportional to the height and is converted to a voltage, which is used to drive a direct reading display or as data for the Flight Management Computer, which will generate a display on the PFD.

Beermat Calculation 5.3

A radio altimeter emits a continuous SHF wave whose frequency is swept in the range 4250–4350 MHz in a cycle that repeats 300 times per second. If a beat frequency of 60 kHz is detected, what is height of the antenna above the ground in feet disregarding the slant distance correction?

If the pattern repeats 300 times per second, that is, 3.33 ms per cycle then it takes 1.67 ms to scan one way over 100 MHz. Thus, if the beat frequency is 60 kHz, the time between the transmitted and reflected signals is:

$$60\,000/1 \times 10^8 \times 1.67\,\text{ms} = 1.002 \times 10^{-6}\,\text{s}$$

(a)

Aircraft in landing configuration and at the flare attitude

Offset added

Ground

(b)

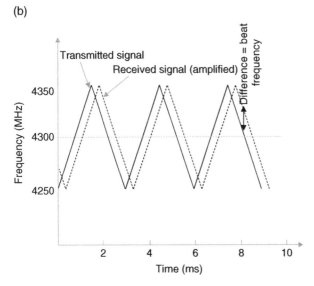

Figure 5.32 (a) Transmitter and receiver geometry in a continuous wave radio altimeter. The system is calibrated to measure the height off the ground of the bottom of the lowest wheel in the landing configuration and in the flared attitude. (b) Frequency variation of transmitted and reflected signals.

Since light travels at 3×10^8 m s^{-1}, this corresponds to a distance of 300.6 m, which is the return distance from the transmitting to the receiving antenna, thus the height off the ground is:

$$153.3\,\text{m} = 493\,\text{ft}.$$

The beat frequency directly measures the slant height, d, of the antenna above the ground and this is converted to a vertical height, h, by

$$h = \sqrt{d^2 - \frac{x^2}{4}} \qquad\qquad (5.36)$$

where x is the distance between the transmitting and receiving antenna. Then a constant factor is added so that the displayed height is measured from the bottom of the lowest wheel in the landing configuration and at the flare pitch attitude as illustrated in Figure 5.32a.

Problems

1   The timing parameters of a pulse train from a ground surveillance radar are shown in the diagram.

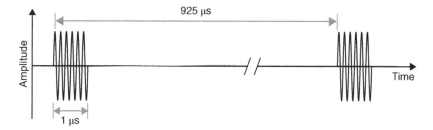

From the timing parameters determine:

a) The theoretical (no dead time) maximum range of the radar in nautical miles.

b) The radial distance resolution of the radar in nautical miles.

c) The theoretical (no dead time) minimum distance at which a target can be detected in meters.

d) Assuming the beam width is 6°, what is the azimuthal resolution in nautical miles of the radar for a target at the theoretical maximum range?

(The speed of light can be taken to be 3×10^8 m s^{-1} or 1.62×10^5 nautical miles per second and 1 nautical mile = 1850 m.)

2   A long-range area surveillance radar has a parabolic antenna of diameter 5 m, rotates at 6 rpm and emits 1 GHz pulses every 4000 μs. If the k-factor for the antenna is 70° determine the dwell time and the number of hits per scan for a target.

3   An ATC radar working at 3 GHz has an average transmitting power of 1 kW and an antenna gain of 4000. Compare the power received at the antenna

from a wide-bodied airliner 20 nautical miles away with that from a helicopter 3 nautical miles away.

4 An AWR is displaying a cloud ahead that contains a rainfall rate of 5 mm h^{-1}. Calculate the relative increase in power received at the antenna if the rainfall rate increases to 20 mm h^{-1}.

References

1 Mätzler, C. (2002). Effects of rain on propagation, absorption and scattering of microwave radiation based on the dielectric model of Liebe. *Research Report 2002-10.* University of Bern. https://www.researchgate.net/publication/303607688

2 Marshall, J.S. and Palmer, W.M.K. (1948). The distribution of raindrops with size. *Journal of Meteorology* **5**: 165–166.

3 Battan, L.J. (1973). *Radar Observation of the Atmosphere.* Chicago: The University of Chicago Press.

4 Uijlenhoet, R. (2001). Raindrop size distributions and radar reflectivity – rain rate relationships for radar hydrology. *Hydrology and Earth Sciences* **5**: 615–627.

5 Marshall, J.S., Hitschfield, W., and Gunn, K.L.S. (1955). Advances in Radar weather. *Advances in Geophysics* **2**: 1–56.

6 Reed, I.S. and Solomon, G. (1960). Polynomial codes over certain finite fields. *Journal of the Society for Industrial and Applied Mathematics* **8**: 300–304.

7 George, T. (2014). AOPA blog, 21 March. https://blog.aopa.org/aopa/2014/03/21/faa-upgrades-alaska-aircraft-to-national-ads-b-standard (accessed 12 June 2018).

8 Kilford, J. and Murphy, P. (2017). UKGA online magazine, 3 August 2017. http://ukga.com/news/view?contentId=41002 (accessed 12 June 2018).

9 Chen, Y.H., Lo, S., Enge, P., and Jan, S.S. (2014). Evaluation & comparison of ranging using Universal Access Transceiver (UAT) and 1090 MHz Mode S Extended Squitter (Mode S ES). In Proceedings of Position, Location and Navigation Symposium (PLANS), Monterey CA, USA, May 2014. doi:10.1109/PLANS.2014.6851456.

10 RTCA (1997). Minimum Operational Performance Standards for Traffic Alert and Collision Avoidance System II (TCAS II) Airborne Equipment, Radio Technical Commission for Aeronautics (RTCA), Document 185A, Washington, DC, 16 December 1997.

6

General Principles of Navigation

6.1 Coordinate Reference System for the Earth

Navigation is the process of moving from one's current position to another specific point on the surface of the Earth that is not in view. Early explorers would follow sequences of topographical features, producing maps as they went but navigation across large featureless areas such as oceans or deserts requires a global coordinate system. It was realized from ancient times that the Earth is a sphere with the earliest recorded proposal around 500 BCE by Pythagoras. When the cause of lunar eclipses was identified around 430 BCE by Anaxagoras, the curved shadow of the Earth on the moon provided direct visual evidence of the spherical shape of the planet. Around 250 BCE, Eratosthenes estimated the circumference of the Earth based on the difference in the elevation of the Sun at the same time on the same day of the year at Syene (modern-day Aswan) and Alexandria in Egypt, which are around 800 km apart. His value was within around 10% of the modern-day accepted value of 40 075 km around the equator.

6.1.1 Latitude and Longitude

Having established the Earth is a sphere, the most convenient coordinate system is spherical polar coordinates (Figure 6.1) and since the radius coordinate is constant, it can be omitted so that just two angles are required to specify any point on the surface. The polar angle ϕ, referred to as latitude is measured relative to the equatorial plane while the azimuthal angle, θ, termed the longitude requires the zero position to be specified as there is no natural reference provided by the Earth. Lines of fixed latitude and longitude are called parallels and meridians, respectively. This coordinate system is also ancient, dating back to at least 600 BCE and the Greek astronomer Hipparchus proposed around 150 BCE the use of a defined reference meridian to be set at zero degrees, which he suggested should pass through Rhodes. This idea continues to the present day with the modern reference 0° meridian, known as the prime meridian, passing through Greenwich in London.

Aircraft Systems: Instruments, Communications, Navigation, and Control,
First Edition. Chris Binns.
© 2019 John Wiley & Sons, Inc. Published 2019 by John Wiley & Sons, Inc.
Companion website: www.wiley.com/go/binns/aircraft_systems_instru_communi_Navi_control

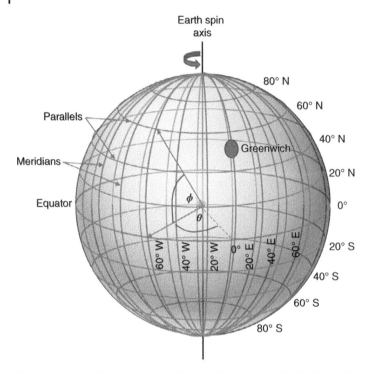

Figure 6.1 Coordinate system used to specify a point on the Earth's surface in terms of two angles, ϕ (latitude) and θ (longitude).

The system differs from the common mathematical definition of spherical polar coordinates with the polar angle, ϕ, measured relative to the equatorial plane rather than the vertical axis and with the values above and below the equator labeled N and S, respectively, instead of + and –. Also, the values of θ do not vary from 0 to 360 but 0 to 180 E and 0 to 180 W from the prime meridian. The nautical mile is defined as the average distance at the surface covered by one minute of latitude along a meridian (the average is used as the Earth is not a perfect sphere – see Section 6.4), thus it corresponds to 40 000 km/(360 × 60) or 1852 m.

The position of a point on the surface is normally specified as the 13-digit plus 2-character code illustrated in Figure 6.2 with latitude and longitude specified as degrees, minutes, and seconds, which provides a precision of 1/60 nautical miles or about 31 m. Sometimes the seconds are omitted and the minutes are specified as a decimal number with two decimal places, giving a position to within 18.5 m. Thus, for example, the center of the runway of the airport where I work (Kavala, Greece) is at N405450 E0243710 while the terminal building is at N405521 E0243721. Sometimes the N(S) or E(W) are placed at the end

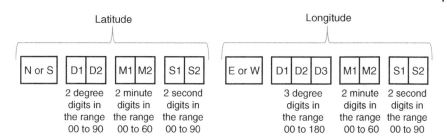

Figure 6.2 Standard format for specifying the coordinates of a point on the surface giving a precision of about 31 m.

of the 6- or 7-digit block, respectively. If greater positional resolution is required, decimal places are added to the seconds digits or minutes digits in the alternative system.

Latitude is relatively easy to measure by the angle of known stars above the horizon at a specific time or, in the Northern hemisphere, their angle from the pole star (Polaris) and could be accurately determined from ancient times. Longitude can be determined by knowing the local time relative to that at Greenwich since the Earth rotates at 15° per hour. Maintaining a Greenwich time reference onboard a ship, however, relied on having marine chronometers that kept accurate time while being exposed to all the rolling, pitching, and yawing moments on a ship as well as the temperature, humidity, and pressure changes in an unregulated environment. Sufficiently, accurate chronometers were not available till after the work of the English clockmaker, John Harrison, in 1735. Following the invention of the atomic clock in 1955, an extremely accurate coordinated universal time (UTC) has been available worldwide and time signals are broadcast by various sources. UTC is used by a number of navigation sources including ADS-B (Section 5.4.10) and GNSS (Chapter 8).

6.1.2 Great Circle Routes, Rhumb Lines, and Departure

A great circle on the Earth is any maximum circumference circle around the globe and is the intersection line between a plane passing through the center of the Earth and the surface as illustrated in Figure 6.3. Examples of great circles include the equator and all meridians. Any other circle is referred to as a minor circle and examples include the parallels that mark fixed latitudes around the globe.

It is straightforward to show that the shortest path between two points on the surface of a sphere is the arc of a great circle that passes through the two points as illustrated in Figure 6.4. Consider two arbitrary points Q and P on the surface at coordinates (θ_1, ϕ_1) and (θ_2, ϕ_2) (Figure 6.4a). Given the spherical symmetry,

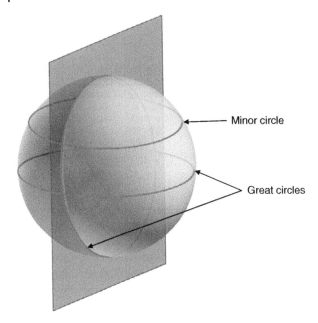

Figure 6.3 Great circles are maximum circumference circles drawn on the globe and correspond to the intersection line between a plane passing through the center and the surface. Examples include the equator and all meridians. Minor circles are smaller circles and examples include parallels.

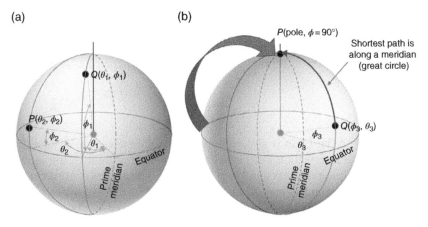

Figure 6.4 Demonstration that the shortest path between two points on the surface of a sphere is a minor arc of a great circle.

the axes of the coordinate system can be rotated so that P is on the pole (Figure 6.4b), so its latitude coordinate is transformed to $\phi = 90°$ (note that at the pole the azimuthal coordinate, θ, is arbitrary). Q will have the transformed coordinates (θ_3, ϕ_3). It is clear that the shortest path is along the meridian that connects Q and P, that is, an arc of a great circle passing through the two points. There is also a formal mathematical proof of this conclusion based on the calculus of variations [1].

Another useful definition of a track in navigation is a *Rhumb line*, which is defined as a path that crosses all meridians at the same angle, for example, all meridians and parallels are Rhumb lines. The significance of a Rhumb line is that it is the path that an aircraft would follow if it maintained a constant track relative to true North. In the case of a track along a cardinal direction, North, South, East, or West, there would be a stable path along the relevant parallel or meridian but for any other track the path would spiral toward the nearest pole as illustrated in Figure 6.5.

In reality, there is not a simple way of maintaining a Rhumb line as magnetic compasses point to magnetic North, which has an angular offset (variation) from true North and as discussed in the next section the variation changes with position on the globe. For distances over which the change in variation is not

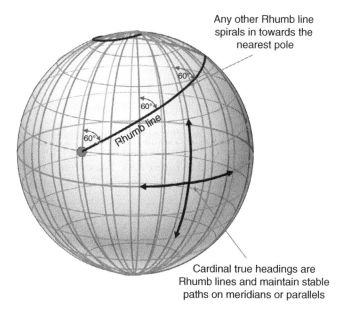

Any other Rhumb line spirals in towards the nearest pole

Rhumb line

Cardinal true headings are Rhumb lines and maintain stable paths on meridians or parallels

Figure 6.5 A Rhumb line is a track that maintains a constant angle with meridians and is the path that an aircraft would follow if it maintained a constant true heading. Rhumb lines along cardinal true headings maintain a stable path along a meridian or parallel but at any other angle the path spirals in toward the nearest pole.

significant, however, an aircraft following a magnetic heading is approximately following a Rhumb line.

The meridians are parallel to each other at the equator but at latitudes away from the equator they converge and make an angle, λ, relative to each other, which depends on how far apart they are and the latitude via the relationship:

$$\lambda = \Delta(\theta)\sin(\phi) \tag{6.1}$$

where $\Delta(\theta)$ is the difference in longitude of the two meridians and ϕ is the latitude at which the angle is required. Departure is defined as the distance between two meridians along a specified parallel (Figure 6.6a). Along a great circle, for example, the equator, each minute of arc corresponds to one nautical mile but along a parallel away from the equator, the convergence of the meridians results in a nautical mile corresponding to more than one minute of arc. As illustrated in Figure 6.6b, the radius of each parallel is given by

$$r = R_{\oplus}\cos\phi \tag{6.2}$$

where $R_{\oplus}$ is the radius of the Earth and ϕ is the angle of the parallel. Thus, the circumference is $\cos\phi$ times the circumference of a great circle, so departure is given by

$$\text{Departure (nm)} = \Delta\text{longitude (minutes)} \times \cos(\text{latitude}). \tag{6.3}$$

For example, the departure between the meridians 10° E and 20° E at a latitude of 60° N is $10 \times 60 \times \cos(60°) = 300$ nautical miles. Equation (6.3) can be rearranged to describe the change in longitude after traveling a given distance along a specific parallel, that is:

$$\Delta\text{longitude (degrees)} = \text{Distance (nm)} \times \sec\phi/60 \tag{6.4}$$

One of the nonintuitive aspects of traveling over the surface of a sphere is that if one moves equal distances along the four cardinal directions in a square, the end point is displaced from the start as illustrated in Figure 6.6a. The difference can be calculated using Equation (6.3) as demonstrated in Beermat Calculation 6.1.

Beermat Calculation 6.1

An aircraft starts at coordinates N200000 E0800000, travels directly North, then East, then South, and finally West with each leg being 2400 nautical miles (this is the path illustrated in Figure 6.6a). What are the coordinates at the end of the journey?

Traveling directly North, the aircraft, which starts at a latitude of 20° N is moving along a great circle where 1° corresponds to 60 nautical miles so after 2400 nautical miles it will be at a latitude of 60°. After turning East it travels along the 60° parallel and according to Equation (6.4), after 2400 nautical miles it will have changed longitude by $2400 \times \sec(60°)/60 = 80°$, that is, it will be on the prime meridian. Moving South it will go back to its starting latitude of 20° along the prime meridian. Finally, turning West it will change longitude by $2400 \times \sec(20°)/60 = 42.568° = 42°34'04''$. So, the final coordinates are N200000 E423404.

(a)

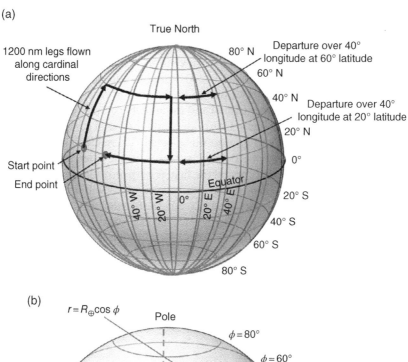

(b)

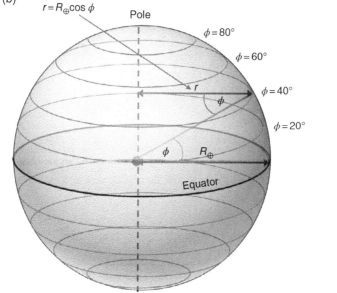

Figure 6.6 (a) Departure is the distance between two meridians along a specified parallel. On a square path of equal length legs along the four cardinal directions, the start and end points are displaced. (b) The radius of a parallel and hence the circumference are a factor of $\cos \phi \times$ the circumference of a great circle where ϕ is the latitude.

6.2 Compass Heading, Variation, and Deviation

The Earth's magnetic field provides a global reference direction and this was recognized from ancient times with the invention of the magnetic compass around 200 BCE in China. From around 1200 CE it revolutionized maritime navigation and has been in continuous use up to the present day. The generation mechanism of the global magnetic field and some of its complexity were introduced briefly in Section 3.3.1 but to fully understand the use of compass heading in navigation it is necessary to go into more detail. There are physical poles, North and South, which can be defined as the position where the angle of dip reaches 90°, that is, the direction of the field is perpendicular to the surface. These are often called "dip poles" and they move chaotically with time, as shown for the case of the North pole in Figure 6.7, due to the stochastic nature of the Earth dynamo that generates the geomagnetic field [2]. The current (2017) position of the North dip pole is at N8630.00 W17236.00 and it is moving at around 55 km per year. The South dip pole is at S6413.80 E13633.40 and moving at about 15 km per year. The poles move independently of each other and the

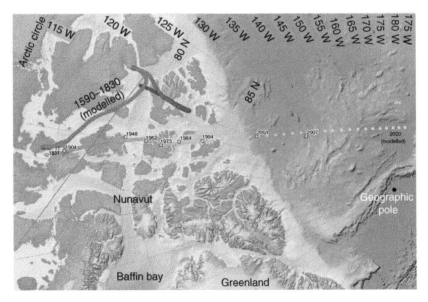

Figure 6.7 Wandering of the North magnetic dip pole between 1590 and 2020 as predicted by a computer model with the time evolution represented by the color changing from blue to yellow. Actual observations available since 1831 are shown by the yellow squares. *Source:* Data from the National Oceanic and Atmospheric Administration (NOAA) [2].

two positions are not antipodal, that is, the line joining them does not pass through the center of the Earth. The field completely reverses around every 250 000 years and during the process it is likely that several poles appear around the globe.

Despite the chaotic nature of the Earth's magnetic field, the wandering of the magnetic poles is sufficiently slow (~0.1° per year) that they serve as useful reference points for navigation and the angular offset between the direction to the magnetic pole and the geographical pole is termed the *variation*. At any time, the value depends on the location on the Earth's surface. For example, the current position of the North dip pole relative to the geographic pole is shown in Figure 6.8 and it is evident that for most of Western Europe the variation is close to zero. As shown in the figure the variation can vary widely and for the extreme

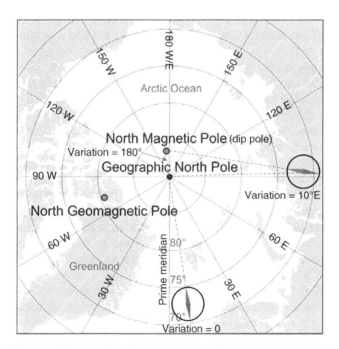

Figure 6.8 The angular offset between the direction to the magnetic pole and geographical pole is termed variation and varies between 0 and 180°, the latter value found on the line directly between the two poles. The sense of the variation is termed East or West and is defined as going from the geographic pole to the magnetic pole. *Source:* Reproduced with permission of https://en.wikipedia.org/wiki/North_Magnetic_Pole#/media/File: North_Magnetic_Poles.svg Licensed under CC BY 4.0 [3].

case of locations directly between the geographic and dip pole it reaches 180°. The sense of the variation is termed East (decrease in heading angle) or West (increase in heading angle) and is defined as moving from the geographic pole to the magnetic pole. For example, to follow a true ground track relative to geographic North of 100 in zero wind with a variation at the current position of 8° East, the compass heading would need to be maintained at 092. The line of zero magnetic variation around the globe is called the *agonic line* while lines of equal magnetic variation are known as *isogonal lines* and are marked on aviation charts.

In addition to the long-term motion of the magnetic dip poles, variation has shorter period fluctuations due to the interaction of the solar wind with the ionosphere. The ionosphere is electrically conducting and currents within it generate a magnetic field in addition to the geomagnetic field. There is thus a diurnal variation due to changes in the ionosphere between day and night (see Section 4.2.3.1). In addition, the 11-year sunspot cycle imposes a cyclic change on the same timescale. These periodic fluctuations are small imposing changes in variation of less than 0.1°, however, solar storms can produce temporary changes by up to 1°.

Since the Earth's magnetic field is approximately dipolar, it is sometimes convenient to represent the real field by the closest fitting pure dipolar field that would be produced by a bar magnet at the center as illustrated in Figure 6.9. This is referred to as the International Geomagnetic Reference Field (IGRF) [4] and is useful when modeling the interaction of the global magnetic field with the interplanetary magnetic field and the solar wind. Note that the bar magnet has the South pole pointing Northward and vice versa for the North pole since in the normal terminology the North magnetic pole of the Earth is the pole toward which the North pole of a magnet would point. The IGRF has two poles that are antipodal at opposite parallels, that is, 80.5° N and 80.5° S and reversed longitudes 72.8° W and 107.2° E (= 180 − 72.8). The position of the North geomagnetic pole in relation to the dip pole and true North is also shown in Figure 6.8. It should be reiterated, however, that compasses point to the nearest dip pole and variation is relative to these points.

As discussed in Section 3.3.2, the compass orientation is also sensitive to local aircraft magnetism, which is either classed as soft or hard. Any misalignment due to aircraft magnetism is termed *deviation* and in small aircraft is given by a table of corrections on a placard near the compass as shown in Figure 6.10. The same terminology is used for the sense of the deviation as for variation, that is, West denotes that the variation has to be added to derive the compass heading and East denotes that it should be subtracted. In all but the simplest aircraft, in addition to the magnetic compass, there is a magnetometer placed on the aircraft wing that senses the Earth's magnetic field direction with negligible deviation (see Section 3.3.3). The correct terminology for the various ways of describing aircraft heading is:

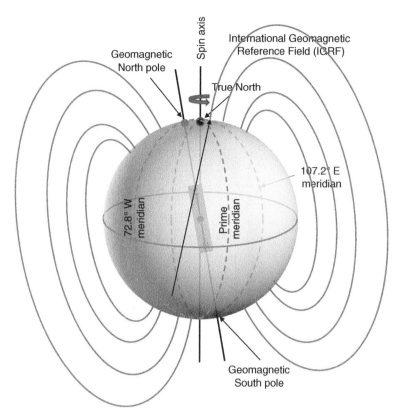

Figure 6.9 The International Geomagnetic Reference Field (IGRF), which represents the actual magnetic field by the best fitting pure dipolar field that would be produced by a bar magnet placed at the center of the Earth.

True heading: Heading with respect to true North (Earth spin axis).
Magnetic heading: Heading with respect to the magnetic North or South dip poles, that is, after variation has been added or subtracted to the true heading.
Compass heading: Magnetic heading corrected for deviation.

6.3   Aviation Charts

6.3.1   General Chart Properties: Chart Scale, Orthomorphism, and Conformality

Chart scale is the ratio of the linear distance between two features on a chart and the distance between the same two features on the Earth. For example, Figure 6.11

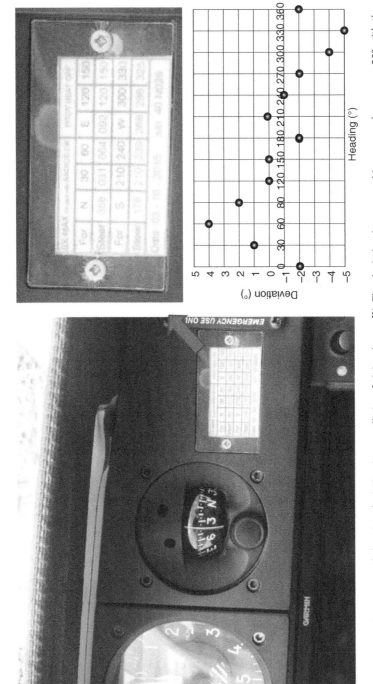

Figure 6.10 Compass card showing deviation in a small aircraft (pitot heat off). The deviation is measured in a ground run every 30° with the engine running and the value is extrapolated for headings in between.

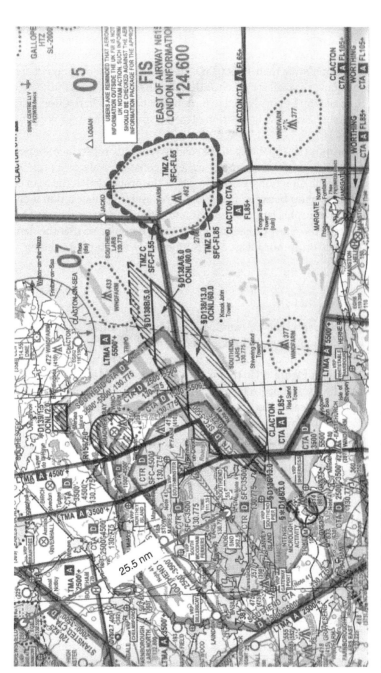

Figure 6.11 Section of a UK CAA 1 : 500 000 chart. The 25.5 nm (47 225 m) between Stansted and Southend airports would be represented by 9.4 cm.

shows a piece of a UK CAA 1 : 500 000 chart and the distance between Stansted and Southend airports, which is 25.5 nautical miles (47 225 m) would be represented by 47 225/500 000 m = 9.4 cm on the chart. On a 1 : 250 000 chart the same distance would be represented by 18.8 cm and here it is worth pointing out that in the normal terminology a "larger" scale means a smaller denominator, that is, 1 : 250 000 is a larger scale than 1 : 500 000. The scale on a chart always changes with position but over a small area such as that shown in Figure 6.10, the change is <0.1% and the scale can be considered to be constant over the chart. Over a larger area the change of scale with position depends on the chart projection as described below.

Orthomorphism is the capability of the map to correctly reproduce shapes of features on the Earth and this requires either a constant scale or a scale that changes equally in all directions. Conformality is the property that the map correctly reproduces angles between features. To be useful for navigation a chart needs to be orthomorphic and conformal and although this is impossible to achieve absolutely, the chart projection needs to be designed so that the departures from perfect orthomorphism and conformality are minimized over the area depicted.

6.3.2 Chart Projections

It is impossible to represent the surface of a sphere on a flat surface without introducing distortions, but various projections are available that minimize the distortions for specific applications. Three of the most common used in navigation are shown in Figure 6.12, that is the cylindrical or Mercator projection (Figure 6.12a), the conical projection (Figure 6.12b), and the gnomic projection (Figure 6.12c). In all cases the chart would display the land areas that would appear if a light source was placed at the center of the globe and the projected shapes were plotted on the corresponding surface, hence the term projection. Since the entire Earth is digitized, the (x,y) coordinates on the map of any feature can be obtained from its longitude and latitude angles (θ, ϕ) by a suitable transform.

The Mercator projection is capable of showing very large areas including the entire globe up to latitudes of about 80° North and South and is the most commonly used chart for general mapping. The Conical projection is limited to regions around the parallel of tangency but will display these with lower distortion than Mercator charts and is the most commonly used projection for aviation charts. The Gnomic projection is normally used to chart regions around the poles. The detailed properties of the different projections are described below.

6.3.2.1 Mercator Projection
The normal Mercator projection starts with a cylindrical projection tangential at the equator (Figure 6.12a) and if no additional scaling is imposed the

(a)
Cylindrical (Mercator) projection

(b)
Conical projection

(c)
Gnomic projection

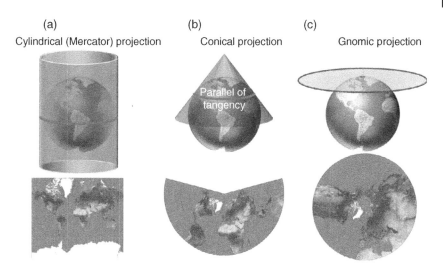

Figure 6.12 (a) Cylindrical (Mercator) projection typically used in marine navigation charts terminated at 82° N and S. (b) Conical projection commonly used for aeronautical navigation charts (after some modification – see Section 6.3.2.2) terminated at 30° S. (c) Gnomic projection normally used for polar navigation charts. The projections at the bottom were reproduced under creative commons 3.0 license [13] from Ref. [5].

North–South scale increases with $\tan(\phi)$, where ϕ is the latitude. The tangent of the latitude becomes infinite at 90°, which makes it impossible to represent the poles. The chart is limited to around 80° latitude before the distortion becomes unacceptable. The meridians are all parallel straight lines, that is, their chart convergence is zero and since the meridian convergence on the Earth is proportional to $\cos(\phi)$, where ϕ is the latitude, the East–West scale on the Mercator chart increases with $\sec(\phi)$. Sometimes an extra scaling is imposed so that the North–South scale also increases with the secant of the latitude. If the chart needs to represent the poles, a transverse Mercator projection can be used where the cylinder is wrapped around the globe in the orthogonal direction.

The mathematical transforms relating (x,y) coordinates on the chart to latitude (ϕ) and longitude (θ) for the Mercator projection are [6]:

$$x = \theta - \theta_0 \tag{6.5}$$

$$y = \ln\left[\tan\left(\frac{\pi}{4} + \frac{\phi}{2}\right)\right] \tag{6.6}$$

Figure 6.13a shows a rectangular region covering the area of the British Isles spanning latitudes 50° N to 60° N and longitudes 2° W to 10° E. After applying the Mercator transformation given by Equations (6.5) and (6.6) to map the region on a chart (Figure 6.13b), the region is rectangular but the vertical elongation is clear.

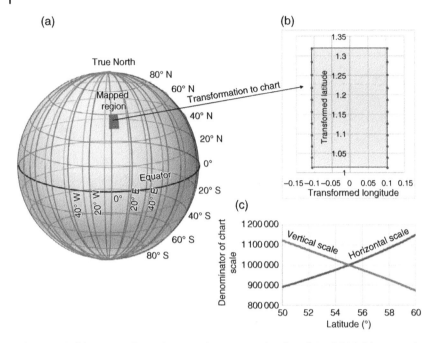

Figure 6.13 (a) Rectangular region covering an area the size of the British Isles spanning latitudes 50° N to 60° N and longitudes 2° W to 10° E. (b) The same region plotted on a Mercator projection chart after using the transformation in Equations (6.5) and (6.6). (c) Variation in the denominators of the horizontal and vertical scales as a function of latitude with the scale set to 1 : 1 000 000 at the central latitude of 55° N.

If the scaling at the central latitude of 55° N is set to be 1 : 1 000 000, which is typical for a chart covering this size of region, the denominators of the horizontal and vertical scales vary with latitude as plotted in Figure 6.13c and it is evident that the variation is ~10–15%.

Summary of Mercator chart properties:

- The scale is correct at the equator and increases with secant (latitude).
- The chart is conformal (angles on the Earth are represented correctly on the chart).
- The chart is orthomorphic over small areas but distortion occurs over large areas, especially at high latitudes.
- The graticule has straight meridians and parallels that cross at right angles.
- The chart convergence is zero everywhere, which is correct only at the equator.
- Rhumb lines are always straight anywhere on the chart.
- The equator and meridians are straight lines but all other great circles are concave toward the equator and convex toward the nearest pole.

6.3.2.2 Conical Projection

The simple conical projection is a cone centered on a pole and tangential to the globe at a chosen latitude, which is known as the parallel of tangency or the parallel of origin. The angular arc, α, defined by the rolled out segment is given by

$$\alpha = \Delta\theta \sin\left(\phi_0\right) \tag{6.7}$$

where $\Delta\theta$ is the range of longitudes spanned by the chart and ϕ_0 is the parallel of origin. For example, for a cone with a parallel of origin, $\phi_0 = 45°$, that spanned the entire 360° of longitude the arc would cover $360 \times \sin(45°) = 255°$. In general, charts are a small section of the cone with the parallel of tangency at the midpoint of the latitude scale of the chart. The quantity $\sin(\phi_0)$ is known as the constant of the cone or convergency factor and is normally denoted by n. The angle of inclination of the meridians, λ, or chart convergence is given by

$$\lambda = \Delta\theta \sin\left(\phi_0\right) \tag{6.8}$$

which is similar to Equation (6.7) except that $\Delta\theta$ now describes the angle between the two meridians in question rather than the whole chart.

A modification to the simple conic projection published by Lambert in 1772 [7] was to move the cone within the globe so that it intersects along two standard parallels as illustrated in Figure 6.14. These are spaced uniformly around the parallel of origin and are typically at the 1/3 and 2/3 positions up the range of latitudes. The resulting projection is known as the Lambert conformal projection and it does not change the distortion or variation of scale but spreads it uniformly around the chart center. Within the standard parallels the scale is decreasing, while outside it is increasing.

If the parallel of origin is ϕ_0, the standard parallels are ϕ_1 and ϕ_2, and the reference longitude is θ_0, the mathematical transforms relating (x,y) coordinates on the chart to latitude (ϕ) and longitude (θ) for the Lambert conformal projection are [8]:

$$x = \rho \sin\left[n(\theta - \theta_0)\right] \tag{6.9}$$

Figure 6.14 Lambert conformal projection.

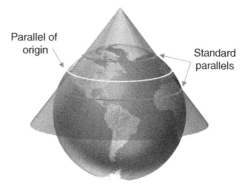

Parallel of origin

Standard parallels

$$y = \rho_0 - \rho \cos\left[n(\theta - \theta_0)\right] \tag{6.10}$$

where

$$n = \frac{\ln(\cos\phi_1 \sec\phi_2)}{\ln\left[\tan\left(\dfrac{\pi}{4} + \dfrac{\phi_2}{2}\right)\cot\left(\dfrac{\pi}{4} + \dfrac{\phi_1}{2}\right)\right]} \tag{6.11}$$

$$\rho = F\cot^n\left(\frac{\pi}{4} + \frac{\phi}{2}\right) \tag{6.12}$$

$$\rho_0 = F\cot^n\left(\frac{\pi}{4} + \frac{\phi_0}{2}\right) \tag{6.13}$$

$$F = \frac{\cos\phi_1 \tan^n\left(\dfrac{\pi}{4} + \dfrac{\phi_1}{2}\right)}{n} \tag{6.14}$$

If the standard parallels are moved to coincide with the parallel of origin, that is, we revert to the simple conical projection, the numerator and denominator for the expression of n (Equation 6.11) are both zero so n is indeterminate, however, taking the limit as $\phi_1 \rightarrow \phi_2$ gives $n = \sin\phi_0$ as before.

Figure 6.15a shows the same rectangular region as in Figure 6.13a, that is, an area covering the British Isles from latitudes 50° N to 60° N and longitudes 2° W to 10° E. The area mapped on the chart after applying the Lambert transformation given by Equations (6.9)–(6.14) is shown in Figure 6.15b with standard parallels at 53° N and 57° N. If the scaling at the standard parallels is set to be 1 : 1 000 000, the denominators of the horizontal and vertical scales vary with latitude as plotted in Figure 6.15c. As predicted, the scale decreases (larger denominator) in the region within the standard parallels and increases outside. It is evident that over the charted region the scale variation is much smaller than for the Mercator projection.

Summary of Lambert conformal chart properties:

- The scale is correct at the standard parallels and decreases within them but increases outside.
- The chart is conformal (angles on the Earth are represented correctly on the chart).
- The chart is orthomorphic over relatively large areas.
- The graticule has straight meridians originating from the pole and curved parallels centered on the pole.
- The chart convergence is constant and is given by Δ(longitude) × sin (latitude).
- Rhumb lines are concave to the pole with the exception of meridians.
- Great circles are slightly concave to the parallel of origin.

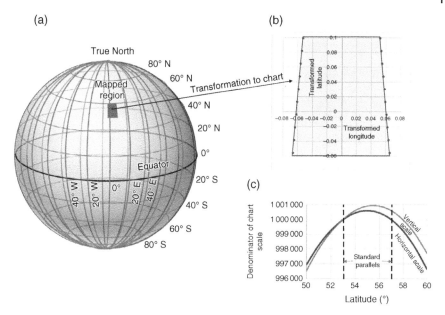

Figure 6.15 (a) Rectangular region covering an area the size of the British Isles spanning latitudes 50° N to 60° N and longitudes 2° W to 10° E. (b) The same region plotted on a Lambert conformal projection chart after using the transformation in Equations (6.9)–(6.14). (c) Variation in the denominators of the horizontal and vertical scales as a function of latitude with the scale set to 1 : 1 000 000 at the standard parallels of 53° N and 57° N.

6.3.2.3 Gnomic and Polar Stereographic Projection

The general Gnomic projection is illustrated in Figure 6.12c, but a specific case used in charts is to have the flat plane contact the globe at one of the poles and to move the light source to the opposite pole as illustrated in Figure 6.16. This is known as a polar stereographic projection and can be used for an entire hemisphere as shown in Figure 6.16, but normally it is used specifically to provide charts for the polar regions.

The mathematical transforms relating (x,y) coordinates on the chart to latitude (ϕ) and longitude (θ) for the stereographic projection are [9]:

$$x = k \cos\phi \sin(\lambda - \lambda_0) \tag{6.15}$$

$$y = k[\cos\phi_0 \sin\phi - \sin\phi_0 \cos\phi \cos(\lambda - \lambda_0)] \tag{6.16}$$

where ϕ_0 and λ_0 are the central latitude and longitude of the mapped region and for a sphere of radius R:

$$k = \frac{2R}{1 + \sin\phi_0 \sin\phi + \cos\phi_0 \cos\phi \cos(\lambda - \lambda_0)} \tag{6.17}$$

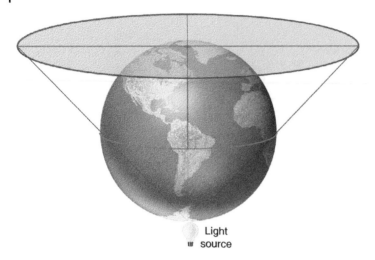

Figure 6.16 Polar stereographic projection of the North pole.

Figure 6.17a shows the same rectangular region as in Figures 6.13a and 6.15a, that is, an area covering the British Isles from latitudes 50° N to 60° N and longitudes 2° W to 10° E. The area mapped on the chart after applying the stereographic transformation given by Equations (6.15)–(6.17) is shown in Figure 6.17b. If the scaling at the standard midpoint is set to be 1 : 1 000 000, the denominators of the horizontal and vertical scales vary with latitude as plotted in Figure 6.17c.

The variation in scale is larger than for the Lambert projection and similar in magnitude to that in the Mercator chart. However, unlike the Mercator projection, the vertical and horizontal scales vary in the same sense so that the chart is orthomorphic over larger areas. In addition, the polar stereographic projection would not normally be used so far away from the pole and at higher latitudes the variation in scale is significantly less, for example, at 80° N, the scale is within 1% of that at the pole.

Summary of polar stereographic conformal chart properties:

- The scale is correct at the pole and increases away from the pole.
- The chart is conformal (angles on the Earth are represented correctly on the chart).
- The chart is orthomorphic over relatively large areas.
- The graticule has straight meridians radiating from the pole and circular parallels centered on the pole.
- The chart convergence is constant, given by Δ(longitude) and is correct only at the pole.
- Rhumb lines are concave to the pole and those along parallels are arcs of circles.
- Great circles are slightly concave to the pole.

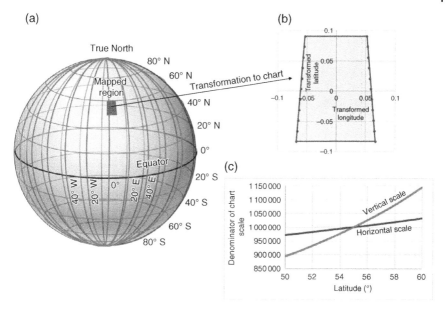

Figure 6.17 (a) Rectangular region covering an area the size of the British Isles spanning latitudes 50° N to 60° N and longitudes 2° W to 10° E. (b) The same region plotted on a polar stereographic chart after using the transformation in Equations (6.15)–(6.17). (c) Variation in the denominators of the horizontal and vertical scales as a function of latitude with the scale set to 1 : 1 000 000 at the midpoint of the region.

6.4 Non-Sphericity of the Earth and the WGS84 Model

The discussion so far has assumed the Earth to be a sphere and although the deviation from perfect sphericity is small, it needs to be taken into account for precision position fixing by satellite-based navigation. The shape of the Earth is most accurately described as an oblate spheroid (ellipsoid), with a slight bulging at the equator. The equatorial radius is 6 378 137 m while the polar radius is 6 356 752.3 m, which is a difference of only 0.3%. The ellipsoidal shape of the globe results in an ambiguity regarding latitude as there are two possible definitions at any point on the surface as illustrated in Figure 6.18. These are the Geocentric latitude whose origin is at the center of the Earth and the Geodetic (or Geographic) latitude, which is the angle between the normal to the meridian of the point in question and the equatorial plane. It is the Geodetic latitude that is used in charts and the maximum angle difference between the two latitudes occurs at 45° N or S and is 11.6′ while at the poles and the equator they are equal.

Position fixing by satellites gives an absolute position in space relative to the center of the Earth and to know where one is relative to the surface requires a

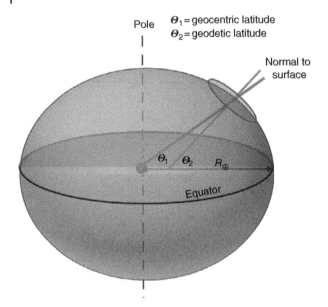

Pole $\Theta_1 = $ geocentric latitude

$\Theta_2 = $ geodetic latitude

Normal to surface

Θ_1 Θ_2 $R_{\oplus}$

Equator

Figure 6.18 Geocentric and Geodetic latitude on an oblate spheroid.

knowledge of precisely where the surface is. So, although the non-sphericity of the Earth is small, assuming the Earth to be a perfect sphere would give an error of up to 12 nautical miles in latitude and up to 10 nautical miles in altitude. There are also some second-order corrections as the true surface deviates slightly from a perfect spheroid but these are not considered here. It is also assumed that the radius does not depend on longitude.

To determine latitude and longitude requires an accurate knowledge of the shape of the Earth spheroid and various measurements including measurements of satellite orbits led to the values of the standard parameters listed in Table 6.1. These are the specifications of the World Geodetic System model published in 1984 [10], which is commonly referred to as the WGS84 model. There have been some refinements since, but this is still the basic standard used in satellite navigation systems.

In addition to latitude and longitude, aviation requires the altitude over sea level to be measured and the WGS84 model provides a mapping of the sea level reference over the whole globe. Sea level is an equipotential surface of the gravity field of the Earth, that is, the average sea surface after removing tides and waves is at the same gravitational potential everywhere. Starting with a spherical body of mass M, the gravitational potential at a distance r_1 from the body is the work done in bringing a unit mass from infinity to r_1 (see Figure 6.19a), which

Table 6.1 WGS84 parameters.

Parameter	Symbol	Value
Universal gravitational constant × the mass of the Earth[a]	$M_e \times G$	$3.986004418 \times 10^{14}$ m^3 s^{-2}
Equatorial radius	a	6378137 m
Polar radius	c	6356752.3 m
Flattening $= (a - c)/a$	$1/f$	1/298.257223560
Rotation rate	ω	7.292115×10^{-5} rad s^{-1}
Dynamic form factor (see below)	J_2	1.081874×10^{-3}

[a] The product $M_e \times G$ is known much more accurately from measurements than M_e or G alone so it is the product that is specified in the WGS84 parameters.

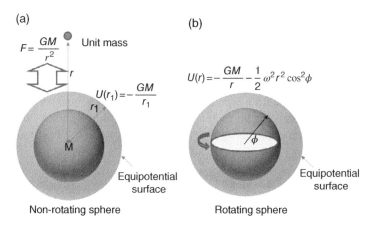

(a)

$$F = \frac{GM}{r^2} \quad \text{Unit mass}$$

$$U(r_1) = -\frac{GM}{r_1}$$

M

Equipotential surface

Non-rotating sphere

(b)

$$U(r) = -\frac{GM}{r} - \frac{1}{2}\omega^2 r^2 \cos^2\phi$$

Equipotential surface

Rotating sphere

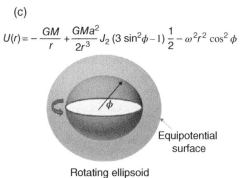

(c)

$$U(r) = -\frac{GM}{r} + \frac{GMa^2}{2r^3} J_2 (3 \sin^2\phi - 1)\frac{1}{2} - \omega^2 r^2 \cos^2\phi$$

Equipotential surface

Rotating ellipsoid

Figure 6.19 (a) Equipotential surface (sea level) on a nonrotating sphere. (b) Equipotential surface on a rotating sphere. (c) Equipotential surface on a rotating ellipsoid.

can be calculated from the gravitational force of attraction, F, between the two masses, given by

$$F = \frac{GM}{r^2} \tag{6.18}$$

The gravitational potential at r_1, $U(r_1)$, is then:

$$U(r_1) = \int_\infty^{r_1} \frac{GM}{r^2} dr = -\frac{GM}{r_1} \tag{6.19}$$

and is negative as negative work is required to bring together attracting bodies.

On a nonrotating sphere the equipotential surfaces are spheres and sea level on such a planet would be at a constant distance from the center as shown in Figure 6.19a.

To determine the shape of the equipotential surface that determines sea level on the Earth, requires several steps. First, the rotation of the Earth needs to be accounted for as the centrifugal acceleration will generate a correction to the gravitational potential given by

$$U(r) = -\frac{GM}{r} - \frac{1}{2}\omega^2 r^2 \cos^2\phi \tag{6.20}$$

where ϕ is the latitude and r is now dependent on ϕ. The equipotential surface in this case develops an equatorial bulge as shown exaggerated in Figure 6.19b. Then, taking into account the spheroidal shape of the Earth modifies the potential to [11]:

$$U(r) = -\frac{GM}{r} + \frac{GMa^2}{2r^3} J_2 (3\sin^2\phi - 1) - \frac{1}{2}\omega^2 r^2 \cos^2\phi \tag{6.21}$$

where J_2 is given in Table 6.1. This causes the equipotential surfaces to move back toward spherical slightly. On a perfectly smooth rotating ellipsoid the equipotential surfaces and thus sea level would also have a smooth spheroidal shape (Figure 6.19c) and although not an ellipsoid the shape is well defined.

Finally, the Earth is not completely covered with water and is not a uniform density, so we need to take account of the departure from the smooth equipotential surface caused by topography, changes in local material density, etc. The real equipotential surface defining sea level taking into account the actual topography of the Earth is known as the Geoid and is illustrated in Figure 6.20a. The departures of the geoid from the smooth ellipsoidal equipotential are mapped over the whole globe so that sea level is known everywhere including land areas that are remote from the sea. Figure 6.20b shows a map of the Earth geoid in gravity units of milligal where 1 milligal is approximately 10^{-6} g. It is clear that the geoid departures from the smooth Earth ellipsoid are very small undulations and in terms of distance range from the extremes

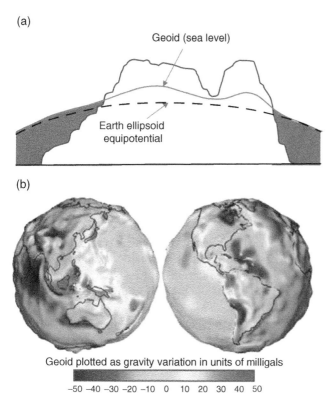

Figure 6.20 (a) Illustration of the geoid departure from the perfect Earth ellipsoid in the vicinity of terrain or other gravitational anomalies. (b) Representation of the Earth geoid in terms of gravity variation in milligals over the whole globe. *Source:* NASA.

−106 to +85 m from the perfect ellipsoid. In the original WGS84 model the resolution of the geoid was about 200 km but the model has been upgraded with a new standard referred to as the Earth Gravitational Model first published in 1996 (EGM96). This has the same Earth ellipsoid but a geoid with a resolution of about 100 km and future revisions are expected to achieve a resolution approaching 20 km.

6.5 Navigation by Dead Reckoning

Navigation by dead reckoning is the process of getting from a current known position to a waypoint or destination by calculating a magnetic heading to fly and time taking into account the winds aloft. It is the most basic form of navigation and during the flight it requires only a compass, a clock, and a chart. It is still an important element of pilot training so that pilots can navigate with basic tools in the event of the loss of radio navigation aids. Prior to radio aids and global navigation systems, dead reckoning was used successfully by aviators to navigate all over the globe.

6.5.1 Calculating the True Airspeed

To fly to the next waypoint during a navigation requires the accurate calculation of two variables, the magnetic heading to fly and the time required to reach the waypoint. To calculate either of these requires a knowledge of the true airspeed (TAS) at the altitude that the aircraft will fly and this must be determined from the indicated airspeed (IAS), which is the only directly measured quantity. The IAS needs to be corrected for instrument errors to give the calibrated airspeed (CAS) and since the difference is usually small, a short table of values at discrete airspeeds is sufficient with intermediate difference values obtained by interpolation. As an example, Figure 6.21 shows the table of values to convert IAS to CAS for a Piper PA28 light aircraft.

Airspeed indicators directly measure dynamic pressure and are calibrated to read the correct airspeed at ISA sea level so the TAS, v_{TAS}, can be obtained from the CAS, v_{CAS}, using the equation:

$$v_{TAS} = v_{CAS} \sqrt{\frac{\rho_0}{\rho}} \tag{6.22}$$

where ρ is the density of the air at which the aircraft is flying and ρ_0 is the density of air at ISA sea level (1.225 kg m^{-3}). At a given altitude, the density can be obtained using Equation (2.12) or (2.13) from Chapter 2 depending on whether the altitude is above or below 11 km (36 089 ft) and from the outside air temperature the density can be corrected for non-ISA conditions. In flying schools, student pilots learn how to obtain the TAS rapidly using a circular slide rule, known as a CRP5, in which the altitude is set against temperature, which sets the TAS against the CAS on the outer scale as illustrated in Figure 6.22. In this example, it is planned to fly the aircraft at 10 000 ft at which it is known the outside air temperature is −2 °C. Then, setting −2 °C against 10 000 ft in the airspeed window sets up the TAS scale correctly against the CAS scale at the outer edge of the slide rule. In this case the CAS is 135 knots corresponding to a TAS at altitude of 155 knots. Having determined the TAS, it becomes possible to calculate both the heading and the ground speed (GS) if the wind velocity (WV) and direction is known at the altitude to be flown.

IAS (knots)	CAS (knots)
89	89
102	100
113	111
125	121
154	148

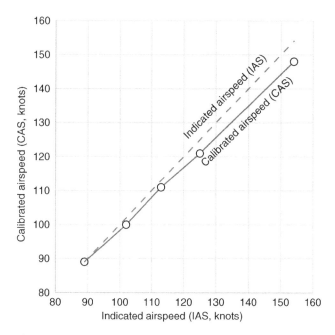

Figure 6.21 Calibrated airspeed (CAS) as a function of indicated airspeed (IAS) for a Piper PA28.

6.5.2 Calculating the Heading and Ground Speed in a Known Wind

The wind is reported as a speed in knots and the direction it is blowing from, which is given as a true bearing in degrees, with the exception of local reports at airports (METAR and TAF) where it is the magnetic direction that is given. For example, a wind specified as 180/20 is blowing from the South at a speed of 20 knots. A simple vector addition of the heading vector, whose magnitude is the TAS and the wind vector gives the ground track vector, whose magnitude

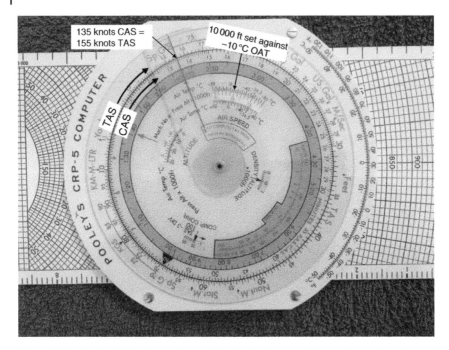

Figure 6.22 Circular slide rule used to rapidly obtain the true airspeed from the calibrated airspeed. In this example −2 °C has been set against 10 000 ft in the airspeed window so that on the outer scales it is seen that 135 knots CAS corresponds at altitude to 155 knots TAS.

is the GS as illustrated in Figure 6.23a. Normally, when planning a navigation, however, the true track between waypoints is known whereas the heading required to fly to maintain that ground track in the presence of the known wind and the corresponding GS are unknowns. As illustrated in Figure 6.23b, given the ground track (TRK), the WV, and the wind direction (WD), the heading (HDG) is:

$$HDG = TRK + \tan^{-1}\left(\frac{WV \times \sin(WD - TRK)}{TAS}\right) \qquad (6.23)$$

where TAS is the true airspeed and the GS is:

$$GS = TAS - WV \times \cos(WD - TRK). \qquad (6.24)$$

The angle between HDG and TRK is known as the wind correction angle (WCA), that is,

$$WCA = \tan^{-1}\left(\frac{WV \times \sin(WD - TRK)}{TAS}\right) \qquad (6.25)$$

(a)

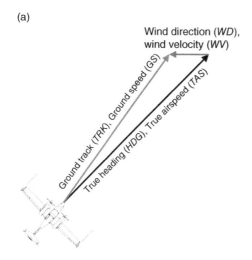

Wind direction (*WD*),
wind velocity (*WV*)

(b)

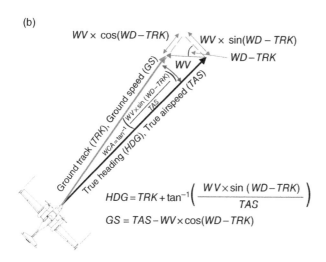

$WV \times \cos(WD-TRK)$ $WV \times \sin(WD - TRK)$

$WD-TRK$

WV

$HDG = TRK + \tan^{-1}\left(\dfrac{WV \times \sin(WD-TRK)}{TAS} \right)$

$GS = TAS - WV \times \cos(WD-TRK)$

Figure 6.23 (a) Ground track followed (TRK) and ground speed (GS) is given by the vector addition of the true heading vector (HDG) with magnitude true airspeed (TAS) and the wind vector with direction WD and magnitude WV. (b) Normally in a navigation planning it is the track that is known and heading and GS that are unknown. The wind correction angle (WCA) between TRK and HDG is shown as are the relationships connecting HDG and GS to TRK, TAS, WD, and WV.

Equation (6.23) is exact while Equation (6.24) is an approximation for small WCAs, which is normally the case.

Beermat Calculation 6.2

An aircraft is required to fly a ground track (TRK) of 039° and has a TAS of 155 knots. The wind at the altitude of the aircraft is 30 knots from 090°. Calculate the heading (HDG) required to maintain the required track and the corresponding GS. This is the situation depicted in Figure 6.23.

From Equation (6.23) the heading is

$$HDG = 39 + \tan^{-1}\left(\frac{30 \times \sin(90-39)}{155}\right) = 048°$$

$$GS = 155 - 30 \times \cos(90-39) = 136\,knots$$

It is clear from Figure 6.23 that the heading would need to be to the right of the ground track so that the wind maintains the aircraft along the required track and also there is a headwind component so the GS will be less than the TAS.

As an alternative to using Equations (6.23) and (6.24), the reverse side of the CRP5 shown in Figure 6.22 has a grid that enables a fast calculation of the heading and GS as illustrated in Figure 6.24 for the same example as in Beermat Calculation 6.2 and Figure 6.23. First, the wind is set up as in Figure 6.24a, in this case from 090 and 30 knots. Then, as shown in Figure 6.24b, the blue central circle is moved up to the TAS (155 knots) and the outer circle rotated to the ground track (039°) rotating the wind vector to its direction relative to the ground track. The end of the wind vector then points to the GS (136 knots) and the WCA is +9°, which makes the heading 048°.

6.5.3 Pilot Log for a Visual Flight Rules (VFR) Navigation

A navigation to be followed using dead reckoning is conducted under visual flight rules (VFR), which are specified in the Enroute section of a state's Aeronautical Information Publication (AIP) [12]. The rules are mostly generic with small changes between different states. Having calculated the true heading, it needs to be corrected for magnetic variation and if the compass and an unslaved direction indicator are to be used for direction control, deviation must also be included to give a magnetic heading to follow. The time between waypoints is simply calculated by dividing the measured distance between them on the chart by the GS. Finally, when all the calculation is done, the two important numbers that emerge are the magnetic heading and time and if the pilot flies accurately the magnetic heading from a waypoint, they will reach the next waypoint at the calculated time. These two numbers are highlighted on a pilot's navigation log (navlog) along with all the other information required for the flight.

(a)   (b)

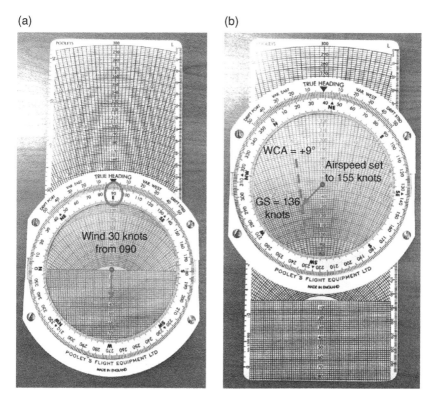

Figure 6.24 (a) Rapid calculation of the heading required to maintain a given ground track in the presence of a known wind and (b) the corresponding ground speed using a CRP5.

Important additional information is the minimum safety altitude (MSA) in each region, which is the highest feature in that region rounded up to the next higher 100 ft plus 1000 ft. On a VFR navigation, it is not necessary to fly above the MSA since it is possible to stay clear of terrain visually, however, it is important to be aware of it so that if visual conditions are lost, the aircraft can be climbed to an altitude that is safely above terrain. The planned altitude, however, needs to comply with the rules for VFR cruising levels, which apply for altitudes above 3000 ft and are illustrated in Figure 6.25. These specify that if flying magnetic tracks in the range 000–179, the altitude is an odd number of 1000s + 500 ft, for example, 3500 ft, 5500 ft, etc. while for magnetic tracks in the range 180–359, the altitude is an even number of 1000s + 500 ft, for example, 4500 ft, 6500 ft, etc.

Figure 6.26 shows a VFR navigation from Kavala airport (LGKV), Greece (where I currently work), to Kefalonia airport (LGKF) following the authorized VFR route around Thessaloniki airport (LGTS) with all the waypoints marked.

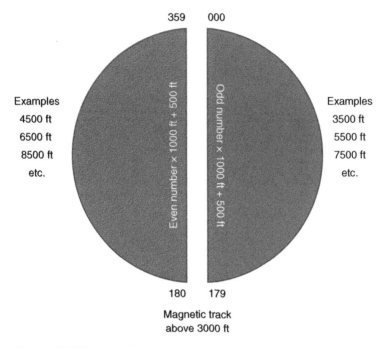

359   000

Examples   Examples
4500 ft   3500 ft
6500 ft   5500 ft
8500 ft   7500 ft
etc.   etc.

Even number × 1000 ft + 500 ft

Odd number × 1000 ft + 500 ft

180   179

Magnetic track
above 3000 ft

Figure 6.25 VFR cruising levels.

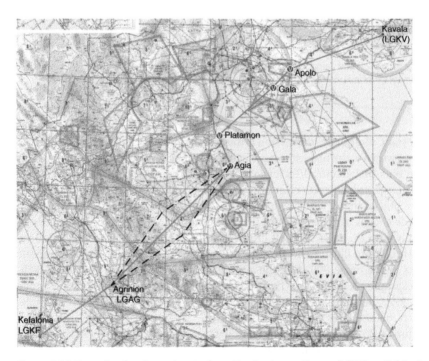

Figure 6.26 Example of a planned route from Kavala airport, Greece (LGKV) to Kefalonia airport (LGKF).

On the leg from Agia to Agrinion airfield are shown 10° deviation lines, which would be drawn on all the longer legs and used to calculate corrections for drifting off track as described in the following section. The route is chosen to follow approved VFR routes, avoid restricted areas, and provide good options for diversion airfields in case of bad weather or emergencies. Having selected the route, the true track from waypoint to waypoint can be measured directly from the map using a protractor or read electronically from a digital map.

The TAS can be calculated at the planned altitude knowing the temperature and the average forecast wind along each leg can be found from various aviation meteorology websites. It is important at this planning stage to get the most reliable information available and often a number of sources are consulted to be confident that the numbers are as close as possible to reality. In addition, it is important to ascertain from meteorological forecasts that VFR conditions can be maintained throughout the flight. On this length of navigation, the magnetic variation only changes by about 1° from the start of the route to the end so it is acceptable to assume an average. Having obtained all the necessary data, the navlog can be prepared and the one corresponding to the navigation in Figure 6.26 is shown in Figure 6.27.

The magnetic headings to fly and the times from waypoint to waypoint, calculated as described in Section 6.5.2, are highlighted as these are the core data for the navigation. The chart also contains all the other information required for the flight including the planned altitudes for each leg, VHF and LF/MF frequencies for communication and navigation aids. Spaces are provided to record departure and arrival information and any enroute clearances. When the takeoff time is known, the estimated time of arrival (ETA) at each waypoint is written in the box indicated. This is an important piece of information as it is an indicator at what time the waypoint should be visible and if not, an indicator that something has gone wrong with the navigation. Above each waypoint, the actual time of arrival (ATA) is noted and any discrepancy between the ETA and ATA is noted somewhere on a blank space on the navlog and this discrepancy (e.g. − 3 minutes, +4 minutes, etc.) is added to the ETA for all future waypoints. Generally, all times in the ETA boxes are just kept at their original values and only the discrepancy is altered overhead each waypoint. This description has been simplified as other factors, including, for example, the time required to climb, which is at a different speed and fuel consumption to the cruise will affect the times but the principle is the same.

6.5.4 Correcting Track Errors

If the wind and temperature information gathered before the flight for each leg is correct, then the magnetic headings and times are correct and, in principle, it should be possible to reach overhead the destination simply by accurately flying the correct heading for each leg for the calculated time. There are several factors

VFR Flight log

		DEPARTURE INFORMATION			ARRIVAL INFORMATION		
		Engine start	Time	Tacho	Engine stop	Time	Tacho
Date	03/11/2017	Runway in use					
Pilot	Binns	Aircraft DA42 SX-BEO	QFE			QFE	
From	Kavala LGKV	QNH				QNH	
To	Kefalonia LGKF	Taxy instr.			Runway in use		
Distance	265	Flight Time	Clearance instr.	97	EN ROUTE INFORMATION		

F FUEL	Fuel Consump.	42 l/hour				Sunset				
R RADIO	Total Required			Var.	68	4	E/W	E		
E ENGINE	Fuel on board	190	2000'	W/V					Temp.	
D DIRECTION	Reserve + conting.	33	5000'	W/V					Temp.	
A ALTIMETER	Total Endurance	4.5 hours	Pl/Alt	W/V					Temp.	

FROM/TO	MSA	PL/ALT	TAS	TR(T)	WD	WV	HDG(T)	HDG(M)	G/S	Time	DIST	E	TIME	ETA	ATA
Kavala LGKV / Apolo	3600	4500	145	250	80	30	248	244	175		55	E	19		
Apolo / Gala	4100	4500	145	226	70	28	222	218	171		14		5		
Gala / Platamon	3500	4500	145	221	60	35	217	213	178		39		13		
Platamon / Agia	7300	7500	152	170	80	38	156	152	152		14		6		
Agia / Agrinion	8400	8500	155	224	100	25	216	212	169		94		33		
Agrinion / Kefalonia LGKF	6200	4500	145	234	150	28	223	219	142		49		21		

Station	SVC	Freq
Kavala	TWR	118.400
Kavala	ATIS	128.150
Thess'lki	APP	120.800
Thess'lki	TWR	118.100
Athens	CTL	TBA
Agrinion	INFO	123.650
Preveza	APP	120.450
Andravida	APP	121.125
Kefalonia	TWR	122.250
Kefalonia	ATIS	126.450
KPL	VOR	108.800
KHR	NDB	327.000
TSL	VOR	112.100
STV	VOR	112.900
STF	NDB	420.000
KFN	VOR	115.500
KEF	NDB	318.000

Figure 6.27 VFR navlog for the navigation flight shown in Figure 6.26.

that can modify the headings and times however. These include pilot inaccuracy in following a heading or altitude, since the wind changes with altitude, incorrectly forecast winds, being asked to maintain a different altitude by air traffic control, etc. It is therefore important to monitor the progress of the flight by comparing with the track on the chart and to establish a procedure for correcting track errors.

In the planning stage, each leg, unless it is very short, has 10° deviation lines marked from the beginning and end as shown for the track between Agia to Agrinion airfield in Figure 6.26. This leg is shown expanded in Figure 6.28 and as an example, suppose the pilot establishes that about half way along the leg, where there is a bow-shaped reservoir (labeled), the aircraft is on the wrong side of the reservoir. It is easy to see from the deviation lines that the aircraft is off track by about 6° and has been following a track from Agia 6° off the intended one. Correcting the heading by 6° will simply alter the track to be parallel to the correct one so, since the aircraft is half way along the leg, a further 6° correction is required to fly to the next waypoint so the aircraft will need to turn right by 12°. In general, up to the half-way point, if the off-track angle is θ, the correction angle, $\Delta(\text{HDG})$, required to reach the next waypoint is

$$\Delta(\text{HDG}) = \theta + \tan^{-1}\left[\frac{x\tan\theta}{d-x}\right] \tag{6.26}$$

where d is the total distance of the leg and x is the distance along the leg. Equation (6.26), however, is not a calculation one would want to do while flying and up to the half-way point it is sufficiently accurate to apply the corrections, 2θ for $x = d/2$, $1.5\,\theta$ for $x = d/4$, 1.7θ for $x = d/3$, etc. Beyond the half-way point applying a correction is simple, the heading just needs to be adjusted by the measured track error defined by the deviation lines to the end of the leg to put it on course to arrive over the waypoint.

Another method to correct track errors is known as the "1 in 60" rule, which relies on the fact that for small angles, $\tan\theta = \theta$, where θ is in radians. If the distance off-track, y, is known after flying a distance x, then the track error in degrees, θ, is $y/x \times (180/\pi) = 57.2y/x$, which is approximated to $60y/x$. In order to then find the closing angle to the next waypoint, the rule is applied to the rest of the leg giving a closing angle of $60y/(d - x)$. Thus, the total heading change, $\Delta(\text{HDG})$, required for an off-track distance error of y is

$$\Delta(\text{HDG}) = 60\left(\frac{y}{x} + \frac{y}{d-x}\right) \tag{6.27}$$

For example, the distance track error of the aircraft shown in Figure 6.28 is 4.5 nautical miles after having traveled a distance of 44 nautical miles along a leg 94 nautical miles long. Applying the 1 in 60 rule gives a heading correction of 11°, which is only 1° different to the method using deviation lines.

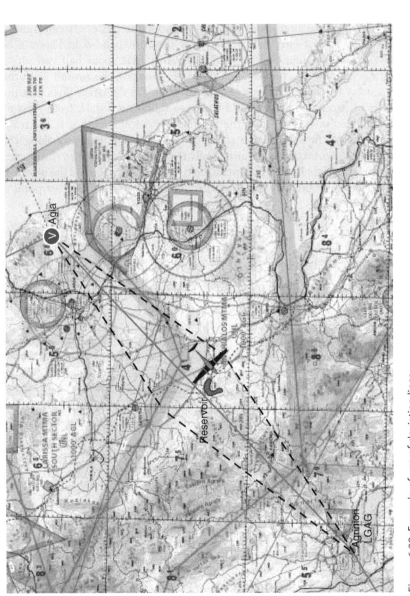

Figure 6.28 Example of use of deviation lines.

Problems

1 Two aircraft set off from the latitudes 30° N and 52° N on the prime meridian and head directly true East. After both have traveled 500 nautical miles, what are the longitudes of the positions of the aircraft?

2 An aircraft flies from Cardiff airport whose coordinates are N512346 W0032031 to Dusseldorf airport with coordinates N511731 E0064553. What is the distance traveled in nautical miles?

3 Currently the distance between the North magnetic dip pole and the true North pole is 210 nautical miles. Southern France at a parallel of 45° N has a magnetic variation of zero. Estimate the distance required to travel East or West along the 45° parallel for the magnetic variation to change by 1°.

4 A Lambert conformal chart is drawn to represent the area between the parallels 40° and 60° N and the meridians 10° E to 10° W. The standard parallels are at 46° N and 54° N at which the scale is 2 000 000 : 1. What is the scale along the parallel 56° N.

5 The Earth's cross-section can be assumed to be an ellipse with semimajor and semiminor axes of 6 378 137 and 6 356 752.3 m, respectively. The eccentricity of an ellipse is given by

$$e^2 = \left(a^2 - b^2\right)/a^2$$

where a is the semimajor axis and b is the semiminor axis. Use this information to estimate the angular difference between the geocentric and geodetic latitudes at a latitude of 45° N.

6 An aircraft is flying at an altitude of 10 000 ft above sea level under ISA conditions. The aircraft has a magnetic heading of 100° with an IAS of 160 knots and the wind is reported as being 20 knots from 030°. The local magnetic variation is 5° E. Calculate the true ground track and the GS of the aircraft.

7 An aircraft is performing a VFR navigation and the current leg is 120 nautical miles long with a magnetic heading of 180. After traveling 45 nautical miles the pilot estimates that the aircraft is 2 nautical miles off track to the West. Calculate the new heading required to fly to the next waypoint.

References

1 Weinstock, R. (1974). Chapter 3. In: *Calculus of Variations*. New York: Dover.

2 National Oceanic and Atmospheric Administration (NOAA). Wandering of the geomagnetic poles. https://www.ngdc.noaa.gov/geomag/GeomagneticPoles. shtml (accessed 12 June 2018).

3 Cavit. https://creativecommons.org/licenses/by/4.0/deed.en. Image reproduced from: https://en.wikipedia.org/wiki/North_Magnetic_Pole#/media/File: North_Magnetic_Poles.svg (accessed 12 June 2018).

4 An excellent description and interactive map of the IGRF has been produced by Data Analysis Center for Geomagnetism and Space Magnetism, Graduate School of Science, Kyoto University, Japan. http://wdc.kugi.kyoto-u.ac.jp/poles/ polesexp.html (accessed 12 June 2018).

5 Strebe. Mercator projection: https://commons.wikimedia.org/wiki/File: Mercator_projection_SW.jpg. Conical projection: https://en.wikipedia.org/ wiki/Lambert_conformal_conic_projection#/media/File:Lambert_conformal_ conic_projection_SW.jpg. Gnomic projection: https://en.wikipedia.org/wiki/ Gnomonic_projection#/media/File:Gnomonic_projection_SW.jpg (accessed 12 June 2018).

6 Weisstein, E.W. Mercator projection. From Mathworld – A Wolfram web resource. http://mathworld.wolfram.com/MercatorProjection.html (accessed 12 June 2018).

7 Lambert, V.J.H. (1772). *Anmerkungen und Zusätze zur Entwerfung der Land- und Himmelscharten* (Eng. trans: *Notes and Additions for Creating Ground and Aerial Charts*). https://quod.lib.umich.edu/cgi/t/text/text-idx?c=umhistmath; idno=ABR2581 (accessed 12 June 2018).

8 Weisstein, E.W. Lambert Conformal Conic Projection. From Mathworld – A Wolfram web resource. http://mathworld.wolfram.com/ LambertConformalConicProjection.html (accessed 12 June 2018).

9 Weisstein, E.W. Stereographic Projection. From Mathworld – A Wolfram web resource. http://mathworld.wolfram.com/StereographicProjection.html (accessed 12 June 2018).

10 National Imagery and Mapping Agency (NIMA) (1997). Technical Report NIMA TR8350.2, 4 July.

11 Heiskanen, W.A. and Moritz, H. (1967). *Physical Geodesy*. San Francisco: W.H. Freeman and Company.

12 UK National Air Traffic Service. For example the UK enroute section of the UK's AIP.

13 Creative commons (1953). http://creativecommons.org/licenses/by-sa/3.0/ legalcode (accessed 12 June 2018).

7

Short-Range Radio Navigation

The navigation techniques discussed in Chapter 6 are for flights under visual meteorological conditions (VMC). Although accurate calculations with good weather forecasts can produce precise navigation without external references, in practice, the ground track has to be monitored to correct for errors in wind predictions. In addition, without external guidance, lining up on a runway and landing requires visibility with the ground. Since the advent of aviation, the aim has been to make air transport weather proof and this requires methods of navigating safely without external visual references, that is, under instrument meteorological conditions (IMC). Apart from occasional local extreme climate phenomena, all-weather global aviation has been available for decades. In this chapter, the radio navigation systems that provide navigational guidance for ranges up to around 100 nautical miles will be described. These are important in the departure, approach, and landing phases of a flight and are integrated with global navigation systems such as inertial guidance and global navigation satellite systems (GNSS) that are used enroute. Thus, GNSS or inertial guidance brings an aircraft to an "initial approach fix," defined by radio navigation beacons close to an airport. These are then used by pilots to follow tracks defined by the local beacons that align the aircraft with the runway and then, in the case of larger airports the aircraft is guided horizontally and vertically to land on the runway. Increasingly, procedures are being defined that use GNSS for the entire journey including departure, approach, and landing, but the majority of instrument approaches and departures are still made using the local beacons described here. These integrated systems and procedures enable flights to be made without external visual references even down to landing in zero visibility in the case of larger airports and suitably equipped aircraft with autopilots.

Aircraft Systems: Instruments, Communications, Navigation, and Control,
First Edition. Chris Binns.
© 2019 John Wiley & Sons, Inc. Published 2019 by John Wiley & Sons, Inc.
Companion website: www.wiley.com/go/binns/aircraft_systems_instru_communi_Navi_control

7.1 Automatic Direction Finder (ADF)

7.1.1 Principle of Operation

Automatic direction finding is based on a nondirectional beacon (NDB) on the ground usually consisting of a simple dipole antenna, which has an isotropic emission intensity with a vertical polarization (see Section 4.4.2). The total operating frequency range of NDBs specified by the ICAO is 190–1750 kHz in the LF and MF band, however, for most aviation NDBs the frequency range is 255–535 kHz. At these frequencies, the propagation is by surface wave (see Section 4.2.2.1) and the range depends on the power of the transmitter. A receiver in the aircraft is tuned to this signal and determines its direction. This is the oldest system in use for radio guidance, dating back to the 1920s and as described in Section 1.7.1, the observation of the basic principle behind it goes back to within a couple of years of the discovery of radio waves. Frank Hertz, who was the first to correctly identify radio waves as electromagnetic waves also noticed in 1888 that a simple wire loop antenna gives a maximum signal when the loop is aligned along the direction of the transmitting antenna and zero signal when the loop is turned face on.

This behavior can be explained in terms of the radiation pattern of a loop antenna, which was described in Section 4.4.6. For a loop whose diameter is much smaller than the wavelength the electric field radiation pattern in polar coordinates has a figure of eight cross-section aligned along the plane of the loop as depicted by the red line in Figure 7.1a. In the corresponding plane a dipole antenna parallel to the loop will have the isotropic radiation pattern shown by the blue line. Recall that the radiation pattern applies to both receiving and transmitting and rotating the loop to the position shown in Figure 7.1b aligns the null in the radiation pattern with the dipole transmitter so that the signal in the loop antenna goes to zero. A problem with this simple arrangement for direction finding is that rotating the loop through 360° produces two nulls and so the transmitter is only defined as being on an axis and not in a specific direction. This can be resolved by coupling a dipole element, called a sense antenna to the loop via a 90° phase-shift as shown in Figure 7.1c. The lower half of Figure 7.1c demonstrates that at any instant the electric polarization of the loop is in opposite directions in the two halves of the radiation pattern while the polarization of the dipole after rotation is parallel to that of the loop, but constant. Thus, when combining the electric fields of the loop and dipole they add on one side and subtract on the other to give a total radiation pattern that has a cardioid shape with a single null. Thus, the uniaxial anisotropy of the loop alone becomes a unidirectional anisotropy of the combined loop and sense antenna. It is evident from Figure 7.1 that the signal variation around the null point is narrower as a function of angle than the maximum, which is why the null is used rather than the maximum signal.

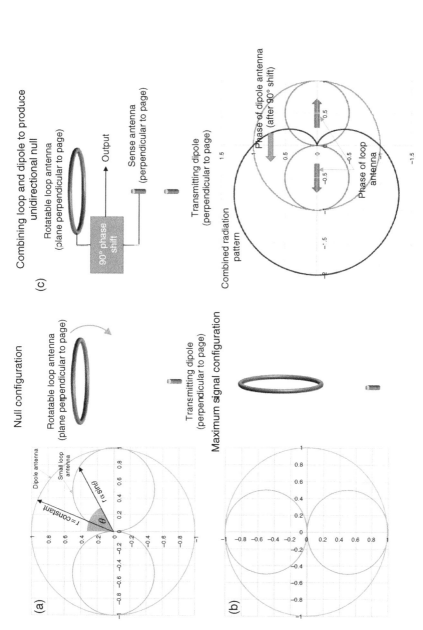

Figure 7.1 (a) Small loop has a figure of eight radiation pattern and when the loop plane is face-on to a dipole transmitting antenna, a null in the radiation pattern is lined up with the transmitter and no signal is detected. There are two nulls 180° apart. (b) Rotating the loop by 90° produces maximum signal. (c) The ambiguity of the direction to the transmitter along the null axis is resolved by combining the loop with a simple dipole (sense) antenna after shifting the phase of the dipole by 90°. The combined radiation pattern is a cardioid shape with a single null.

(a)　　　　　　　　(b)　　　　　　　　(c)

Figure 7.2 (a) Orthogonal loops wound on a ferrite core used in a modern ADF antenna. (b) Aerodynamic housing containing loop and sense antennas on the bottom of the fuselage of a light aircraft. (c) Twin ADF antennas on the roof of an airbus A380. *Source:* Adapted from Ref. [1] and reprinted under Creative Commons license 2.0 [2]. Reproduced with permission of https://en.wikipedia.org/wiki/Airbus_A380#/media/File:Airbus_A380_blue_sky.jpg. Licensed under CC BY-SA 2.0.

Early direction-finding loops had an air core and were rotated by hand or by electric motors to find the null, but modern ADF antennas consist of a stationary arrangement of perpendicular loops wound on a ferrite core to boost the signal (Figure 7.2a). This is combined with the sense antenna in an integral aerodynamic housing, which also contains the preamplifiers and other electronics as shown in Figure 7.2b and c.

Originally, the perpendicular loop arrangement was used to generate orthogonal oscillating magnetic fields within a separate set of coils in the receiver known as a radiogoniometer or just goniometer (see Section 1.7.1). In this arrangement, the null is sensed by a rotatable search coil which produces a direct drive to the display within the cockpit. Current ADF systems have no moving parts anywhere and detect the null in the received radio signal by processing the difference signal in the two orthogonal coils mixed with the sense antenna in software [3] and provide a digital output to a glass cockpit display.

7.1.2   ADF Cockpit Instrumentation

The most basic instrumentation found in a legacy cockpit comprises the LF/MF receiver tuner and the display (Figure 7.3a), which is the ADF needle pointing in the direction of the station. The needle alone shows the relative bearing (i.e. relative to the aircraft longitudinal axis) of the transmitter antenna. In all but the most basic instruments, however, a compass card around the circumference of the display can be rotated till the direction displayed at the top is the heading of the aircraft so that the needle shows the magnetic bearing to the station. To fly straight to the station ("homing") the aircraft simply needs to be turned till the needle is vertical and then maintain that heading with a wind correction (see Section 6.5.2).

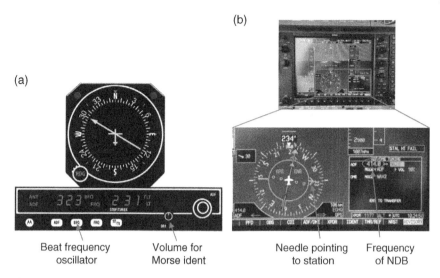

(b)

(a)

Beat frequency
oscillator

Volume for
Morse ident

Needle pointing
to station

Frequency
of NDB

Figure 7.3 (a) ADF receiver and display found in a light aircraft. *Source:* Adapted from Ref. [4] and reprinted under Creative Commons license 2.0 [2]. Reproduced with permission of https://upload.wikimedia.org/wikipedia/commons/thumb/6/67/Adf_mdi.svg/2000px-Adf_mdi.svg.png. Licensed under CC BY-SA 2.0. (b) Radio magnetic indicator display in a glass cockpit with the needle pointing to a nearby NDB at Bristol airport, United Kingdom.

The NDB modulates the carrier with a three-letter Morse code identification sequence for the beacon either by switching the carrier on and off, that is, on–off keying (OOK) or by amplitude modulation. The former method has the ITU code A1A (see Section 4.3.8) and is used on older long-range NDBs, which have mostly been switched off. In order to hear the code, the received signal needs to be mixed with a local signal whose frequency is slightly offset from the carrier so that the beat frequency is an audible tone as illustrated in Figure 4.11. This is the purpose of the beat frequency oscillator (BFO) button on the receiver shown in Figure 7.3a. While the identification sequence is being transmitted, direction finding is interrupted so that the code is transmitted relatively sparsely. Amplitude-modulated Morse code identification sequences have the ITU code A2A and in this case direction finding continues as normal while the code is transmitted. To hear the identification sequence the pilot simply needs to turn up the volume control indicated in Figure 7.3a. The same onboard equipment can tune into AM radio stations transmitting in the MF band for direction finding (as long as the location of the transmitter is known) and in this case turning up the volume passes the voice and music broadcast by the station to the pilot's headphones.

The simple setup shown in Figure 7.3a is only found in older light aircraft and in more modern cockpits the ADF needle is superimposed on the direction finder as shown in Figure 7.3b. In this type of display the ADF needle is referred to as a radio magnetic indicator (RMI) and always shows the magnetic bearing to the station. The heading to fly in zero wind conditions to reach the station is referred to as "QDM" following the old "Q-codes" developed in the early days of commercial radiotelegraph communications. These were originally developed to reduce the length of messages in Morse code and despite their age, survive into modern aeronautical voice communications. Conversely, the tail of the needle shows the magnetic bearing from the station, also referred to as QDR. For example, Figure 7.3b shows an aircraft flying on heading 234 with a nearby NDB at Bristol airport, United Kingdom transmitting on 414 kHz at a magnetic bearing 327 to the station (QDM) and a magnetic bearing 147 (QDR) from the station.

Setting up the ADF needle as an RMI facilitates more sophisticated use of the NDB other than simple homing. For example, it is straightforward to intercept and track a specific bearing to or from a station, as illustrated in Figure 7.4. The example shown is an aircraft initially traveling North that will approach the NDB inbound on the 240 radial (QDR = 240, QDM = 060). The Northerly heading can be continued until the ADF needle points to 060 and then the aircraft can be turned onto 060 (with a wind correction) to maintain a track along the desired radial.

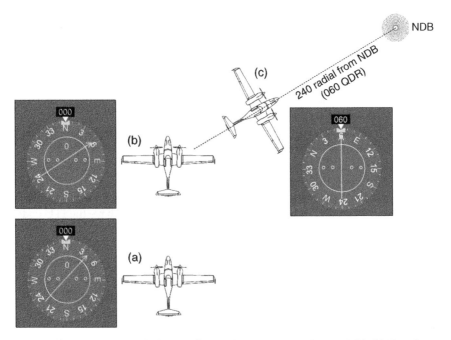

Figure 7.4 ADF used to track the specific radial 240 (QDR = 060) to an NDB. (a) Aircraft traveling North on an intercept track continues till the needle points to 060 at (b). (c) The aircraft turns onto 060 plus a wind correction to maintain the track along the required radial.

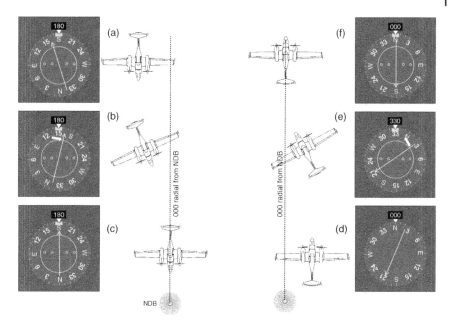

Figure 7.5 Correcting track errors using an NDB. (a) Aircraft traveling South has drifted off the 000 radial from the NDB. (b) To correct the error the heading is changed till it is on the opposite side of the needle from the required bearing ("push the head") till the needle moves on to the bearing. (c) The track is recovered. (d) Aircraft traveling North has drifted off the 000 radial from the NDB. (e) The aircraft is turned until the heading is further away from the tail of the needle in the same direction as the required bearing ("pull the tail"). (f) The track is recovered.

Once established on a radial, the NDB can be used to maintain that radial by applying appropriate corrections as illustrated in Figure 7.5, which shows an aircraft tracking inbound and outbound along the 000 radial from an NDB. In Figure 7.5a, the aircraft is tracking inbound (QDM = 180) and has wandered off track so that the ADF needle has moved away from the required bearing of 180. To correct the track, the aircraft is turned until the heading is on the opposite side of the needle from the required bearing (Figure 7.5b), which in pilots parlance is referred to as "push the head." The heading is then maintained till the needle moves back onto the required bearing at which point the aircraft is turned back onto the required track with a wind correction if necessary. For an aircraft tracking outbound, the needle is pointing backward and it is easier to manipulate the tail. In Figure 7.5d the aircraft is off track and the aim is to bring the tail of the needle back to the QDR of 000. In fact, the needle will slowly move in the right direction without a heading change but the intercept will occur at infinity. To move onto the desired track, the aircraft is turned until the heading is further away from the tail of the needle in the same direction as the required bearing ("pull the tail"). This heading is then maintained till the needle moves onto the required bearing.

Many regional airports have an NDB either as the sole radio navigation aid or as a backup for a VOR (see Section 7.2) to provide approach guidance onto the runway. Figure 7.6a shows an example of a published approach onto runway 07 at Alexandroupolis airport, Greece (LGAL) using the NDB with Morse identification code "ALP" transmitting at 351 kHz. For a light aircraft (category A or B), the guidance is to follow an outbound track of 221 from the NDB while descending to 2000 ft for three minutes, then turn and follow an inbound track of 053. The descent is then continued to the minimum descent altitude (MDA) of 640 ft, which must be maintained if the runway is not in sight. If the runway is still not in sight at the position of the NDB, a go-around is initiated. Figure 7.6b shows a photograph taken from inside the cockpit of a light aircraft with an RMI (light blue needle) as it tracks outbound from the beacon during a night approach with the tail of the needle close to 221. Figure 7.6c shows a similar photograph during the inbound leg with a QDR of 053. Note that the green arrow and course deviation of a VOR display (see Section 7.2) is also displayed on the direction indicator.

Single needle tracking is a common technique in radio navigation and in a modern avionics display such as that shown in Figure 7.3b, the pilot can select the RMI needle to point to an NDB, a VOR beacon (see Section 7.2), or a GPS waypoint. In the latter case the waypoint can be defined by the user as any desired position on the Earth so the basic method for track guidance described above can be used to track to or from any point.

Despite the age of NDB/ADF technology, which is now approaching a century, it remains as part of the radio navigation infrastructure and is likely to do so for some time. It is still a significant part of the training curriculum for pilots obtaining their instrument rating. In order to use the technology, however, it is important to be aware of some of the shortcomings, which include:

- Thunderstorm effect: Discharges during thunderstorms often emit more radio energy in the NDB frequency band than the transmitters so that the ADF needle points toward thunderstorm rather than the beacon.
- Night effect: The D-band of the ionosphere absorbs LF/MF radio waves but it disappears at night allowing skywave contamination of the surface wave used by the ADF (see Section 4.2.3.2).
- Mountain effect: In mountainous areas, reflection and diffraction of the transmitted wave produces bearing errors.
- Coastal refraction: The ground wave from the NDB refracts if it crosses the land/sea boundary at an angle causing bearing errors (see Section 4.2.2.1).
- Quadrantal error: The fuselage reflects and reradiates the radio waves causing bearing errors (see Section 4.4.3).

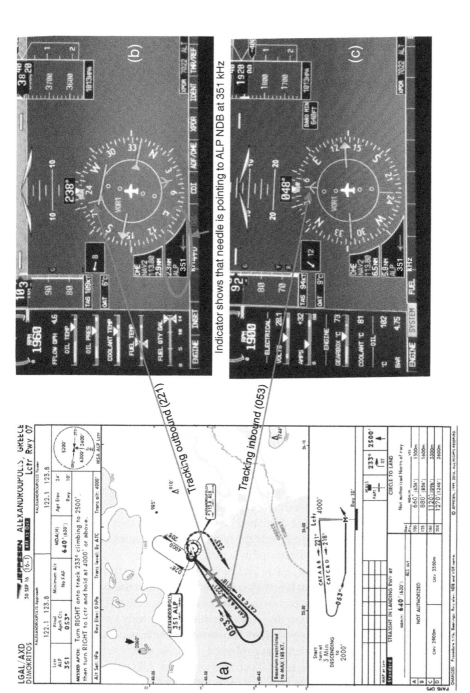

Figure 7.6 (a) Instrument approach plate onto runway 07 at Alexandroupolis airport, Greece (LGAL) using the NDB with Morse identification "ALP" transmitting at 351 kHz. (b) Photograph of the RMI (light blue needle) of a light aircraft doing a night approach onto runway 07 on the outbound track 221. (c) The RMI display on the same approach following the inbound track 053.

- Bank error: The loop aerial is designed to use a vertically polarized beam and any horizontal component of polarization generates a signal in the horizontal loop antenna elements causing bearing errors.

7.2 VHF Omnidirectional Range (VOR)

By the late 1930s, it was recognized that many of the problems associated with LF/MF beacons were due to the low-frequency radio waves and the ground wave propagation mode. It was decided to develop a new type of radio beacon operating in the VHF frequency band and this became the VHF omnidirectional range (VOR), which was selected by the ICAO in 1949 as the international civil navigation standard. Here, the word "range" in the title is carried over from the name "radio range" given to the old LF navigation system in the United States and is not related to distance, which the VOR on its own is unable to determine.

7.2.1 Principle of Operation

Each VOR operates at a frequency in the range 108–117.95 MHz with a channel spacing of 50 kHz, but some channels within the range are allocated to ILS localizers (see Section 7.4). The VOR beacon generates two 30 Hz sine waves modulated onto the VHF carrier, one of which is an omnidirectional reference signal with the same phase in all directions and the other is a variable signal whose phase varies continuously around the circle from 0 to 360° relative to the reference signal. The two signals are in phase along magnetic North, so comparing the phase angle between them gives the angle of the bearing to or from the station as illustrated in Figure 7.7.

The earliest VORs produced the reference signal by a rotating antenna, but by the early 1960s these had been replaced by systems that generated the variable phase as a function of angle electronically. There are two basic types in service today that use different methods to produce the variable and reference 30 Hz modulation and these are labeled the conventional VOR (CVOR) and the Doppler VOR (DVOR). The two types are described in detail below but they both use the same design of VHF antenna to generate the carrier, which is known as the Alford loop, patented in 1942 [5] and illustrated in Figure 7.8a. It consists of a pair of resonant 90° dipoles arranged in a square and each fed in antiphase from the oscillator. The result is a current that is in the same direction around the loop so in the far-field a receiving antenna detects an in-phase signal, irrespective of the bearing from the Alford loop. Other desirable characteristics are a nearly isotropic radiation pattern and horizontally polarized emission, which is important for the implementation of VORs in aviation as explained below. In practice, the antenna elements are made of sheet metal as illustrated in Figure 7.8b, which gives the antenna a broader band performance and makes it more mechanically stable.

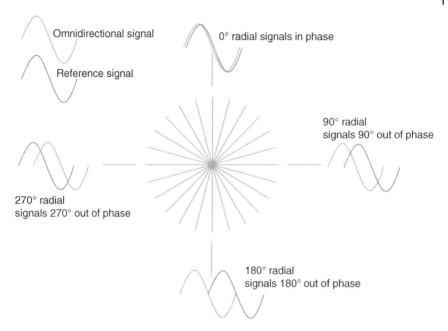

Figure 7.7 Relative phase of the 30 Hz variable and reference signals from a VOR as a function of bearing from the station.

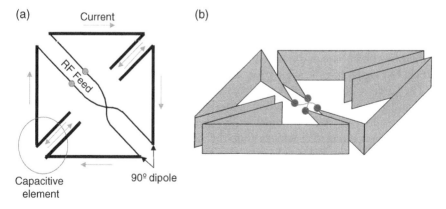

Figure 7.8 (a) Plan schematic of Alford antenna consisting of two 90° dipoles fed in antiphase and capacitive end elements. The emission is nearly isotropic and horizontally polarized. (b) Alford antennas in VORs are normally made of sheet metal to give them a broader band performance and greater mechanical stability.

7.2.2 Conventional VOR (CVOR)

The CVOR was the first solid-state implementation of the VOR and by the early 1960s had replaced the first generation of VORs that used mechanically rotating antennas. The antenna system consists of four Alford loops arranged and labeled as shown in Figure 7.8a and placed within a weatherproof radome. The radome also contains a set of parasitic antennas that are energized by any vertically polarized emission from the Alford loops and reradiate with a horizontal polarization. Any vertically polarized emission will cause bearing errors while an aircraft is banking as described in the previous section on the NDB. The normal installation has the radome placed on a circular metal mesh that forms a well-defined ground plane and is normally placed on the roof of the shack containing the electronics. An example is shown in Figure 7.9b, which is the CVOR at Megas Alexandros airport, Kavala, Greece. Also shown in the photograph is a UHF band dipole antenna used for the distance measuring equipment (DME), which is normally colocated with the VOR and used to determine the distance of an aircraft from the beacon (see Section 7.3).

The North reference is a 30 Hz signal, which is frequency modulated onto a 9960 Hz subcarrier with a frequency variation of ±480 Hz, that is, a modulation index of 0.05 (see Equation 4.28). As described in Section 4.3.4, the subcarrier is generated by amplitude modulation of the main carrier at the desired frequency offset, in this case 9960 Hz and this subcarrier will contain the 30 Hz FM signal as illustrated in the block diagram of Figure 7.10a. This also shows how the Morse code identification is added to the reference signal. The end result is a double sideband full carrier (DSBFC or conventional AM) signal with the 30 Hz FM signal on a 9960 Hz subcarrier and additional sidebands at 1020 Hz carrying the Morse identification. This is normally a three-letter code, for example, the identification for the CVOR at Kavala airport shown in Figure 7.9b is "KPL" (.-. .--. --). The various components of the reference signal in the frequency domain are shown in Figure 7.10b. The North reference signal is passed to all four Alford loops in phase and is broadcast as an omnidirectional beam in which the 30 Hz FM signal has the same phase in all directions.

The variable 30 Hz signal is amplitude modulated directly onto the VHF carrier and the directional radiation pattern is rotated electronically at 30 Hz, which produces a space-modulated 30 Hz signal in which the phase is fixed along a given radial. To achieve this, the same 30 Hz generator that is used for the reference signal is split into a sine and cosine component using a 90° phase-shifter and each of the components is passed into a balanced modulator that is also fed by the VHF carrier. The output of the modulators is a double side band suppressed carrier (DSBSC) signal that is passed to the RF distributor (Figure 7.11b), which drives the Alford loop pairs in antiphase as illustrated in the figure. A single Alford loop pair transmitting in antiphase will produce a static directional figure of eight pattern but the two pairs fed in quadrature

(a)

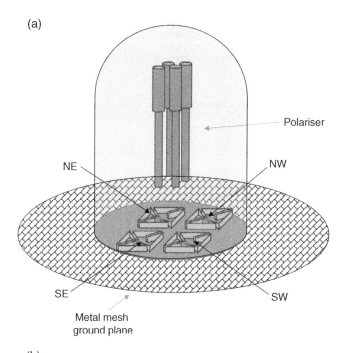

Polariser

NE

NW

Metal mesh
ground plane

SE

SW

(b)

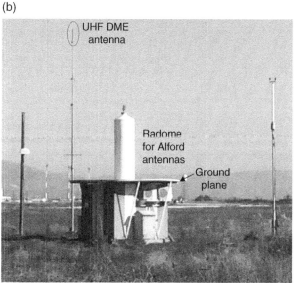

UHF DME
antenna

Radome
for Alford
antennas

Ground
plane

Figure 7.9 (a) Arrangement of Alford loops and the polarizer within the radome of a conventional VOR (CVOR), which is placed on a wire mesh ground plane. (b) The CVOR at Megas Alexandros airport, Kavala, Greece with the radome and ground plane on the roof of the electronics shack. Also shown is the UHF antenna used for the distance measuring equipment (DME).

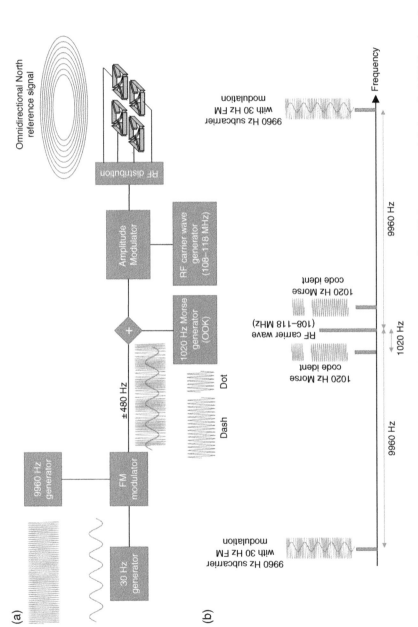

Figure 7.10 (a) Generation of the North reference signal in a CVOR consisting of a 30 Hz frequency modulation of a 9960 Hz subcarrier. There is also a 1020 Hz ident signal modulated onto the carrier by on–off keying (OOK). (b) The components of the double sideband full carrier (DSBFC or conventional AM) reference signal in the frequency domain.

with the sine and cosine components of the 30 Hz modulation produce the rotating figure of eight pattern shown in Figure 7.11a, which is driven clockwise. Since the modulation and the rotation are exactly synchronized, there will be a 30 Hz AM signal of constant phase each specific radial. One complication is that the figure of eight pattern will go through two maxima per cycle, so a receiver picking up the variable signal alone would record a 60 Hz modulation. In the far field, the rotating variable signal mixes with the synchronized reference signal with its isotropic radiation pattern and in a similar manner to the case of the NDB (see Figure 7.1) the result is a radiation pattern with a single maximum and minimum per rotation as illustrated in Figure 7.11c. In this case the pattern is a Pascal's Limaçon, from the French for snail. Also shown in Figure 7.11a is a phase-shifter between the 30 Hz signal generator and the reference signal generator, which is used to calibrate the reference for magnetic North. The receiver in the aircraft is tuned to the VHF carrier of the VOR and obtains the 30 Hz FM

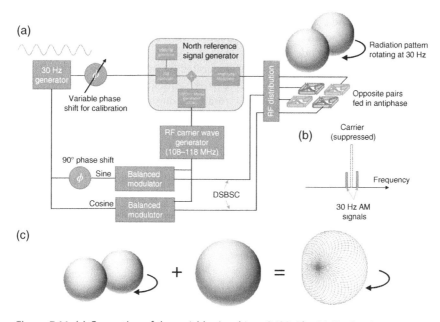

Figure 7.11 (a) Generation of the variable signal in a CVOR. The 30 Hz signal generator output is split into quadrature components and passed to two balanced modulators along with the VHF carrier. The DSBSC output from the balanced modulators is passed into the two opposing pairs of Alford loops in antiphase. The result is a clockwise rotating directional beam whose phase is fixed along a given radial. (b) The output from each balanced modulator in the frequency domain. (c) In the far field the combination of the figure of eight radiation pattern of the variable signal and the isotopic pattern from the reference signal produces a Limaçon-shaped radiation pattern with a single maximum and minimum per cycle. *Source:* Reproduced with permission of Robert FERREOL.

reference signal from the 9960 Hz subcarrier. It then compares this to the phase of the 30 Hz AM signal and the phase angle is a direct measurement of the bearing from magnetic North. The cockpit instrumentation driven by the VOR receiver is described in detail below.

7.2.3 Doppler VOR (DVOR)

Errors in the measured bearing from a VOR caused by signals reflected from terrain and man-made objects, referred to as multipath (see Section 7.2.5), are a particular problem in the CVOR making the choice of location very important. In order to minimize these errors and make the beacon less sensitive to siting a different design of VOR was developed in the mid-1960s known as the DVOR. The antenna consists of 48 Alford loops, each housed in a weatherproof glassfiber box and arranged in a ring as shown in Figure 7.12a. The variable signal is produced by the ring and a single Alford loop at the center is used to generate the North reference signal. The assembly is placed over a wire mesh that acts as a ground plane and mounted on the roof of the instrument shack in a similar fashion to the CVOR. A photograph of the Daventry VOR in the United Kingdom with Morse code identifier "DTY" (-.. - -.--) is shown in Figure 7.12b.

In this beacon, the variable signal is generated by switching an unmodulated 9960 Hz subcarrier of the main VHF carrier sequentially around the ring of Alford loops to simulate an antenna rotating counterclockwise at 30 Hz. In the far field, the result of this is that the frequency of the subcarrier has a sinusoidal modulation due to the Doppler shift whose phase depends on the bearing from the VOR as illustrated in Figure 7.13. In order to maintain compatibility with the CVOR, the modulation index is made the same, that is, the frequency variation, Δf, is ±480 Hz, though it is important to point out that here the frequency modulation is on the variable signal and not on the North reference signal as in the CVOR.

The modulation index and carrier frequency define the radius of the circle of Alford loops since:

$$\frac{\Delta f}{f_0} = \frac{v}{c} \tag{7.1}$$

where f_0 is the carrier frequency, v is the effective speed of the switching around the circle, and c is the speed of light. If r is the radius of the circle, the effective speed around the circle is:

$$v = 2\pi r \times 30 \, \text{ms}^{-1} \tag{7.2}$$

So, from Equation (7.1)

$$r = \frac{c\Delta f}{60\pi f_0} = \frac{8c}{\pi f_0} \tag{7.3}$$

(a)

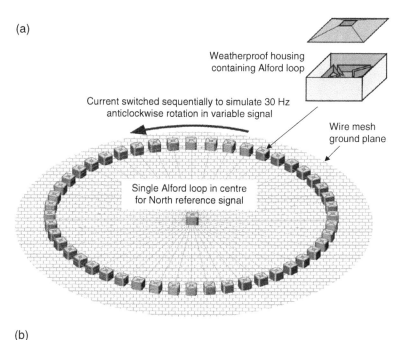

Weatherproof housing
containing Alford loop

Current switched sequentially to simulate 30 Hz
anticlockwise rotation in variable signal

Wire mesh
ground plane

Single Alford loop in centre
for North reference signal

(b)

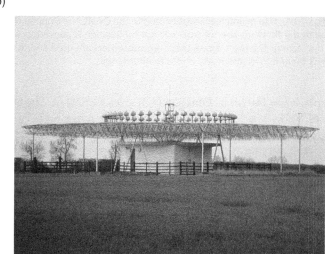

Figure 7.12 (a) Construction of a Doppler VOR (DVOR) consisting of 48 Alford loops in a ring that generate the variable signal and a single Alford loop at the center that produces the North Reference signal. The assembly is placed on a wire mesh ground plane and mounted on the roof of the instrument shack. (b) The DVOR at Daventry (DTY) in the United Kingdom. *Source:* Photo reproduced with permission from c/o Trevor Diamond td@trevord.com.

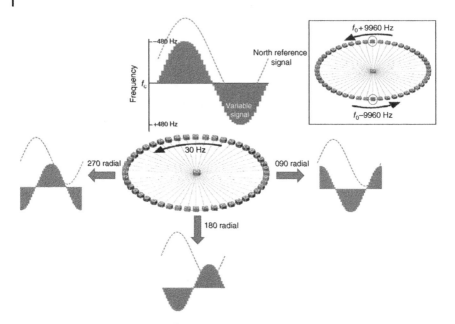

Figure 7.13 The 30 Hz rotation of the signal around the ring produces a sinusoidal FM modulation of the subcarrier due to the Doppler shift and the radius of the circle is calculated to produce a ±480 Hz variation to maintain compatibility with the CVOR. The Doppler shift is in discrete steps due to the finite number of antennas but modulation is converted to a smooth sine wave by filtering. The inset shows that both subcarriers are used and the switching driver feeds them into diametrically opposite Alford loops.

Thus, for example, if f_0 = 110 MHz, Equation (7.3) gives r = 6.94 m. The entire frequency range of VORs is encompassed by radii in the range 6.47–7.07 m. Typically, there are 48 Alford loops around the circle so the frequency is stepped as opposed to smoothly varying as shown in Figure 7.13, however, a simple filter will recover a smooth sinusoidal wave from the digitized frequency variation. An additional detail in most VORs is that both subcarriers at ±9960 Hz are used and the switching driver feeds them into diametrically opposed Alford loops as shown in the inset in Figure 7.13.

The North reference signal is generated by a single Alford loop that is fed by the VHF carrier amplitude modulated by a 30 Hz signal as shown in Figure 7.14a. As in the case of the CVOR, the reference signal also has mixed in a 1020 Hz subcarrier with the Morse code identification. The block diagram for the generation of the variable and reference signals is shown in Figure 7.14a and the constitution of the signals in the frequency domain is illustrated in Figure 7.14b.

Note that there is a direct swap of the frequencies and modulations of the variable and reference signals in the CVOR and DVOR. In the case of the CVOR,

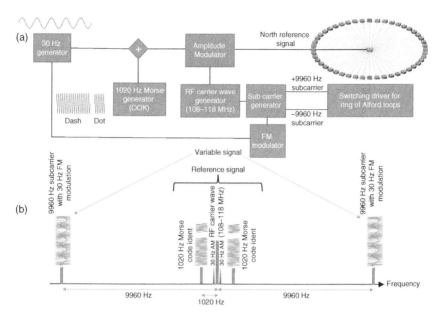

Figure 7.14 (a) Block diagram of the generation of reference and variable signals for the DVOR. (b) The variable and reference signal components in the frequency domain.

the reference signal is FM on the 9960 Hz subcarriers and the variable signal is AM on the main carrier, while for the DVOR the reference signal is AM on the main carrier and the variable signal is FM on the 9960 Hz subcarriers. Thus, the receiver on the aircraft will decode and measure the phase between the signals in exactly the same way for both types of VOR except the reference and variable signal phases with be transposed for the two types. The space modulation of the variable signals rotate in opposite directions for the DVOR and CVOR so that the returned bearing is the same irrespective of which type of VOR signal is being received. Table 7.1 shows a summary of the differences between the two types of VOR.

In this brief description, many details have been omitted, for example, how opposite Alford loops are fed in antiphase in the case of the CVOR and how the complex switching mechanism works in the DVOR. An excellent tutorial that goes into more detail is available on Youtube [6].

7.2.4 VOR Cockpit Instrumentation

The signals from VORs are received at the aircraft normally by a quarter-wave Marconi antenna oriented horizontally to match the polarization of the VOR signals. In the cockpit, the relative phase of the 30 Hz reference and variable

Table 7.1 Comparison of reference and variable signals in the CVOR and DVOR.

Type of VOR	Reference signal	Variable signal	Rotation direction
CVOR	30 Hz FM on 9960 Hz subcarrier	30 Hz AM on main carrier	Clockwise
DVOR	30 Hz AM on main carrier	30 Hz FM on 9960 Hz subcarrier	Counterclockwise

signals from the VOR is determined and the bearing information conveyed to the pilot by various types of cockpit instruments. The most basic type of display is a stand-alone electromechanical instrument similar to the one shown in Figure 7.15a commonly found in light aircraft. In this type of instrument the phase difference is converted to an analog voltage, which is used to drive the deflection of a needle labeled the course deviation indicator (CDI) in Figure 7.15a. In order to determine the bearing to or from a beacon the omni bearing selector (OBS) knob is rotated, which moves the rotating course card around the outside of the instrument and when the pointer or omni bearing indicator (OBI) is aligned with the bearing to or from the VOR, the CDI

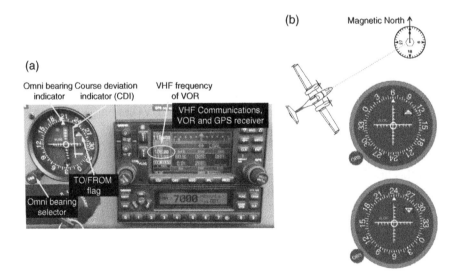

Figure 7.15 (a) Cockpit instrumentation for a VOR display in a light aircraft consisting of the VHF receiver and electromechanical display. (b) An aircraft on the 240 radial from a VOR and the cockpit display with the OBS rotated to the 060 bearing selected (CDI centered and "TO" flag showing) and the 240 bearing selected (CDI centered and "FROM" flag showing).

deflection is zero. Whether the bearing is to or from the station is indicated by a TO/FROM flag on the instrument and Figure 7.15b shows an example of the two zero CDI deflection bearings for an aircraft on the 240 radial from a VOR. These will be obtained when rotating the OBS knob so that the OBI shows 060 (TO flag displayed) and 240 (FROM flag displayed).

The divisions on the CDI needle scale are 2° apart (the edge of the central circle also indicates 2°) and the deflection of the needle from center determines the number of degrees the aircraft is off the selected bearing up to a maximum of 10°. The reading of the VOR display is independent of the heading of the aircraft, which is treated as a point object as illustrated in Figure 7.16a for three aircraft having different headings on the 210 radial from a VOR. In order to join and track a specific radial using this type of display, the card is rotated till the OBI reads the correct bearing and the needle then indicates whether to fly a heading to the right or left of the selected bearing to approach the required radial. For example, Figure 7.15b shows three aircraft around the 210 radial from a VOR with aircraft 1 on radial 218 and aircraft 2 on radial 202. The course TO the beacon is 030 and for aircraft one the CDI needle is deflected to the right showing that a heading to the right of 030 must be flown to intercept the 210 radial. The opposite indication is shown for aircraft 2, which is to the right of the radial and so needs to choose a heading to the left of 030 to intercept. Aircraft 3

(a)   (b)

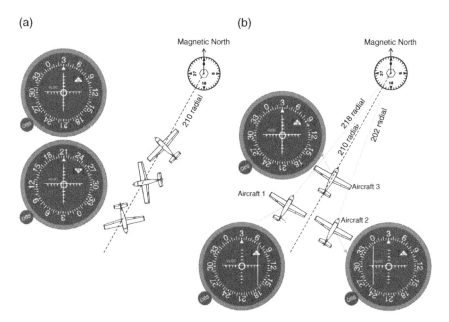

Figure 7.16 (a) VOR display is unchanged by the heading of an aircraft on a given radial. (b) CDI deflection shows which way to fly to intercept the radial selected on the OBI.

is on the radial so the CDI is centered and the heading needs to be maintained at 030 plus a wind correction to track the radial.

The problem with the simple instrument shown in Figure 7.15 is that the pilot needs to consider both the selected radial on the OBI and the current heading to develop a situational awareness and then decide a new heading to fly that will intercept the selected radial. This can be a heavy workload, especially if the CDI is off-scale and the aircraft is flying a heading in a semicircle opposite to the one containing the required bearing. The instrument does not directly provide a situational awareness of the relationship between heading and bearing. To provide a more ergonomic display, more sophisticated cockpits combine the CDI and OBI display with the direction indicator so that the relationship between the required bearing and heading is clear. The combined instrument is referred to as a horizontal situation indicator (HSI) and an example of an older stand-alone display with its separate VHF receiver is shown in Figure 7.17a. The OBI is moved around the same display as the direction indicator and always shows the bearing relative to the current heading and when the aircraft turns, the OBI rotates with the direction indicator. Thus, the direction the aircraft is going relative to the selected bearing is always clear and easy to read. In a more modern cockpit the HSI becomes part of the primary flight display (PFD) as shown in Figure 7.17b. Note that the HSI also displays up to two RMI needles and in the example shown, one is pointing to an NDB and the other to a GPS waypoint.

Figure 7.18a shows an example of an approach to runway 23 at Kavala airport, Greece (LGKV) using the VOR KPL sited next to the runway. The guidance is to follow an outbound track of 067 from the VOR to a distance of 10 nautical miles while descending to 2400 ft. Note that the distance to the VOR, which is highlighted in the cockpit photograph shown in Figure 7.18b, is obtained using DME described in Section 7.3. At 10 nautical miles, the aircraft turns to intercept radial 052 (232 inbound) from the VOR and track this while descending to the runway using the descent profile shown in the bottom half of Figure 7.18a. If the runway is not visible by 600 ft altitude a go-around must be initiated. Figure 7.18b and c shows photographs of the PFD from inside the cockpit of a light aircraft during the outbound and inbound leg of this approach.

7.2.5 VOR Track Errors

The major cause of track errors in VOR signals is interference between the direct wave from the beacon and any reflected signal arriving at the aircraft as illustrated in Figure 7.19. This is known as multipath error and is further distinguished as lateral multipath, due to horizontal reflections (Figure 7.19a) or longitudinal multipath due to vertical reflection (Figure 7.19b). Lateral multipath can be due to reflections from terrain, tree lines, power cables, buildings,

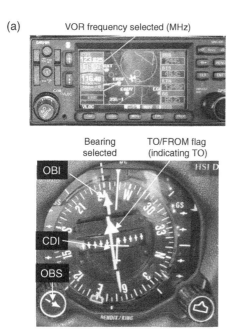

(a) VOR frequency selected (MHz)

Bearing selected   TO/FROM flag (indicating TO)

OBI

CDI

OBS

(b) VOR frequency selected (MHz)

OBS

CDI   173°   OBI

RMI needle 1 (GPS waypoint)

RMI needle 2 (NDB)

TO/FROM flag (indicating TO)

Figure 7.17 (a) Stand-alone HSI and separate receiver/tuner. (b) HSI within a primary flight display (PFD) in a glass cockpit. The display can also show up to two RMI needles. In this example one is pointing to a GPS waypoint and the other to an NDB.

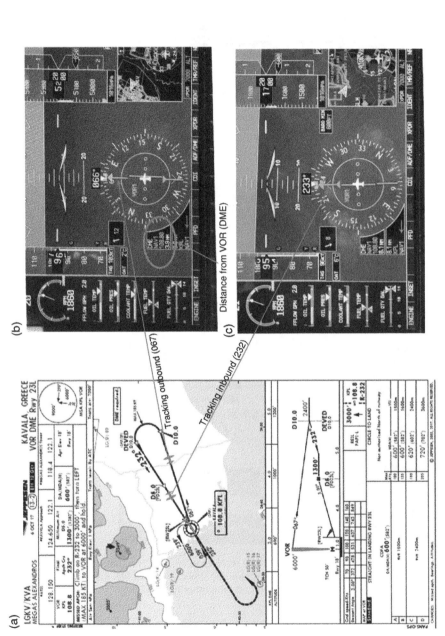

(b)

(c) Distance from VOR (DME)

Tracking outbound (067)

Tracking inbound (232)

Figure 7.18 (a) Approach plate for runway 23L at Kavala airport, Greece (LGKV) using the VOR KPL transmitting at 108.8 MHz. (b) Photograph of the primary flight display (PFD) of a light aircraft on the outbound leg of the approach tracking radial 067 (QDR). Also highlighted is the distance from the VOR obtained using distance measuring equipment (DME – see Section 7.3). (c) Photograph of the PFD of a light aircraft on the inbound leg of the approach tracking radial 052 /232 QDM.

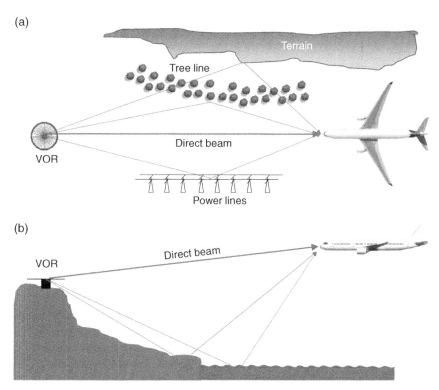

Figure 7.19 (a) Lateral multipath can be due to reflections from terrain, powerlines, tree lines, etc. (b) Longitudinal multipath occurs for VORs placed on hill tops that can result in reflections from terrain below or sea water.

metal fences, etc. Longitudinal multipath arises from VORs placed on hilltops that can result in reflections off terrain below or the sea surface, which is quite reflective for 120 MHz (~2.5 m) radio waves. Although the reflected signals are much weaker than the direct beam they can cause some enhancement or cancelation of the direct signal when they recombine at the aircraft, which leads to an error in the determination of the phase-shift between the reference and the variable signals. The result is that an aircraft that is flying accurately along a specific selected radial as determined by, for example, GNSS will display changes in the CDI deflection indicating an apparent drift off the selected track. Lateral multipath is the most common form of interference and track errors are more likely to occur when following a CVOR due to the fact that the reflected signal will be present for many cycles of the 30 Hz variable signal. In the case of the DVOR, due to the moving source of the emission, it is likely that the reflected signal will be present only for a part of each cycle leading to a smaller probability of the cockpit instrument measuring a change in phase.

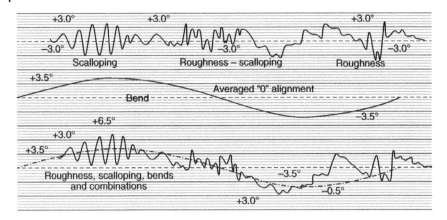

Figure 7.20 Illustration of scalloping, roughness, and bend deflections on CDI needle due to multipath errors and the tolerance required for VORs as specified by the ICAO document 8071 vol. 1 [7].

The ICAO has produced a comprehensive manual on the testing of radio navigation aids [7] that defines the required accuracy of VORs and how to test them. It subdivides irregularities in the cockpit display into three different types, that is, roughness, scalloping, and bends. Roughness is irregular rapid movements of the CDI needle and can be due to irregular reflecting surfaces such as tree lines. Scalloping is regular rapid movements of the CDI needle and can be caused by straight reflecting features such as power lines or metal fences. Both of these types of deflections are too fast to follow with the aircraft and it is possible to maintain an average track through them. Bends are slower undulations of the CDI needle due to large-scale terrain that can be followed by the pilot or autopilot to keep the CDI needle centered while flying an undulating track around the required one. The ICAO specification for all VORs is that combined roughness and scalloping should produce deviations no more than 3° from the average track while bends should not exceed more than 3.5° away from the required radial. The types of errors and the tolerances are illustrated in Figure 7.20.

7.2.6 Airways System Defined by VORs

Over land masses VORs define the nodes of the airways system that is still used by ATC to control traffic and provide separation between flights. Figure 7.21 shows the network of airways in an area of sky over Northern Greece with the VORs at the intersections highlighted by red circles. Two possible routes from Kavala airport (LGKV) to nearby airports along the airways system are marked by red lines with relevant reporting points marked by red triangles. An aircraft flying from Kavala to Thessaloniki airport (LGTS) under instrument

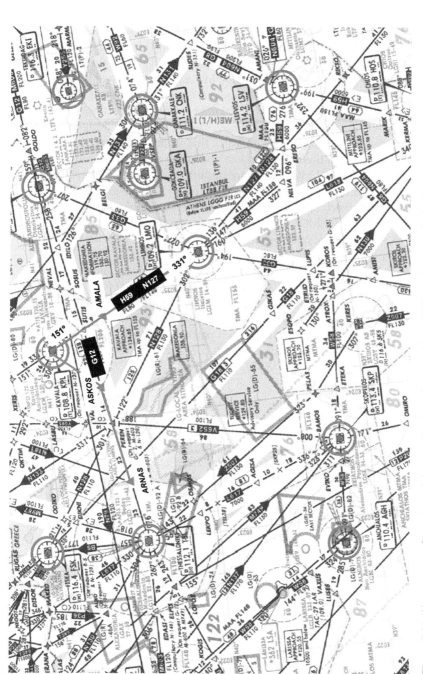

Figure 7.21 Airways map of a section of sky over Northern Greece with the VORs acting as nodes highlighted by red circles. Airways routes from Kavala airport (LGKV) to neighboring airports a: Thessaloniki (LGTS) and Limnos (LGLM) are highlighted in red.

flight rules (IFR) after takeoff would follow a departure procedure using the Kavala airport VOR ("KPL") that would take it to the reporting point ASKOS. There it would join the airway G12 that in this region is defined as running between the VORs "ALX" at Alexandroupolis airport (LGAL) and "TSL" close to Thessaloniki airport. From there the aircraft would track the G12 airway using the "TSL" VOR along radial 076 (QDM = 256) till it reached the reporting point ARNAS. From there it would be radar vectored to be aligned with the ILS system on one of the runways at Thessaloniki airport (see Section 7.4).

An aircraft flying from Kavala to Limnos airport (LGLM) would again follow a departure procedure from the VOR "KPL" that would take it to the reporting point AMALA. From there it would join one of the airways that runs between the VORs "KPL" and "LMO" at Limnos airport. These are designated H59 for low-altitude flights (down to flight level 80) or N127 for high-altitude flights (down to flight level 140). Having tracked the airway along the 331 radial (QDM = 151) from "LMO" to arrive overhead, the aircraft would then use an approach similar to the one shown in Figure 7.18 to line up with the runway by tracking radials from and to "LMO."

Many VORs are located at airports but there are also a number of enroute beacons that act as nodes in the airways system, an example is the VOR "FSK" on the upper left side of Figure 7.21. In addition, the VOR "TSL" is not at Thessaloniki airport but about 4 nautical miles away. There is a low-power VOR on the airport site that is not part of the airways map that is used just for local traffic carrying out approach procedures. This type of beacon is referred to a terminal VOR (TVOR).

7.2.7 Area Navigation (RNAV)

Having aircraft fly along tracks defined by beacons is a traffic management system that dates back to the low-frequency radio range (LFR) system developed in the late 1920s in the United States and described in Section 1.7.2. Restricting flights to airways that join beacons does not normally provide the shortest route between departure and destination points. With increased emphasis on saving fuel it is rational to utilize the flexibility of the aeroplane to be able to go anywhere and fly routes that minimize the distance traveled, thus in the late 1960s the concept of area navigation (as opposed to linear tracking between beacons) was developed and the first implementation of this was by "virtual beacons." With increased miniaturization of electronics allowing microprocessors to be incorporated into VHF navigation receivers, it became possible to relocate VORs electronically within the receiver to any desired position as shown schematically in Figure 7.22. Consider an aircraft on radial 1 at a distance x from a real VOR (it is assumed that the VOR has a colocated DME – see Section 7.3). The pilot can enter a radial (radial 2) and distance (y) required to shift that VOR to create a virtual VOR over a desired destination, for example, an airport without a VOR.

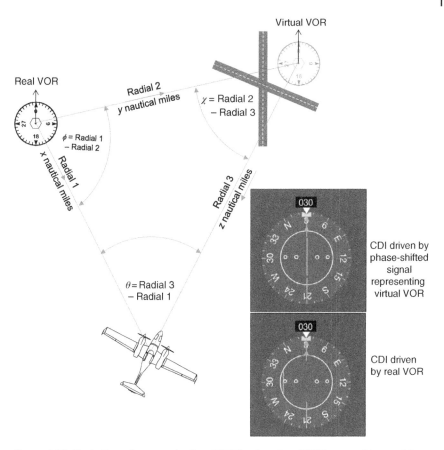

Figure 7.22 Illustration of area navigation (RNAV) using virtual VORs created by applying a radial and a distance offset to a real VOR.

The distance to the real VOR (x) and the radial (radial 1) are measured and since the shift (y) and the radial (radial 2) to the virtual VOR are known, the distance, z, to the virtual VOR can be determined using the law of cosines, that is:

$$z^2 = x^2 + y^2 - 2xy\cos\phi \qquad (7.4)$$

where ϕ = radial 1 − radial 2. Having determined z, the angle, θ can be calculated using the law of sines, that is:

$$\theta = \sin^{-1}\left(\frac{y\sin\phi}{z}\right) \qquad (7.5)$$

and θ is the phase-shift that has to be applied to the variable signal from the real VOR to obtain the radial from the virtual VOR (radial 3). These phase-shift and distance z are constantly updated as the aircraft flies so that the cockpit instrumentation can be used as normal for tracking as if there was a real VOR (and colocated DME) stationed at the position of the virtual VOR. This is illustrated by the schematics of the cockpit displays in Figure 7.22, which show the readings if the display is switched to the virtual VOR showing the aircraft is on track along the 210 radial (QDM = 030) or the real VOR, in which the CDI indicates an off-scale "fly left" deflection. The ability to randomly place a virtual VOR anywhere led to the term RNAV (the "R" stands for random).

The first commercially available RNAV receivers appeared in the late 1970s when single-chip microprocessors became available and an example is the Bendix-King KNS80 shown in Figure 7.23. This incorporated a DME readout and a ground speed calculated from the rate of change of distance. From a given real VOR that the receiver was tuned to, the pilot could generate up to four virtual VORs by specifying a radial and distance of each of them from the real VOR. The system would drive an electromechanical display of the type shown in Figure 7.15a and pressing the VOR button would drive the CDI from the real VOR as in a normal VOR receiver and the distance read out would be from the real DME. Pressing the RNAV button would get the currently selected virtual VOR to drive the CDI and the distance readout would be the one calculated to the virtual VOR using Equation 7.4.

Inertial navigation and satellite navigation systems provide true area navigation capability and the routing of air traffic is starting to move to a system known as required navigational performance (RNP) that was introduced in Section 1.8.4. This combines all available navigation systems to follow optimized curved paths in three dimensions. Currently, over densely populated land masses, the airways system defined by VORs is still an important part of the

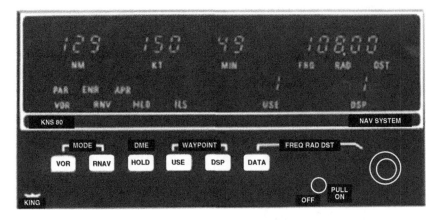

Figure 7.23 Bendix-King KNS80. The first commercially available RNAV system.

traffic control system and flight plans are submitted using airways but wherever possible controllers will clear flights direct to destinations knowing that the necessary primary and backup area navigation systems are on board.

Interesting Diversion 7.1: Visual-Aural Radio Range (VAR)

The VOR and the instrument landing system (ILS) described in Sections 7.2 and 7.4, respectively, which are still standard radio navigation tools used today, are both evolved from an earlier system first developed in the 1930s known as the visual-aural radio range (VAR). The VAR in turn emerged from an earlier network of LF beacons described in Section 1.7.2 (see Figure 1.18) called the LFR. This operated at around 500 kHz and suffered from the interference problems occurring in the LF band described in Section 7.1, so by the 1930s it was decided to replace it by a network of beacons transmitting in the VHF band. The network was designed to be similar to the LFR with each beacon defining four tracks that aircraft would follow as shown in Figure ID7.1.1 for the Guthrie range beacon on the Dallas network. In one pair of arms the signals providing guidance for the track were the same as for the LFR, that is, the Morse code letters "A" and "N," which are the inverse of each other (. _ and _. respectively). Thus, on each side of the track "N" or "A" is heard in the pilot headphones while along the track there is a continuous tone. For the other pair of arms the signal was amplitude modulated so that on one side of the track there was predominantly a 90 Hz tone (the yellow sector) while on the other there was predominantly a 150 Hz tone (the blue sector). Along the track the modulation at the two frequencies was of equal amplitude. This was the forerunner of the localizer in the ILS still in use today (see Section 7.4). The modulations would be used to drive a cockpit instrument that had a needle that would be centered if the aircraft was on track (equal amplitude tones) and would give a fly right or fly left indication if the aircraft was off track as illustrated in Figure ID7.1.2a, thus this was the "visual" arm of the beacon. In some displays the needle would track vertically across while in others it was hinged at the top. The displays would also have a second, horizontal needle that would indicate whether the aircraft was on the glide slope on an approach to a runway in early ILS systems. Figure ID7.1.2b shows a VAR transmitting station and antenna at Mangalore airfield in Australia taken in 1954.

VAR networks were installed to define airways across the United States and other countries throughout the 1940s and 1950s but they had a relatively brief life as they were rapidly overtaken by the more advanced VOR that allowed the aircraft to track any desired radial to or from a beacon. Indeed, this is why the term "omnidirectional" was used to distinguish it from the LFR and VAR networks that defined just four courses from each beacon. However, much of the technology used on VORs and ILSs was developed with the VAR.

(Continued)

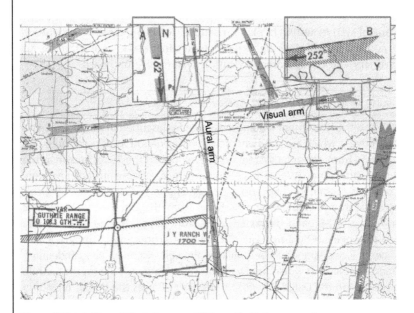

Figure ID7.1.1 Map of Guthrie range VAR on the Dallas network.

(a)

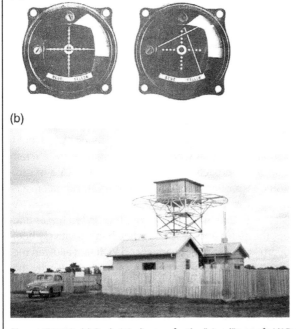

(b)

Figure ID7.1.2 (a) Cockpit indicators for the "visual" arm of a VAR. (b) VAR transmitter station and antenna at Mangalore, Australia in 1954. *Source:* Reproduced with permission of CAHS.

7.3 Distance Measuring Equipment (DME)

Most VORs have an additional facility to measure the distance of an aircraft from the beacon, known as the DME. The VOR determines the radial an aircraft is on from the beacon, so combining this with a distance provides a position fix. DME works on the principle of secondary radar (see Section 5.4), that is, a signal sent by the aircraft triggers a response from a transponder on the ground colocated with the VOR. The system works in the band 962–1213 MHz (UHF) with a 1 MHz spacing providing 252 channels and in the aircraft receiver the DME UHF frequency for a specific VOR is tuned automatically when the VOR frequency is selected on the VHF navigation receiver. For example, the Daventry VOR shown in Figure 7.12b transmits at the VHF frequency of 116.4 MHz and the DME receives aircraft transmissions at 1073 MHz, but this is selected automatically when 116.4 MHz is selected on the receiver. Both interrogation and return signals are transmitted from omnidirectional antennas and the ground station transmits at a frequency 63 MHz lower than the interrogation signal from the aircraft to avoid directly reflected pulses being processed.

The operation of the system is illustrated in Figure 7.24. The aircraft transmits pairs of pulses of width 3.5 μs separated nominally by 12 or 36 μs but the precise spacing is randomized (jitter). When these are received a gate is set to only continue to accept pairs with the same randomized interval. The ground transponder retransmits the pulse pair with the same spacing as those received and on a carrier wave with a 63 MHz lower frequency following a 50 μs delay. Randomizing the pulse spacing in this way means that each aircraft only accepts the pulses from the transponder that have been triggered by its own interrogator and thus each aircraft is uniquely identified. The timing between the transmitted and received pulses gives the distance to an accuracy of 0.25 nm or 1.25% of the range (whichever is the greater error). The DME provides the slant distance, so the height of the aircraft must be taken into account if a ground distance is required. ILS installations (see Section 7.4) also have a paired DME facility so that the distance to the runway threshold is known.

During the initial range search an aircraft transmits at 150 pulse pairs/s (pps), which reduces to 60 pps after 15 000 pulse pairs (100 s). This is further reduced to about 25 pps at "lock on" (tracking mode), following which a lock on gate is activated to only receive pulses after the expected time interval, which is adjusted for changing range. A beacon saturates at 27 000 pps (~100 a/c) and in response the receiver gain (or threshold) is reduced (increased) so that the beacon responds only to the strongest signals. The ground station also transmits a Morse code identifier for the DME, which has the same three-letter code as the VOR but the dots and dashes of the identification are heard at a higher pitch to distinguish them from the VOR identification.

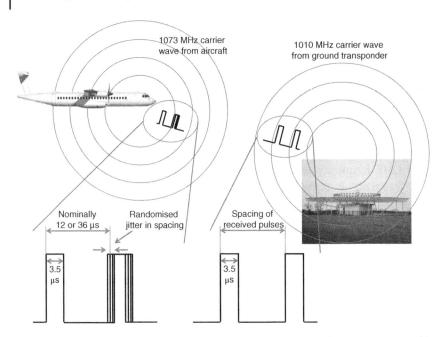

Figure 7.24 DME finds the range by timing the interval between pulse pairs transmitted by the aircraft interrogator and the received pulse pairs from the ground transponder. Each aircraft is uniquely identified by randomizing the spacing between the pulses.

7.4 Instrument Landing System (ILS)

VORs (or sometimes NDBs) can be used to line up the aircraft with a runway in IMC but the final part of the approach requires high-precision lateral and vertical guidance to bring an aircraft to a few meters above a runway with widths typically in the range 45–80 m. The first-ever landing of a commercial aircraft under IMC conditions guided by radio was by a Boeing 247D (see Figure 1.37b) in the United States in 1938 using a space-modulated system originally designed in the 1920s. The international standard adopted by the ICAO in 1949 is a space-modulated VHF/UHF installation on the runway called the ILS, which is still in use today and is likely to remain in service for the foreseeable future. It was scheduled to be replaced by a newer method known as the microwave landing system (MLS) (see Section 7.5) starting in the 1990s, however, development of MLS was shelved as it became clear that a better replacement of ILS would be to a GNSS for approach guidance. Since satellite-based systems are not ready to replace all categories of ILS, it has won a new lease of life.

The basic principle of ILS was outlined in Section 1.7.3 (see Figure 1.23) and consists of two components, that is, the localizer and glideslope that give lateral and vertical guidance, respectively. The localizer transmits space-modulated VHF signals in the frequency range 108.1–111.95 MHz, which is part of the spectrum reserved for radio navigation (see Figure 4.16) with some of the 50 kHz-spaced channels reserved for ILS. The glideslope works in the UHF part of the spectrum with frequencies in the range 329.15–335.0 MHz though this is invisible to the pilot as the aircraft receiver automatically tunes in the glideslope frequency when the VHF frequency of the localizer is selected. Each component will be described separately though the fundamental technique for achieving space modulation is the same.

7.4.1 ILS Localizer

The basic principle of the ILS localizer is illustrated in Figure 7.25. At the far end of the runway, two banks of directional antennas, for example, Yagi–Uda or log-normal arrays emit overlapping beams with a VHF carrier having a horizontal polarization. One is modulated at 150 Hz and the other at 90 Hz so that the overlapping beams produce a space modulation where the relative amplitude of the modulations varies with angle away from the centerline. Along the centerline the amplitudes are equal and a fan with width 2.5° either side of the extended runway centerline is described as the course signal. The cockpit instrumentation is calibrated to give full-scale deflection off track at the edge of the course signal. A simple implementation of this system would be to feed one half of the antenna bank with a 150 Hz amplitude-modulated signal and the other half with a 90 Hz amplitude-modulated signal. This simple method, however, requires that the transmitted power from the two antenna banks being identical, which in turn requires the output power of the transmitters and antenna gains to be identical. If this is not the case the course signal will not follow the centerline and, in practice, the transmitted powers cannot be guaranteed to be equal to the required accuracy.

Figure 7.25 Basic operating principle of the ILS localizer.

To generate a space modulation where the path of the course signal is insensitive to transmitter power and antenna gain, the transmitted signal emitted by each antenna bank combines both AM signals. If $S_1(f_1)$ is the 90 Hz signal and $S_2(f_2)$ is the 150 Hz signal, one combination is $S_1 + S_2$, referred to as the carrier with side bands (CSB) signal and the other is $S_1 - S_2$, referred to as the side bands only (SBO) signal. The subtraction can be achieved by inverting the phase of S_2 before adding it to S_1 and both signals can be generated coherently as shown in Figure 7.26a. It is worth noting at this stage that if S_1 and S_2 are identical, then taking the sum of the CSB and SBO signals reproduces a pure 90 Hz modulation while taking the difference reproduces a pure 150 Hz modulation as shown in Figure 7.26b. Allowing the signals to vary by up to 30% produces relatively little change in the sum and difference as demonstrated in Figure 7.26c, which shows the sum and difference of the CSB and SBO signals when $S_1 = 0.7S_2$.

Although the $S_1 + S_2$ and $S_1 - S_2$ have been labeled CSB and SBO, the format of the transmitted signal is set up by the type of modulation. Normally a balanced modulator is used, which transmits the SBO and for the CSB signal the carrier is added back in as illustrated in Figure 7.27. In a similar fashion to the VOR output, a Morse code identification is added into the CSB signal

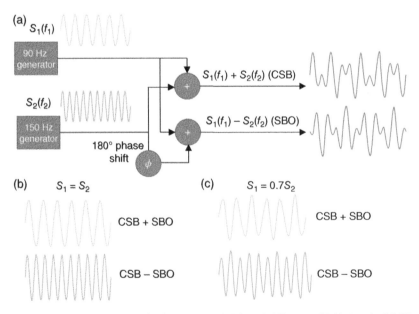

Figure 7.26 (a) Generation of coherent sum (CSB) and difference (SBO) signals. (b) When $S_1 = S_2$ taking the sum CSB + SBO reproduces the pure 90 Hz signal while taking the difference CSB − SBO reproduces the pure 150 Hz signal. (c) Changing the amplitude of one of the signals by 30% produces relatively little change to the modulations.

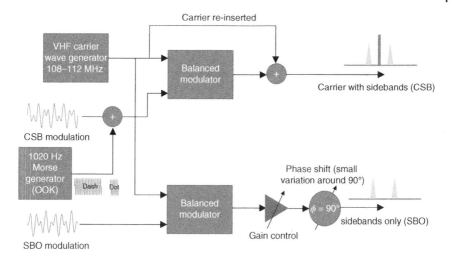

Figure 7.27 Generation of the CSB and SBO signals in the ILS localizer. The CSB modulation is passed into a balanced modulator and the carrier is added back in to produce the carrier with side bands (CSB) output. The SBO signal and carrier passed into the balanced modulator to directly produce the side bands only signal (SBO). The Morse code identification is added to the CSB signal by on–off keying (OOK) of a 1020 Hz signal.

by OOK of a 1020 Hz signal. On the SBO side the balanced modulator output is taken directly but a gain control that enables the relative strength of the SBO signal to be varied is fitted, which can be used to change the width of the course signal as explained below. In addition, a phase-shift of 90° is added, which is required for the addition and subtraction of the CSB and SBO signals in the correct phase as shown below. This includes a limited variability around 90° to compensate for differences in cable lengths.

The CSB signal is fed to both antenna banks in phase while the SBO signal is fed with 180° phase-shift between the two antenna banks. This is illustrated in Figure 7.28, where for simplicity just one antenna is used to represent each bank. Figure 7.28a shows two Yagi–Uda antennas fed in phase by the CSB signal along with the calculated radiation pattern using CocoaNEC [8]. The result is the normal directional lobe with a higher directivity than a single antenna and side lobes attenuated by at least 10 dB. The SBO signal fed in antiphase to the two antennas (Figure 7.28b) has a radiation pattern with two lobes in antiphase and a null signal along the centerline. In the far field (Figure 7.28c) the SCB and SBO radiation patterns combine so that right of the centerline a CSB – SBO signal is received and left of the centerline a CSB + SBO signal is received. With reference to Figure 7.26, these are 150 and 90 Hz signals, respectively, so the space modulation has been achieved and the direction of the centerline is insensitive to the relative transmitter power. This is because the null along the

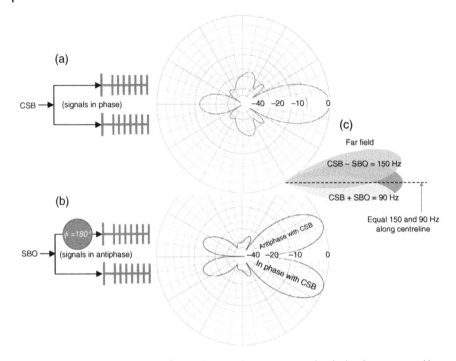

Figure 7.28 (a) The CSB signal is fed in phase to the two antenna banks (each represented by a single Yagi–Uda antenna. The radiation pattern calculated for the geometry shown using cocaoNEC [8] is plotted and has a single directional lobe. (b) The SBO signal is fed in antiphase to the two antenna banks and the radiation pattern, calculated by cocoaNEC has two lobes in antiphase and a null along the centerline. (c) In the far field the combined patterns are CSB − SBO = 150 Hz right of the centerline and CSB +SBO = 90 Hz left of the centerline.

centerline has been produced by splitting the signal from the same transmitter between the two antenna banks. Along the centerline the null in the SBO signal results in just the CSB signal being received, which consists of equal 90 and 150 Hz modulations. The system modulated as described provides a simple method to alter the width of the course signal by changing the output power of the SBO signal using the gain control shown in Figure 7.27. For example, making the SBO signal stronger increases the width of the SBO radiation lobes so that the course signal becomes narrower.

A complication is that the in-phase and antiphase signals are phase-shifted by 90° with respect to each other as demonstrated in Figure 7.29. Consider a point, P, a distance d away from the antenna and an angle, θ, right of the centerline. Two antenna elements a distance, a, apart are shown to represent the two antenna banks. For $a \ll d$, the path length difference between the two beams reaching P is $2a\cos\theta$, so there is a phase-shift between the two signals given by:

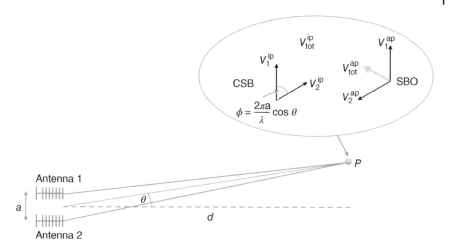

Figure 7.29 Demonstration that the CSB and SBO signals are phase-shifted by 90° with respect to each other so that a 90° phase-shift has to be added to the SBO signal (see Figure 7.27) in order for the CSB and SBO signals to be added and subtracted in the correct phase.

$$\phi = \frac{2\pi a}{\lambda} \cos\theta \qquad (7.6)$$

where λ is the wavelength of the carrier wave (in the range 2.68–2.78 m for ILS signals). When adding the in-phase (CSB) voltages from the two antennas, the total voltage vector shown in yellow in Figure 7.29 is obtained. Adding the anti-phase (SBO) voltages from the two antennas produces the total voltage vector shown in blue and it is easy to show that the blue and yellow vectors are phase-shifted by 90° with respect to each other. This result is independent of the angle, θ, off the centerline and the distance, a, between the antennas. Thus, in order to add and subtract the CSB and SBO signals in the correct phase, 90° has to be added to the phase of the SBO signal and this is the purpose of the phase-shifter shown at the SBO output in Figure 7.27.

The sidelobes transmitted by the directional antennas have the potential to produce false course signals, especially in the case of the CSB radiation pattern, which is shown in Figure 7.30 for the CSB emission from the two Yagi–Uda antennas in more detail. Although the sidelobes have an intensity that is 15–20 dB smaller than the main lobe, in the case of the CSB signal they have equal amplitude 90 and 150 Hz signals and the cockpit instruments will give an on-centerline indication.

A remedy is to transmit a separate CSB–SBO signal at lower power from a subset of the antennas in the two banks as illustrated in Figure 7.31a. Remembering that fewer antennas transmitting will produce a broader beam, this signal

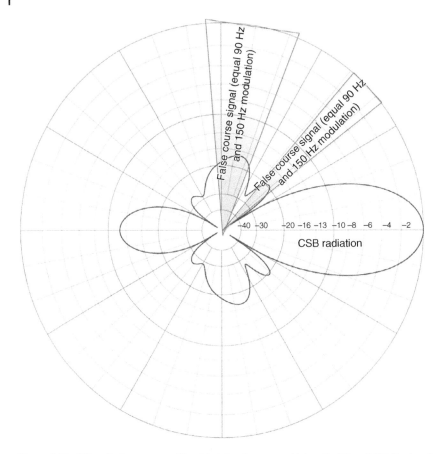

Figure 7.30 CSB radiation pattern. The side lobes have equal intensity 90 and 150 Hz signals and will produce false course signals.

will be wider than the main beam and will be picked up at angles where the side lobes are present. If it is broadcast at a signal strength about 10 dB lower than the main beam it will still overwhelm the side lobe signal strength and will produce the required full-scale fly right or fly left indication in the cockpit instruments. The low-power transmission broadcast for side lobe suppression is referred to as the clearance signal. Sometimes is has a slight frequency offset of around 8 kHz from the course signal and in some ILS systems it is transmitted by a separate antenna array. The ICAO specification of range at which the localizer signals can be picked up is shown in Figure 7.31b. The main course signal should be received and provide a cockpit indication up to 25 nm away within ±10° of the centerline and up to 17 nm within ±35° of the centerline. Note that beyond ±10° of the centerline the signal received will be the clearance signal.

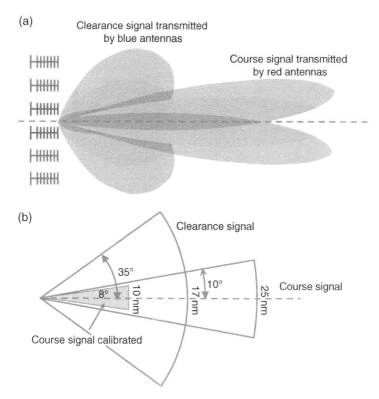

Figure 7.31 (a) A low-power CSB–SBO broadcast referred to as the clearance signal is transmitted at about 10 dB lower power from a subset of the antenna array (shown in blue). It is wider than the main course signal and is received at high angles at which the side lobes from the main transmission are present. It has a higher power than the side lobes and provides a correct fly left or fly right indication in the cockpit instruments. (b) The ICAO specification for the ranges at which ILS signals should be received, that is out to 25 nm ±10° of the centerline (course signal) and out to 17 nm ±35° of the centerline (clearance signal). Out to 10 nm and ±8° of the centerline the signal should be calibrated.

The course signal should be calibrated and accurate out to 10 nm away and up to ±8° either side of the centerline.

The higher the number of elements in the antenna array the greater the directivity of the transmitted beams and generally higher categories of ILS system (see Section 7.4.4) have antennas with more elements. Figure 7.32a and b shows calculations published by the United States Federal Aviation Authority (FAA) [9] of the transmitted power as a function of angle away from the centerline of an 8-element and a 14-element localizer antenna. It is clear that the 14-element array produces beams with a higher directivity. Also shown is the depth of modulation in (%) of the side bands. Figure 7.32c shows a 14-element ILS

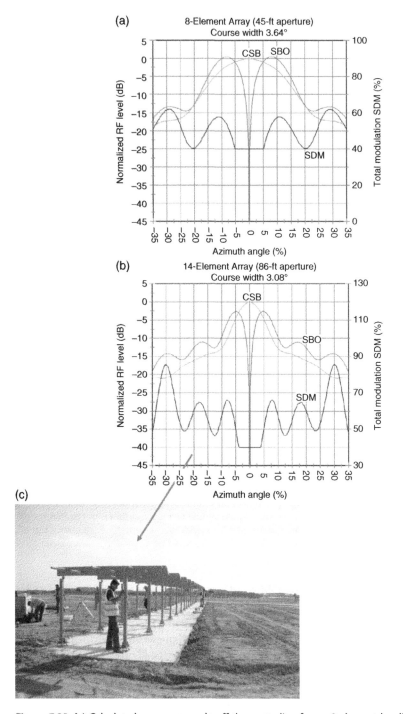

Figure 7.32 (a) Calculated power vs. angle off the centerline for an 8-element localizer antenna. (b) Calculated power vs. angle off centerline for a 14-element localizer antenna. The blue line in both curves shows the depth of modulation (%) in the side bands. (c) A 14-element ILS antenna consisting of log-periodic individual elements being installed and checked for alignment. *Source:* Reproduced with permission of US Air Force.

localizer antenna being installed and checked for alignment at Kirkuk airbase, Iraq. In this array, each individual antenna is a log-periodic directional type (see Section 4.4.7.2).

7.4.2 ILS Glide Slope

The glide slope uses the same CSB–SBO method to produce the space-modulated 90 and 150 Hz signals, but in this case the CSB and SBO signals are separated in the vertical plane as illustrated in Figure 7.33a. The null in the SBO signal and thus the course signal is at an angle in the range 2°–4° with respect to the horizontal, which is normally taken as the surface of the runway landing area. The angle of the course signal is set according to local obstructions but the majority of glide slope course signals are in the range 2.5°–3°. The width of the course signal in the vertical plane is 1.4°, that is, ±0.7° around the glide slope and at the edges of the course width the cockpit instrumentation will show full-scale deflection from the on-glide slope indication. The height at which the glide slope crosses the runway threshold is known as the reference datum height (RDH). As with the localizer signal, in the horizontal plane, the glide slope signal is calibrated within ±8° of the extended runway centerline out to 10 nm from the touchdown point (Figure 7.33b).

The carrier wave for the CSB and SBO modulations is in the UHF frequency range 329.15–335 MHz with 40 discrete channels made available in this band. The glide slope antenna is placed at the side of the runway about one-third of the runway length from the approach end, the correct distance being

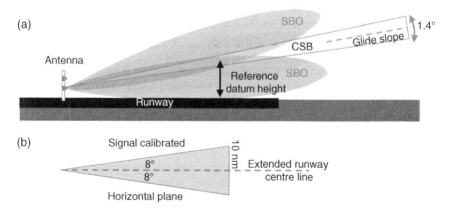

Figure 7.33 (a) The glide slope course signal defined by the null in the SBO transmission is angled up at around 3° from the horizontal with width ±0.7° around the glide slope. (b) In the horizontal plane the calibrated area is ±8° around the runway centerline out to 10 nautical miles (same as for the localizer).

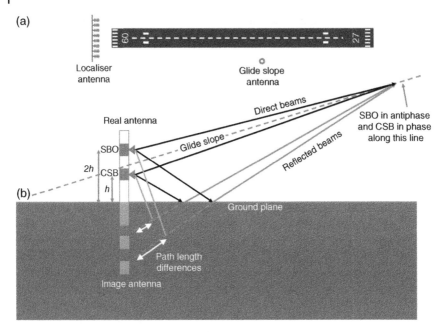

Figure 7.34 (a) Indication of the siting of the localizer and glide slope antennas around a runway. (b) An "image" type glide slope antenna forms the radiation pattern by combining direct and reflected beams from the ground, which is the equivalent of combining emission from an image antenna the same distance below ground. Due to the path length difference, the direct and reflected beams from the CSB and SBO antenna elements combine in phase and antiphase, respectively, along a line (the glide slope).

determined by the required RDH. The most common types of antennas are typically 130 m from the runway centerline and an indication of the positioning of antennas in a normal ILS is shown in Figure 7.34a.

In the most common glide slope antennas, referred to as image systems, the radiation lobes are formed by combining the direct beams from the antennas with the reflected beams from the ground. This gives the same pattern as the emission from an antenna array consisting of the real antennas plus "image" antennas the same distance below ground as illustrated in Figure 7.34b. If the CSB and SBO antennas are placed at different heights, the reflected and direct beams from the upper and lower antennas will recombine with different phase-shifts due to the unequal path length differences. This effect can be exploited to ensure that the SBO direct and reflected beams are in antiphase giving the required null along the glide slope while the CSB signals are in phase, which requires that the upper antenna is at twice the height of the lower one. The in-phase antiphase combination will occur along a locus of points, which form a straight line (the glide slope) and the angle of the glide slope for a given carrier frequency is fixed by the heights of the antennas. For example, for a 335 MHz carrier, the antenna heights need to be at 4.25 and 8.5 m. An undesirable aspect of this design is that a second null in the SBO signal occurs at about twice the angle of the primary reference producing a false glide slope. Since there are no false glideslopes underneath the main one this problem can be circumvented by the simple expedient of always intercepting the glideslope from below, which is the case in all published approach procedures.

The system described is the so-called null reference antenna but there are more complex designs involving three antenna elements, that is, the Marray type (Figure 7.35b) but still use the image technique. One problem with using reflection is that it demands quite a large area of stable ground around the antenna tower and deep snow can cause multiple reflections that nearly cancel the reflected signal. For airports where a large flat area is not available, a different design referred to as an end-fire antenna can be used. This uses horizontally spaced antennas to provide the correct phasing between the CSB and SBO

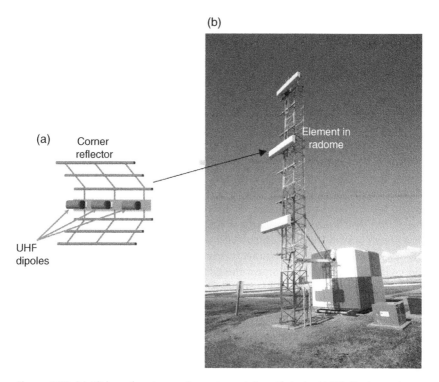

Figure 7.35 (a) Glide path antenna element consisting of stacked UHF dipoles and corner reflector emitting a directed beam with horizontal polarization. Antenna (b) Marray-type glide path antenna. *Source:* Reproduced with permission of NAV CANADA.

signals. A typical antenna element for an image system is shown in Figure 7.35a and consists of a number of UHF dipoles mounted horizontally and a corner reflector providing a directed beam with horizontal polarization. A three-element Marray type antenna is shown in Figure 7.35b.

7.4.3 ILS Cockpit Instrumentation

Since ILS localizers emit horizontally polarized radiation in the same band as VORs, it is possible to use the VOR antenna to receive the localizer signals and this is done on light aircraft, whereas in commercial aircraft the antennas are normally separate for redundancy. The UHF glide slope antenna is normally a horizontal quarter-wave Marconi type. In some airliners, the ILS and glide slope antennas are at the front inside the radome of the weather radar (see Figure 4.24).

In the cockpit, the same receiver as for the VOR is used and it recognizes when tuned to an ILS localizer frequency. It then automatically tunes in the corresponding UHF frequency for the glide slope, measures the difference in the 90 and 150 Hz modulation amplitudes in the VHF and UHF signals, and drives the cockpit display accordingly. In the case of electromechanical displays, the vertical needle shows deviation from the runway centerline with full-scale deflection corresponding to 2.5° off center, that is, a factor of 4 increase in sensitivity compared to the VOR reading as illustrated in Figure 7.36a. When an ILS frequency is chosen, the display also activates a second, horizontal needle, which indicates above or below the glideslope with full-scale deflection in either direction corresponding to 0.7° off the glide slope as illustrated in Figure 7.36b. In this mode, the OBS knob is ineffective and turning it to rotate the OBI has no effect on the needle deflection, though it is generally good piloting practice to rotate the OBI to read runway direction. Note how for both the localizer and the glide slope a given angle off course corresponds to a smaller distance error closer to the runway so that progressively smaller corrective inputs by the pilot are required as the aircraft approaches the runway.

In a glass cockpit, the CDI on the HSI display normally used for indicating VOR or GPS courses and deviations is driven to indicate the angle off the centerline with full-scale deflection on the CDI corresponding to 2.5°. When the ILS frequency is selected, the PFD activates an additional marker and scale next to the altitude moving tape display that shows the deviation off the glide slope with full-scale deflection corresponding to 0.7° above or below. Figure 7.37a shows the instrument approach plate for the ILS approach into Newquay airport, United Kingdom (EGHQ) and Figure 7.37b shows a photograph of the Garmin G1000 glass cockpit during an approach by a light aircraft with the localizer and glide slope displays indicated. The frequency on the radio navigation tuner is set to the ILS frequency at Newquay (110.5 MHz) as shown. The centered green CDI needle on the localizer display and the centered green diamond

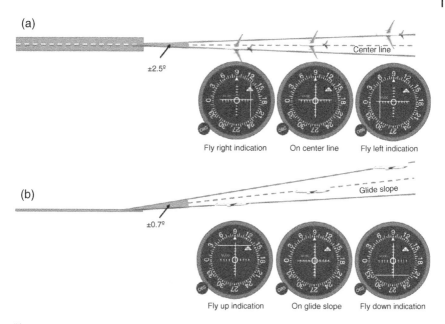

Figure 7.36 Electromechanical cockpit displays when the receiver is tuned to an ILS frequency. (a) Full-scale deflection of the localizer needle corresponds to 2.5° off the centerline. (b) Full-scale deflection of the glide slope needle corresponds to 0.7° off the glide slope.

on the glide slope display show that the aircraft is stabilized along the center of the approach path horizontally and vertically at the positions indicated by the red aircraft symbols on the plate. Note the large wind correction angle required to maintain the course to the runway due to the strong cross wind indicated on the PFD.

7.4.4 Categories of ILS

The ILS is an example of a *precision* approach aid, that is, it provides guidance in both azimuth and height. This is distinct from a *non-precision* approach such as the VOR procedure shown in Figure 7.18, which provides guidance in azimuth only. Other examples of precision approach aids are the MLS described in the following section, guidance by GNSS (see Chapter 8) and precision approach radar. There are strict rules set by the ICAO regarding the minimum altitude an aircraft can descend to following precision and non-precision aids under IMC conditions. Thus, for example, if the glide slope transmission fails on an ILS system it changes status from a precision to a non-precision landing aid and the corresponding minimum approach altitude increases accordingly.

Fully operational ILS installations are classified according to the certified minimum approach altitude and visibility at which they can be used. In the case of precision approach aids, the minimum altitude is specified by a decision height

(a)

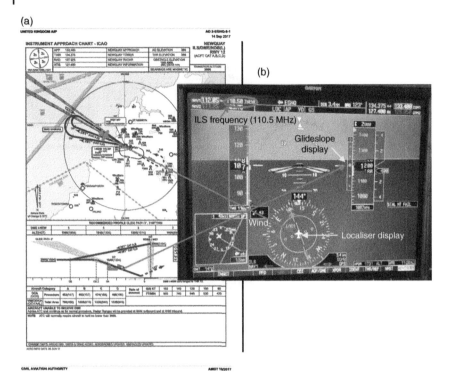

(b)

Figure 7.37 (a) ILS approach plate into Newquay airport, United Kingdom (EGHQ). (b) Photograph of the PFD in a light aircraft during an ILS approach into Newquay at the positions of the red aircraft symbols. The navigation radio is tuned into the VHF frequency of the localizer (110.5 MHz) (highlighted). Tuning to an ILS frequency causes the green CDI needle to show deviation off the centerline with full-scale deflection corresponding to 2.5°. In addition, a glide slope display has been activated next to the altitude moving tape with the green diamond showing deviation away from the glide slope with full-scale deflection corresponding to 0.7°.

above the runway threshold (DH), which is the height that an aircraft can descend to under IMC conditions following the approach before the pilot becomes visual with the runway. If the runway is not visible by the DH an immediate go-around must be initiated. Table 7.2 shows the decision heights and visibilities that pertain to different categories of ILS. The visibility is specified in

Table 7.2 ILS categories DH and RVR.

Category	I	II	IIIA	IIIB	IIIC
DH	≥200 ft	≥100 ft	<100 ft if any	<50 ft if any	0
RVR minimum	≥550 m	≥350 m	200–300 m	200–75 m	0

terms of the runway visual range (RVR), which is the distance over which a pilot of an aircraft on the centerline of the runway can see the surface markings delineating the runway or identifying its centerline.

Normally in commercial flights, landing under IMC conditions is carried out by the autopilot driven by the ILS. It is seen that the highest category of ILS (IIIC) enables landing under zero visibility conditions (dense fog) but very few airports have this facility. In addition to use it under zero visibility conditions the aircraft and the pilot have to be certified as well as the airport installation. Even if the landing can be carried out safely there remains a problem of how to taxi the aircraft to the stand so, in practice, dense fog will even close down airports with Class IIIC ILS installations.

7.5 Microwave Landing System (MLS)

ILS has been the dominant precision landing aid for nearly 70 years but it suffers from a number of shortcomings. These include its inflexibility in changing approach angles and paths, interference with the transmitted radio beams by taxying aircraft, and distortion of beams by aircraft on approach making it necessary to separate approaching aircraft, which limits the traffic density. In order to overcome some of these problems a precision landing system operating in the SHF band frequency range 5031–5090 MHz (similar to radar) was developed in the late 1970s and is known as the MLS. It is far more flexible, able to accommodate larger traffic volumes, and less sensitive to interference than ILS. Installations at civil airports accelerated after 2000 and it was assumed that MLS would be the replacement for ILS. However, the technology only had a brief life span, at least in civil aviation, as GNSS with augmentation (see Chapter 8) provides all the advantages of MLS but without the need for large-scale installations at airports. Most systems have now been decommissioned and here the basic principles are described briefly as MLS is largely an obsolete technology.

There is a 300 kHz channel spacing in the MLS band providing 200 channels (compared to 40 for ILS). The transmissions from the airport installation consist of two scanned fan-shaped beams, one for positioning in azimuth and one for elevation as shown in Figure 7.38. Guidance for the runway centerline is provided by a 15° wide beam transmitted by an antenna at the stop end of the runway scanning backward and forward horizontally through ±40° labeled the TO and FRO beam (Figure 7.38a). The other beam, which is 80° wide, is transmitted by an antenna near the runway threshold and scans up and down through 0–15°, providing guidance for the glide slope (Figure 7.38b). Both beams are usable out to a distance of 20 nm and a height of 20 000 ft, which is further and higher than ILS. The transmitted signals are digital, modulated by binary phase-shift keying (BPSK – see Section 4.3.7.2), and contain timing signals

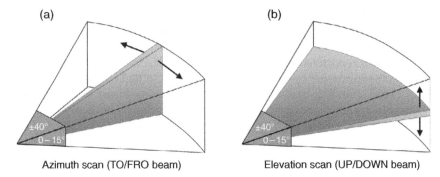

(a)

(b)

Azimuth scan (TO/FRO beam)　　　　Elevation scan (UP/DOWN beam)

Figure 7.38 (a) Azimuthal fan-shaped beam providing runway centerline guidance in the MLS. (b) Elevation fan-shaped beam providing glide slope guidance in the MLS.

among other data. Typical scan times are 15 ms for one sweep of the azimuth beam and 5 ms for the elevation beam.

The principle of operation is illustrated for the azimuth beam in Figure 7.39. If the aircraft is on the runway centerline there will be a specific time between the TO beam and FRO beam detected at the aircraft. If the time is shorter than this standard time the aircraft is on one side of the centerline and if it is longer the aircraft is on the other side. Simply timing the gap between pulses arriving at the aircraft receiver antenna gives the angle off center with no other reference. The glideslope information is provided using a similar principle with an UP and DOWN beam. This general method is known as time-referenced scanning beam (TRSB) location.

The instrumentation in the cockpit consists of the tuner/controller, which is set to the required channel in the 5031–5090 MHz band and the pilot can also select the required glideslope angle and the desired approach azimuth. That is, the approach does not have to be along the runway centerline with a 3° glideslope. Once these are chosen, the electronics drives the same displays as used in ILS. The system also has a built-in DME, which operates on a similar principle to that described in Section 7.3, but provides a higher precision and gives distances to within 30 m, which is required for CAT II/CATIII approaches. The digital message embedded in the transmitted beams includes the station identification, the condition of the system, the runway condition, and weather information.

MLS can circumvent interference problems that plague ILS as the transmissions can be interrupted to avoid reflections from buildings, terrain, and vehicles. Due to the lack of interference from aircraft on the approach it has been demonstrated to enable a higher traffic density. In addition, each aircraft can choose its approach angle and glide path from a distance of 20 nautical miles and a height of 10 000 ft, which maximizes utilization of runways. This makes

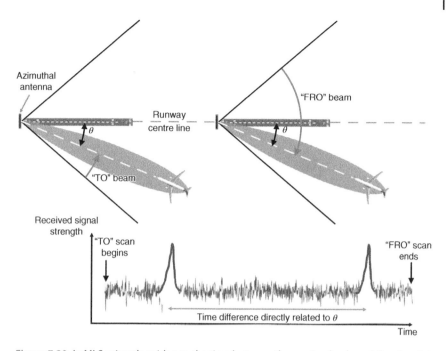

Figure 7.39 In MLS azimuth guidance, the time between the received pulses at the aircraft is used to calculate the angle off the centerline.

MLS particularly suitable for airports at which traffic requiring different approaches is operating, for example, helicopters mixed in with fixed wing traffic that includes short takeoff and landing (STOL) aircraft. This is the situation at many military airports and the few remaining MLS installations are to be found at this type of airport.

Problems

1 Calculate the size of the rectangle that an aircraft has to pass through to be within half-scale deflection of the localizer and glide slope cockpit indicators at the decision height in a class II ILS approach.

2 There is a VOR sited at Daventry with coordinates N521049 W0010650. Calculate the radial and distance one would need to enter into an RNAV receiver to place a virtual VOR on the runway at Leicester airport with coordinates N523628 E0010151. The magnetic variation in the region is 2°W.

3   An aircraft is flying West at 120 knots and is waiting to pick up the 000 (Northerly) radial from a VOR. When the DME tuned to the same VOR reads 29 nautical miles the VOR display in the cockpit is as shown in the figure. How long must the aircraft continue to fly before crossing the 000 radial?

4   The figure shows the VOR cockpit displays in three aircraft flying approximately toward a beacon. In all three aircraft, the pilot in each has set the OBS to 045 with the TO flag showing so the intention is to track 045 to the VOR. Aircraft A is exactly on the 225 radial so the CDI is centered.

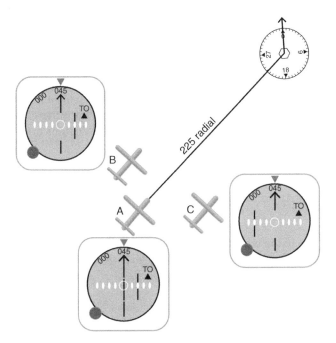

a) On which radial FROM the VOR is aircraft B?

b) On which radial FROM the VOR is aircraft C?

References

1 www.airbattle.co.uk/b_research_3.html (accessed 12 June 2018).
2 http://www.aps.org/publications/apsnews/200604/history.cfm (accessed 12 June 2018).
3 Morgan, D.J. (1987). Electronic Null-Seeking Goniometer for ADF. US patent 4654664, 31 March. http://www.freepatentsonline.com/4654664.pdf (accessed 12 June 2018).
4 Fred the Oyster. https://en.wikipedia.org/wiki/Non-directional_beacon (accessed 12 June 2018).
5 Andrew Alford (1942). Antenna system. US Patent 2283897A, 26 May.
6 Miet, P.L. (2016). *The technical wizardry of VORs – how they work.* https://youtu.be/tZ2gG1v9Xg8 (accessed 12 June 2018).
7 ICAO (2000). Manual on the Testing of Radio Aids, 4, vol. **1**, Doc 8071. www.caa.lv/file/935/280 (accessed 12 June 2018).
8 Chen, K. (2012). cocoaNEC2.0 software.
9 Siting criteria for instrument landing systems. Federal Aviation Authority Document 6750.16E, 10 April 2014. https://www.faa.gov/documentLibrary/media/Order/FINAL_SIGNED_Order_6750_16E_ILS_Siting_Criteria_06-09-2014_for_Web_posting[1].pdf (accessed 12 June 2018).

8

Global Navigation Satellite System (GNSS)

Satellite navigation has become the mainstay of all global navigation systems and is an indispensable tool in modern aviation. The convenience of use and relatively low cost of the user equipment has led to the installation of GNSS navigation receivers in virtually all aircraft from gliders to commercial jet transport. Increasingly, procedures are being implemented for aircraft to approach and land using GNSS as well as navigate enroute so that in the near future it is likely that most flights can be conducted under IMC conditions using only satellite navigation. The important requirement for redundancy, however, means that the onboard equipment and infrastructure for all other navigation systems will be maintained including inertial navigation and the radio beacons described in Chapter 7. In practice, in a commercial airliner, the navigational information from all sources is combined using Kalman filtering to produce an instantaneous three-dimensional position fix as described in Chapter 9. In this chapter the infrastructure, signals, and cockpit equipment required for navigation by satellite will be described.

8.1 Basic Principle of Satellite Navigation

A GNSS system employs a constellation of satellites orbiting the Earth so that a vehicle always has line of sight with several of them. Each satellite has an atomic clock on board that is synchronized to the clocks on the other satellites and to atomic clocks on the ground that define "GPS time," which is also synchronized to coordinated universal time (UTC – see Section 8.3.1). Each satellite continuously broadcasts its current orbital position (ephemeris) with each broadcast starting at a precise time referenced to GPS time in a similar manner to ADS-B and UTC (see Section 5.4.11). Imagine for a moment that there was also an atomic clock in the receiver synchronized to that of the satellite. Then knowing the time that the message was sent and the position of the satellite at that time,

Aircraft Systems: Instruments, Communications, Navigation, and Control,
First Edition. Chris Binns.
© 2019 John Wiley & Sons, Inc. Published 2019 by John Wiley & Sons, Inc.
Companion website: www.wiley.com/go/binns/aircraft_systems_instru_communi_Navi_control

both of which are contained in the message, the time for the signal to reach the vehicle and thus the distance to the satellite would be known. A single satellite would define the position of the vehicle as somewhere on the surface of a sphere (Figure 8.1a). Two satellites would fix the position as somewhere on the circle where the spheres intersect (Figure 8.1b) and three would place the vehicle at one of two points (Figure 8.1c), one near the surface of the Earth and the other out in space, which can be discarded.

The aircraft, however, does not have an atomic clock synchronized to GPS time and although modern quartz clocks can be extremely accurate for normal purposes each microsecond offset from GPS time will produce a 300 m error in the calculated distance. The inaccurate distances obtained without synchronizing the aircraft clock to GPS time are termed *pseudo-ranges*. Bearing in mind that that all the satellite clocks are synchronized, the time offset between the aircraft and satellite clocks is the same for all satellites. Thus, receiving the

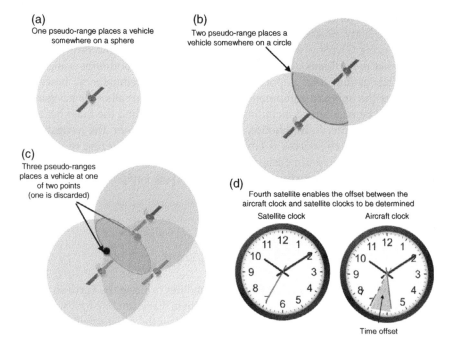

(a) One pseudo-range places a vehicle somewhere on a sphere

(b) Two pseudo-range places a vehicle somewhere on a circle

(c) Three pseudo-ranges places a vehicle at one of two points (one is discarded)

(d) Fourth satellite enables the offset between the aircraft clock and satellite clocks to be determined

Satellite clock   Aircraft clock

Time offset

Figure 8.1 Principle of position determination by GNSS. Pseudo-ranges are determined by timing the signals from a known satellite position (contained in the message) with an unsynchronized clock on the aircraft. (a) One pseudo-range places the vehicle somewhere on a sphere. (b) Two pseudo-ranges place the vehicle somewhere on the circle where the spheres intersect. (c) Three pseudo-ranges place the vehicle at one of two points, one of which is out in space and can be discarded. (d) Receiving a fourth satellite enables the aircraft clock to be synchronized to the satellite clock and convert the pseudo-ranges to absolute distances.

signals from a fourth satellite enables the construction of four simultaneous equations for four variables, that is, the three pseudo-ranges and the time offset between the satellite and aircraft clocks. Knowing this fourth variable allows the aircraft clock to be synchronized to the satellite clocks (Figure 8.1d) and thus the pseudo-ranges can be converted to absolute distances with high accuracy. A by-product of the process is that the GNSS receiver is synchronized to GPS time and thus to UTC with atomic clock accuracy. It is not feasible to solve the four simultaneous equations in real time at the rate at which the signals are received from the satellites and in practice preprogrammed filters are used to stream out positional data in response to the input data arriving from the satellites.

The x,y,z coordinates determined by GNSS define a point in space but to relate this to a geometrical position on the Earth's surface in terms of latitude and longitude requires an accurate knowledge of the shape of the Earth such as the WGS84 model described in Section 6.4. The difference between mean sea level and the WGS84 model is up to 50 m, which is the reason why raw GNSS signals cannot be used for precision landing approaches and augmentation is required (see Section 8.7). A GNSS system is comprised of three segments, that is, the space segment, which is the constellation of satellites in orbit; the control segment, which is the ground infrastructure of control stations, monitoring stations, etc.; and the user segment, which are the GNSS receivers in vehicles.

8.2 The Constellation of Space Vehicles (SVs)

The group of satellites or space vehicles (SVs) that are at the heart of a satellite navigation system are described as a constellation and they transmit in the UHF band, so reception will only occur along the line of sight. Thus, the first aspect of the design that needs to be addressed is how many satellites are required and in which orbits to ensure that at least four are visible at all times from anywhere on the Earth's surface. As will be discussed below the more satellites that are in view, the more accurate the position fix and in addition, having more than four in view enables error checking and monitoring of the integrity of the system. The constellation described in detail here is the US Navstar system that is more commonly known as the global positioning system (GPS).

8.2.1 Orbital Radius of the GPS Constellation

The basic GPS system uses 24 satellites that are placed in an orbit such that each satellite makes exactly two rotations about the Earth for one rotation of the Earth about its axis. The relevant Earth rotation period is the time taken to turn once relative to the distant stars, that is, the sidereal day, which is 23 hours, 56 minutes, and 4 seconds (86 164 seconds). It is slightly shorter than the mean

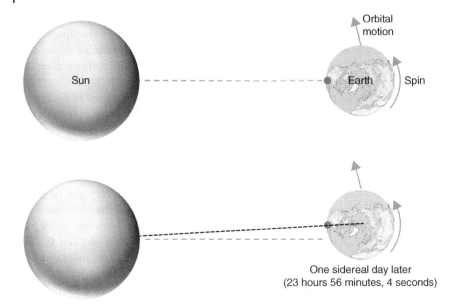

Figure 8.2 Illustration of sidereal day. After rotating once on its axis relative to the distant stars, the Earth has moved in its orbit around the Sun and so has to rotate further for the Sun to appear overhead. Thus, the mean solar day is slightly longer than the sidereal day.

solar day, which is the period between the Sun appearing overhead on consecutive days. The difference is due to the fact that after completing one sidereal rotation, the Earth has moved in its orbit around the Sun so it has to rotate a little further for the Sun to appear overhead as illustrated in Figure 8.2. The Earth is in an elliptical orbit about the Sun so the length of the solar day changes slightly due to the speed of the Earth varying in its orbit, which is why the mean solar day needs to be specified and is the period in common use. Thus, the GPS satellite orbital period is half a sidereal day, which is 11 hours, 58 minutes, and 2 seconds (43 082 seconds).

It is straightforward to calculate the radius of the orbit to achieve the required orbital period. A stable obit occurs when the centripetal force equals the gravitational attraction between the Earth and the satellite, that is:

$$\frac{GM_\oplus m_s}{r^2} = \frac{m_s v^2}{r} \tag{8.1}$$

where $M_\oplus$ is the mass of the Earth, m_s is the mass of the satellite, v is its velocity, r is the radius of the orbit, and G is the gravitational constant ($= 6.67408 \times 10^{-11}$ m^3 kg^{-1} s^{-2}). Equation (8.1) simplifies to:

$$v = \sqrt{\frac{GM_\oplus}{r}} \tag{8.2}$$

The length of the orbit is $2\pi r$ and thus the period is $\tau = 2\pi r/v$, that is:

$$\tau = \frac{2\pi r^{3/2}}{\sqrt{GM_{\oplus}}} \tag{8.3}$$

so, the radius required to achieve a given period is:

$$r = \left(\frac{\tau^2 GM_{\oplus}}{4\pi^2}\right)^{1/3} \tag{8.4}$$

Thus, to achieve an orbital period of 43 082 seconds requires an orbital radius of 26 559 km. This is the distance from the center of the Earth but the mean radius of the Earth is 6367 km (see Table 6.1) so the distance of the satellites above ground is 20 192 km. This analysis is simplified as the satellites have a slight eccentricity in their basic orbit, which is also perturbed by the non-sphericity of the Earth, the Lunar and Solar gravitational fields, and radiation pressure from the Sun. All these factors have to be accounted for in order to get accurate positional information from satellites. In addition, the effects of special and general relativity need to be corrected for but these are known very precisely from Einstein's equations.

8.2.2 Orbital Arrangement for Optimal Coverage by the GPS Constellation

Given the orbital radius of the GPS SVs, it is straightforward to calculate the percentage of the Earth's surface that is in view of a single satellite as illustrated in Figure 8.3. Tangents drawn from the satellite to the Earth define the illuminated spherical cap for line of sight transmissions (Figure 8.3a). As illustrated in Figure 8.3b, the perimeter of the cap and Earth radii define a cone with a spherical cap, which subtends a solid angle at the center of the Earth, Ω, given by:

$$\Omega = 2\pi(1 - \cos\theta) \text{ steradians} \tag{8.5}$$

where 2θ is the linear angle of the cone apex. If R_O is the radius of the satellite orbit and $R_{\oplus}$ is the radius of the Earth, then:

$$\cos\theta = \frac{R_{\oplus}}{R_O} \tag{8.6}$$

and since the solid angle of a hemisphere is 2π steradians, the proportion, P, of the hemisphere illuminated by a satellite is $\Omega/2\pi$, that is:

$$P = 1 - \frac{R_{\oplus}}{R_O} \tag{8.7}$$

For the GPS system, $R_O = 26\,559$ km and we know that $R_{\oplus} = 6367$ km (average), so $P = 76\%$. From any point on the Earth, the maximum amount of time that a satellite is visible each day is given by half the length of a sidereal day

(a)

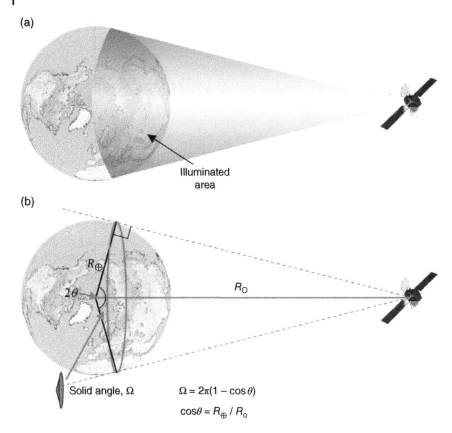

(b)

$$\Omega = 2\pi(1 - \cos\theta)$$

$$\cos\theta = R_\oplus / R_0$$

Figure 8.3 (a) Portion of the Earth's surface illuminated by line of sight satellite transmissions. (b) Solid angle subtended by the cone with a spherical cap defined by the circular perimeter of the illuminated area and Earth radii.

multiplied by $P/2$, that is, $0.25 \times 0.76 \times$ (23 hours, 56 minutes, and 4 seconds) or 4.6 hours. In practice, satellites very close to the horizon may be unusable due to interference by terrain so an average of about four hours a day is a more realistic value.

Beermat Calculation 8.1

The orbital height above the ground of the International Space Station (ISS) is 408 km. Calculate the proportion, P, of the Earth's hemisphere in view from the space station.

Applying Equation (8.7) gives:

$$P = 1 - \frac{6367}{6367 + 408} = 1 - 0.94$$

That is, $P = 6\%$.

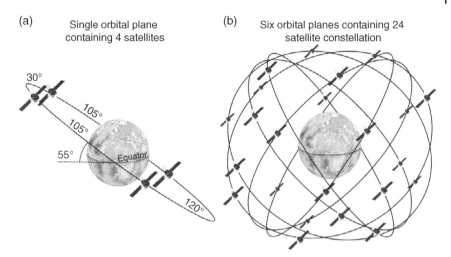

(a) Single orbital plane containing 4 satellites

(b) Six orbital planes containing 24 satellite constellation

Figure 8.4 (a) Each orbital plane is aligned at 55° to the equator and contains four satellites with the angular spacings shown. (b) There are six orbital planes separated by 60° containing the constellation of 24 satellites.

The arrangement of satellite orbits that provides the optimum number of satellites visible at one time from any point on the Earth is shown in Figure 8.4. There are six orbital planes spaced 60° apart and each orbital plane, containing four satellites, is oriented at 55° with respect to the equator. The distribution of satellites is not even within each plane and angles between them are shown in Figure 8.4a, that is, 30°, 105°, 120°, and 105°. The average number of satellites visible at any time from a point on the surface will be 12 times the illuminated proportion of a hemisphere by a single satellite (P), that is, $12 \times 0.76 = 9$ satellites. Due to various factors the actual number visible within the constellation will vary between 6 and 12 so that the minimum of four required for a position fix should be permanently available with some redundancy.

The Navstar GPS system, built by the United States, was the first operational GNSS but there are now several other GNSS constellations put into operation by other countries and these are summarized in the table in Figure 8.5 along with their orbital period and radius. The diagram in Figure 8.5 shows the orbits of the various constellations, which are all mostly in medium Earth orbit (MEO) and the Earth drawn to scale. The Russian system, GLONASS, which started launching in 1982 is complete and provides global coverage. The Chinese system, BEI DOU, had its first launch in 2000 and is complete but provides regional coverage around China. Galileo is a European system that had its first launch in 2011 and was declared operational at the end of 2016 but the full constellation will not be complete till mid-2018 so there are still holes in the coverage. India

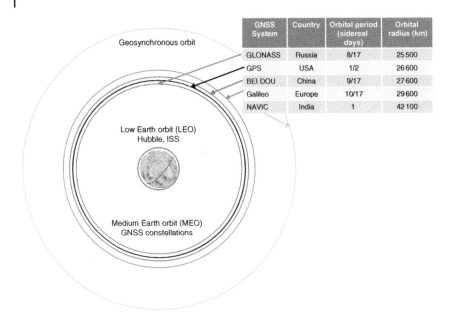

GNSS System	Country	Orbital period (sidereal days)	Orbital radius (km)
GLONASS	Russia	8/17	25 500
GPS	USA	1/2	26 600
BEI DOU	China	9/17	27 600
Galileo	Europe	10/17	29 600
NAVIC	India	1	42 100

Figure 8.5 The various GNSS constellations that are operational and provide global or regional coverage.

also put a regional system called NAVIC (Sanskrit for sailor) into operation with the final satellite launched in August 2017. There are other partial systems such as the Japanese Quasi-Zenith 4-satellite constellation, which provides regional augmentation to other GNSS constellations. Modern GNSS receivers are designed to interpret signals from multiple constellations so that as many satellites as possible are made available for position fixing.

A satellite that has an orbital period of exactly one sidereal day, which requires an orbital radius of 42 100 km is said to be in a geosynchronous orbit. In general, when viewed from the Earth's surface it will trace out a figure of eight pattern in the sky and return to a fixed position overhead at the same time each day. A special case of a geosynchronous orbit is when the plane of the orbit is aligned with the Earth's equator in which case the satellite will appear to stay at one fixed point in the sky so that an Earth-based receiver dish can point to one position without having to track. Some of the constellations use more than one type of orbit, for example, the NAVIC system, which has three satellites in a geostationary orbit and the remaining four in geosynchronous orbits.

The maintenance of a GNSS system requires the commitment to keep launching satellites indefinitely as they have a finite lifetime. This is estimated to be about 7–8 years, due to various possible fail conditions and the exhaustion

Block	Launch period	Successful	Failed	Currently in operation	Channels
I	1978–1985	10	1	0	L1,L2
II	1989–1990	9	0	0	L1,L2
II A	1990–1997	19	0	2	L1,L2
II R	1997–2004	12	1	12	L1,L2
II RM	2005–2009	8	0	7	L1,L2C
II F	2010–present	11	0	11	L1, L2C, L5
III A	scheduled	—	—	—	L1C, L2C, L5
Totals		69	2	32	

Figure 8.6 Table showing the satellites launched in the GPS constellation over the years. The image shows the launch of 24th Navstar SV of the GPS system, one of the Block II A series, on top of a Delta II rocket in 1994.

of the fuel required to adjust the satellite orbit. The first trenche of Navstar satellites, now called Block I, was launched from 1978 to 1985 and none of these are operational. Since then, several more trenches or blocks have been launched and are summarized in the table in Figure 8.6, which shows that since 1978 a total of 69 launches has been required to maintain a viable constellation, which currently comprises 32 satellites. The image in Figure 8.6 shows the launch of the 24th satellite in 1994, which was part of the Block II A series. The GPS system originally transmitted navigation information at two frequencies (channels) labeled L1 and L2, but over the years improvements have been made and capabilities enhanced and in the latest generation of satellites a third channel (L5) has been introduced. These are described in more detail below.

8.3 Transmissions by the GPS SVs

In this section, the format of the transmissions sent by SVs that enable position fixing by the user segment will be explained in detail for the GPS satellites but other constellations use a similar system. In all cases the modulation of the carrier is by BPSK but the information is contained in a complex hierarchy. The full description of the satellite signals is given in Ref. [1].

8.3.1 GPS Time and UTC

The atomic clocks on the satellites are referenced to UTC but are offset from it by an integer number of seconds. GPS time is maintained by the control segment and is continuous while UTC is monitored by the US Naval Observatory who from time to time add a leap second to UTC to compensate for small changes on the rate of rotation of the Earth. There is a very gradual reduction in the rate of rotation over geological timescales due to tidal friction with the Moon but larger fluctuations on smaller timescales occur and are thought to be due to changes in the internal structure of the Earth. GPS time has remained unchanged since it was synchronized to UTC at the midnight changeover from 5 January to 6 January in 1980 but since then 18 leap seconds have been added to UTC. Although all timing is done in terms of GPS time, the navigation message transmitted by the satellites contains the information to synchronize UTC to GPS time. The point of synchronization at the above date is time zero in GPS time, which has been incrementing steadily since then. The units used for operational reasons are the week number (WN) and units 1.5 s long, whose count since the start of the current week is called the time of week (TOW). Thus, at midnight on Saturday, 5 January 1980 the time was (WN = 0, TOW = 0) and since then the TOW has been incremented to a count of 403 200 before being reset to zero at midnight on the following Saturday at which point the WN is incremented by 1. Both the TOW and WN are transmitted in the navigation message as described in Section 8.3.5.

8.3.2 Transmission Channels

The timing on an SV is provided by four exceptionally accurate atomic clocks, two using cesium as the standard and two using rubidium. The international standard for the definition of one second is 9 192 631 770 cycles of the photons in the microwave region emitted by the transition of an electron between the hyperfine split six seconds levels of a Cs-133 atom. The frequency of an atomic transition such as this provides an extremely stable time reference as it is not influenced by environmental factors such as pressure and temperature. A practical clock is produced by locking the frequency of a microwave cavity to 9.19263177 GHz (λ = 3.26 cm) using this transition. This is done by using a magnetic field to separate the excited Cs atoms from those in the ground state and optimizing the flux of the excited atoms by tuning the frequency of the microwave generator. The locked frequency is then used to control a quartz oscillator to compensate for environmental factors. Cs clocks have a demonstrated stability of two parts in 10^{14} (or one second in 1.4 million years). The clocks on all the satellites in a constellation are synchronized to each other and to atomic clocks on the surface via the ground stations in the control segment (see below). The entire system is thus keeping GPS time with atomic clock precision and is referenced to UTC with the same accuracy.

The clocks are used to produce a fundamental frequency, f_0 = 10.23 MHz from the quartz oscillator with the exceptionally high stability described above and the fundamental is used to generate the carrier frequencies of the three navigation channels currently in use. These are:

L1 channel:	$1.57542\,\text{GHz} = 154\,f_0$.
L2 channel:	$1.22760\,\text{GHz} = 120\,f_0$.
L5 channel (since 2010):	$1.17645\,\text{GHz} = 115\,f_0$.

The L5 channel, also called the safety of life channel, was introduced relatively recently and has been available from block II F satellites, beginning in 2010. The L1 and L2 channels are modulated by two so-called pseudo-random noise (PRN) binary codes labeled as the coarse acquisition (C/A) and precision (P) codes mixed with a navigation message as illustrated in Figure 8.7. The PRN codes are not truly random and are deterministic repeating bit streams but they have acquired the label because they appear random and do not carry digital information in the bit pattern. The bit pattern can be used, however, to identify the satellite transmitting the PRN code. In the original scheme, the C/A code was freely available for public use while the precision code was reserved for military use.

The C/A code is transmitted at a rate of 1.023 Mbps and is a repeating binary sequence of 1023 bits, which is different for each satellite, transmitted every millisecond. This code is mixed with a navigation message transmitted at 50 bps by modulo-2 addition and then modulated onto the L1 carrier by BPSK.

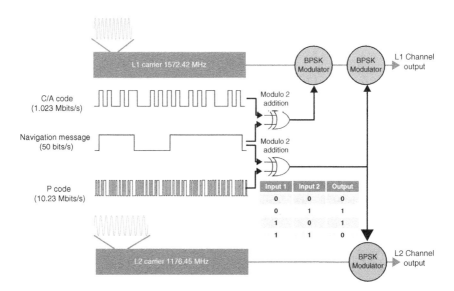

Figure 8.7 Modulation of the L1 and L2 channels by the C/A and P PRN codes and the navigation message in the original scheme for GPS. Modulo 2 addition is achieved using an EX-OR gate whose truth table is shown.

Modulo-2 addition can be achieved by an exclusive-OR (EX-OR) gate whose truth table is shown in Figure 8.7 and reveals that if adding the inputs produces an even number (00, 10) the output is "0" while if the inputs are an odd number (01) the output is "1." The precision code is transmitted at a rate of 10.23 Mbps and is a much longer binary sequence taking a week to transmit. This is also mixed with the navigation message by modulo-2 addition and the result modulated onto both the L1 and L2 channels. Thus, a receiver designed to use the P code can receive the same navigation message at two different frequencies. This is important for removing errors due to ionospheric propagation as explained below and it also provides some protection against interference so that if one frequency is blocked the other is still available. There is now the capability to encrypt the P code, in which case it is called the Y code so it is generally referred to as the P(Y) code.

Since the C/A and P(Y) PRN codes carry no information apart from a means of identifying a specific satellite, the bits are often relabeled "chips" to signify that they are a binary stream whose values do not represent data. Since 2005, a separate civilian code has been added to the L2 channel (known as the L2C code) so that general users can take advantage of position fixing at two different frequencies. A separate L5 channel at a different frequency was added in 2010 carrying additional information. The description below will focus on the legacy system in which the C/A code and navigation message is transmitted only on the L1 channel as this is still the main source of civilian navigation information. The additional GPS signals added to enhance the system will be described in Section 8.3.7.

8.3.3 Construction of the C/A Code

The C/A PRN code is a sequence of chips generated by a process originally invented by Robert Gold [2] for multiplexed bit streams, which have special properties useful for extracting bit patterns from noisy data and distinguishing codes from different sources. In the case of the GPS C/A code, the generation is by a two-step process, whose first stage is the same for all satellites and is illustrated in Figure 8.8. At the start of a 1023-chip sequence taking exactly 1 ms, a 10-bit shift register is first loaded with "1"s and then at each clock cycle $(1/1.023 \times 10^6 = 0.977517\ \mu s)$, the EX-OR result (modulo-2 addition) of chips 3 and 10 is placed into position 1 and all other bits move one place to the right. The chip that is then in position 10 is fed into the second stage of the C/A code generator. Each stage is often described as a polynomial and within this notation the polynomial for stage 1 is defined as $1 + x^3 + x^{10}$ though the mathematical manipulation is not a polynomial in the normal sense but rather the exponent defines which bit is summed. The result of the first 12 iterations of the stage 1

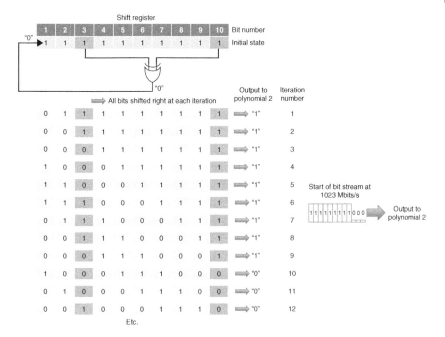

Figure 8.8 Generation of chip stream using polynomial 1, which is independent of satellite number.

polynomial are shown in Figure 8.8, which reveals that the start of the chip stream output to polynomial 2 is 11111111000 at a rate of 1.023 Mbps.

The second-stage polynomial is dependent on the satellite number and as an example, the C/A code generation by satellite 1 will be described. In this case the polynomial is defined as $1 + x^2 + x^3 + x^6 + x^8 + x^9 + x^{10}$, that is bits, 2, 3, 6, 8, 9, and 10 are summed by modulo-2 addition. The bit or chip that results from this is fed into the first position of a 10-bit shift register with all other chips moving one place to the right as illustrated in Figure 8.9. As with the shift register using polynomial 1, all chips are set to "1" at the start of a 1023-chip cycle and then on each iteration the EX-OR result from positions 2 and 6 of the shift register is EX-ORed with the output from polynomial one at the same iteration to produce an output chip to the PRN that is broadcast. The result for the first 10 iterations is shown and the first 10 chips for the 1023-chip PRN code from satellite 1 is $1100100000_2 = 1440_8$.

After exactly 1 ms when 1023 chips of the PRN code have been transmitted, the two shift registers are reset to all "1" and the process repeats. By changing which two bits of the polynomial 2 shift register are EX-ORed together, a different repeating code will be generated and this is how the unique code for each

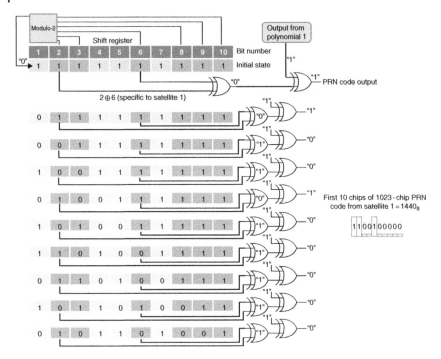

Figure 8.9 Generation of the first 10 chips of the PRN code broadcast from satellite 1.

satellite is produced. Table 8.1 shows the EX-OR combination from shift register 2 for each satellite and the resulting first 10 bits of its PRN code. The PRN codes, although appearing similar to a random stream of chips are deterministic sequences, unique to each satellite. They are the heart of the timing for GPS and can be regarded as a precisely ticking clock with a 1 ms beat in which there is a means of identifying which satellite is broadcasting the time sequence. At the speed of light, a single 1 ms cycle corresponds to 300 km and a single chip within a PRN code to 293 m, which shows that the time measurement has to be performed over a period significantly shorter than a chip.

8.3.4 Multiplexed Decoding of the Navigation Message

The Gold code scheme used to generate the PRN codes may seem overly complex but it is carefully designed to extract multiple signals with high fidelity from noisy data, which is exactly what is required for the reception of GPS signals. All the PRN codes are transmitted at the same L1 frequency and have to be extracted from the carrier. The satellites transmit about 25 W into the L1 band with an antenna gain of 13 dB (approximately ×20). This means that a

Table 8.1 Satellite-dependent EX-OR combinations of bits in register 2 to produce unique PRN code for each satellite and the first 10 chips of the code.

SV No.	Shift register 2 bits combined	First 10 chips output (octal)
1	$2 \oplus 6$	1440
2	$3 \oplus 7$	1620
3	$4 \oplus 8$	1710
4	$5 \oplus 9$	1744
5	$1 \oplus 9$	1133
6	$2 \oplus 10$	1455
7	$1 \oplus 8$	1131
8	$2 \oplus 9$	1454
9	$3 \oplus 10$	1626
10	$2 \oplus 3$	1504
11	$3 \oplus 4$	1642
12	$5 \oplus 6$	1750
13	$7 \oplus 8$	1764
14	$2 \oplus 6$	1772
15	$8 \oplus 9$	1775
16	$9 \oplus 10$	1776
17	$1 \oplus 4$	1156
18	$2 \oplus 5$	1467
19	$3 \oplus 6$	1633
20	$4 \oplus 7$	1715
21	$5 \oplus 8$	1746
22	$6 \oplus 9$	1763
23	$1 \oplus 3$	1063
24	$4 \oplus 6$	1706
25	$5 \oplus 7$	1743
26	$6 \oplus 8$	1761
27	$7 \oplus 9$	1770
28	$8 \oplus 10$	1774
29	$1 \oplus 6$	1127
30	$2 \oplus 7$	1453
31	$3 \oplus 8$	1625
32	$4 \oplus 9$	1712

vehicle at the closest possible distance (directly below) receives a power density of about 10^{-13} W m^{-2} (see Beermat Calculation 8.2) and in most cases the signal will be significantly weaker than this.

Beermat Calculation 8.2

Calculate the received power density in Wm^{-2} of the signal received from a GPS satellite directly above if the transmitter power is 25 W and the transmitter antenna gain is 13 dB.

A satellite directly above will be at a distance, $d = 26\,559 - 6367$ km $= 20\,192\,000$ m.

The power density received, P_D, is thus given by: $P_D = \dfrac{GP_{TX}}{4\pi d^2}$ where P_{TX} is the transmitter power and G is the antenna gain. The linear antenna gain, G, is given by

$$\log_{10} G = 1.3$$

Thus, $G = 19.95$, so putting all values into the equation for P_D gives $P_D = 9.74 \times 10^{-14}$ W m^{-2}.

Gold codes have excellent autocorrelation properties, which means that if we pass a PRN Gold code and the same code with a time offset into an EX-OR NOT combination gate as illustrated in Figure 8.10, a permanent "1" will appear at the output only when the offset is set to zero. If the offset is incremented, the output will remain at "0" for all other times till the offset reaches 1 ms and correlation occurs again. Thus, if we generate within the receiver a template PRN code that matches one of the satellite PRN codes we can lock onto the signal by time shifting the template to achieve a permanent high output. The Gold codes are also

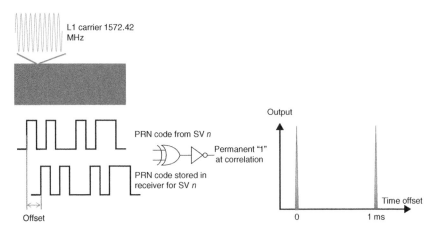

Figure 8.10 Autocorrelation of time-offset PRN codes.

designed to achieve minimum correlation between different PRN codes so there is very little probability that a given PRN template will find correlation with the wrong satellite.

The time offset between the template and the incoming signal to achieve a match is the basis of time measurement in the GPS system. Having locked onto each satellite, the receiver is constantly scanning each template offset by a small amount back and forth to stay locked-on and measure the changing offset due to the motion of the vehicle and the satellite. If the internal clock in the receiver was synchronized to GPS time with atomic clock accuracy, the time offset would simply be the travel time of the signal from the satellite. So, for example, for a typical satellite distance of 30 000 km, this would be 0.1 s and given a satellite orbital speed of about 3900 ms^{-1}, the offset would be changing by up to 12 μs s^{-1}. Having achieved an accurate timing via the PRN code, we need to know the position of the satellite synchronized to the start of the code, which is contained in the navigation message.

When the PRN code is modulated as shown in Figure 8.7 by the 50 bits s^{-1} navigation message, while the navigation code is at "0" the PRN code is unaffected but while the navigation code is at "1" the PRN code is inverted as illustrated in Figure 8.11a. When this signal is received and synchronized to the internally generated PRN code template in the receiver (Figure 8.11b), during a "0" of

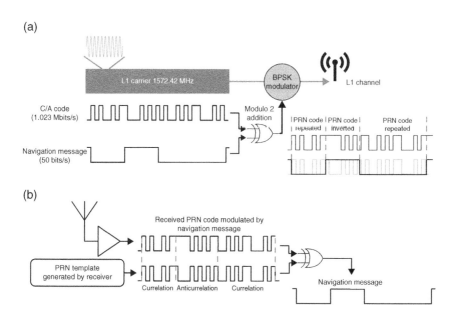

Figure 8.11 (a) Mixing the navigation message with the PRN code as shown produces an output that is the PRN code during a "0" of the navigation message and the inverse of the PRN code during a "1." (b) At the receiver when the PRN template is synchronized with the received-modulated PRN code, the navigation message appears at the output.

the navigation message there is correlation and a "0" is output from the EX-OR sum of the signal and template (note the removal of the NOT gate shown Figure 8.10). Similarly, while the navigation message has value "1," there is anti-correlation and a "1" is output, in other words at synchronization the navigation message appears at the output, otherwise the output remains at "0." Given that a single bit of the navigation message corresponds to 20 complete cycles of the PRN code, it is clear that the navigation message can be extracted with high fidelity. Note that the "1"s and "0"s of the navigation message are labeled bits as they constitute actual data as described below.

The synchronization of the PRN templates generated by the receiver- and the received-modulated PRN codes can be achieved by trial and error. Given the excellent autocorrelation properties of the Gold code and the lack of correlation between different PRN codes, the template for each can be shifted independently until the navigation data stream appears at the output, thus all satellite navigation messages can be received in parallel as illustrated in Figure 8.12. This shows just four satellites, that is, the minimum number for a position fix but clearly the method can be extended to any number. The navigation message contains all the parameters required to determine the position of the satellite at the reference time that the message was sent as described below.

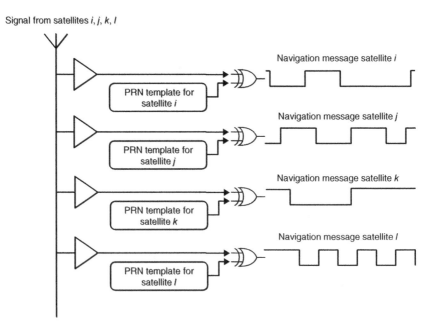

Figure 8.12 Parallel multichannel decoding of the navigation message from a number of satellites.

8.3.5 Format of the Navigation Message

The entire navigation message has a total length of 37 500 bits and is transmitted at a rate of 50 bits s^{-1}, thus taking 12.5 minutes to send. Each message is synchronized to GPS time with atomic clock precision and is internally structured as shown in Figure 8.13. There are 25 frames, each containing 1500 bits, taking 30 seconds to transmit and the frames are further subdivided into five subframes, each containing 300 bits, taking six seconds to transmit. The first three subframes of each frame contain satellite-specific timing and orbital position information while the final two subframes contain almanac parameters for the whole constellation. The data for the almanac is spread through the entire navigation message and thus is refreshed only every 12.5 minutes but generally these parameters do not change except at much longer timescales. The almanac parameters are stored locally on the receiver and unless this data is lost, position fixing can continue frame by frame on a 30-second timescale with the local receiver interpolating if necessary to provide positional updates more frequently. Every piece of the message is precisely synchronized to GPS time and via the time measurement using the PRN code, any specific point in the message, for example, the start of the transmission of the satellite ephemeris data, can be timed to a fraction of a chip in the PRN code.

Each subframe is further subdivided into 10 30-bit words with the most significant bit (MSB) transmitted first and the least significant bit (LSB) last.

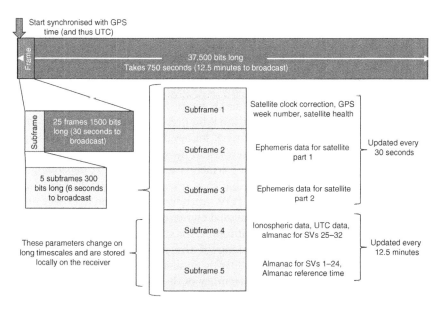

Figure 8.13 Frame and subframe structure within the navigation message.

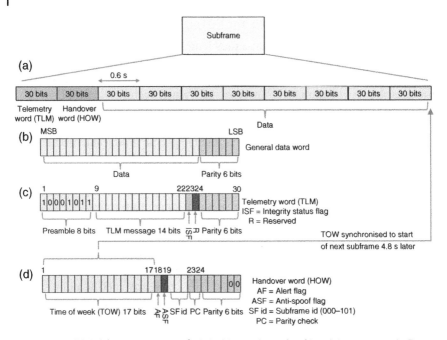

Figure 8.14 (a) Subframe structure of 10 30-bit words, each taking 0.6 s to transmit. Every subframe starts with a telemetry (TLM) word and a handover (HOW) word. (b) The bit structure of a general data word consisting of 24 bits of data and a 6-bit parity block. (c) Bit structure of the TLM word. (d) Bit structure of the handover word.

The first two words are the same for all subframes and are known as the telemetry word (TLM) and the handover word (HOW) as illustrated in Figure 8.14a. The other eight words in each subframe carry data in bits 1–24 and all 10 words have a 6-bit parity block at the end as shown in Figure 8.14b.

The TLM (Figure 8.14c) starts with an 8-bit preamble (which may be inverted depending on the phase of the carrier), followed by a 14-bit TLM message, which is required by authorized users using the precision code. Bit 23 of the TLM is the integrity status flag (ISF), which is set to "0" or "1" depending on the expected level of accuracy and bit 24 is reserved. The first 17 bits of the HOW (Figure 8.14d) represent the TOW introduced in Section 8.3.1, but since it needs 19 bits to fully represent the total count of 1.5 seconds increments in a week (403 200) only the most significant 17 bits are stored in this word. The TOW is referenced to the beginning of the next subframe, eight words (4.8 seconds) later. Bit 18 in the HOW is an alert flag, which if set to "1" is an alert to standard users that the accuracy of the data from the SV is suspect and is used at the consumers' risk. Bit 19 is an anti-spoofing flag, which if set to "1" indicates to authorized users that the precision code is encrypted. Bits 20–22

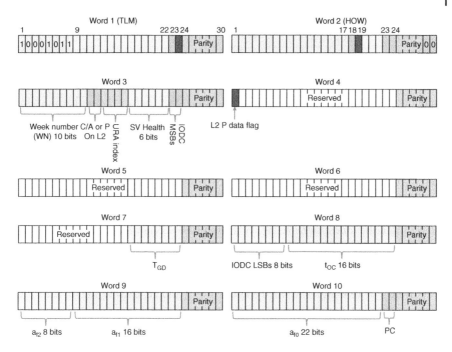

Figure 8.15 Word structure for subframe 1 of the navigation message.

contain the number of the subframe directly in binary, that is, 000 to 101 and the following two bits are set to provide a correct parity check with the last two parity bits 29 and 30 set to zero.

The data contained in subframes 1–3 provides the necessary information to produce a position fix from a specific SV and this is described in detail below. The almanac data distributed through subframes 4 and 5 goes beyond the scope of this book but the details are published in Ref. [1]. The word structure in subframe 1 is shown in Figure 8.15 and as for all subframes, the first two words are the TLM and HOW described in detail above. All words end with a 6-bit parity block and in addition the tenth word has two bits at positions 23 and 24, before the parity block for parity computation, which leaves 190 bits for data. The first 10 bits of word 3 are assigned to the current WN, the next two bits flag whether the C/A code or P(Y) code is being transmitted on the L2 channel (either is possible), and the next four bits provide the user range accuracy (URA) index, N, which varies from 0 to 15 (0000_2–1111_2). If $N \leq 6$, the accuracy is given nominally by $2^{(1 + N/2)}$ meters and if $6 < N < 15$, is given by $2^{(N-2)}$ meters. Thus, for example, if $N = 0$, the accuracy is predicted to be about 2 m whereas if $N = 14$, it is predicted to be about 4 km. A URA index, $N = 15$, indicates the absence of an accuracy prediction and the SV should only be used at the user's discretion.

The next 6 bits describe the health of the SV and if the MSB is set to "0" all navigation data are valid while if it is set to "1" there is a problem and the following 5 bits are set to indicate the nature of the problem. For example, if they are set to "11100" this reports that the SV is temporarily out of service and should not be used on the current pass. This specific code is set during the adjustment of satellite orbits by the control segment. The final 2 bits before the parity block in word 3 are the MSBs of the issue number of the clock correction parameters (IODC), which is described in the detailed description of word 8.

The first bit in word 4 is a flag that is set to "1" if the navigation data has been switched off from the precision code on the L2 channel but apart from this one bit, the data sections of words 4, 5, and 6 are reserved for future capabilities. The first 16 bits of word 7 are also reserved while the final 8 data bits are the group delay time, T_{GD}, which is a parameter that enables users of one frequency only to estimate the ionospheric delay correction (see Section 8.5.2). The first 8 bits of word 8 are the LSBs of the IODC, which with two bits from word 3 make a 10-bit block. The IODC is the issue number of the clock correction parameters and provides a user with a quick determination whether a change in values has occurred. The remainder of word 8 and words 9 and 10 are assigned to the four clock correction parameters, t_{OC} (16 bits), a_{f2} (8 bits), a_{f1} (16 bits), and a_{f0} (22 bits), which are critical to obtaining an accurate determination of the signal transit time. The atomic clocks have a time stability of the order of 1 part in 10^{13} but they are not synchronized with this level of accuracy to GPS time due to a number of factors. If t is the true system GPS time and t_{SV} is the SV time, then we can write:

$$t = t_{SV} - \Delta t_{SV} \tag{8.8}$$

where Δt_{SV} is the correction in seconds required to synchronize the SV clock with system GPS time and varies from SV to SV. The value of Δt_{SV} is modeled as a polynomial of time using the correction parameters in words 8–10 and is written as:

$$\Delta t_{SV} = a_{f0} + a_{f1}(t - t_{OC}) + a_{f2}(t - t_{OC})^2 + \Delta t_r \tag{8.9}$$

where Δt_r is the correction for special and general relativity, which has to be evaluated from the orbital parameters as described in Section 8.6. The clock correction parameters are named clock data reference time (t_{OC}), bias (a_{f0}), drift (a_{f1}), and aging (a_{f2}).

Figures 8.16 and 8.17 show the word structure for subframes 2 and 3, which both start with the usual TLM and HOW words. Word 3 of subframe 2 contains the issue number of the ephemeris data (IODE), which is a convenient way for a user to determine if the ephemeris parameters have changed in a similar manner to the IODC number for the clock parameters in subframe 1. The majority of the rest of the data specifies the ephemeris parameters and any correction factors required to accurately determine the position of the SV at the reference time.

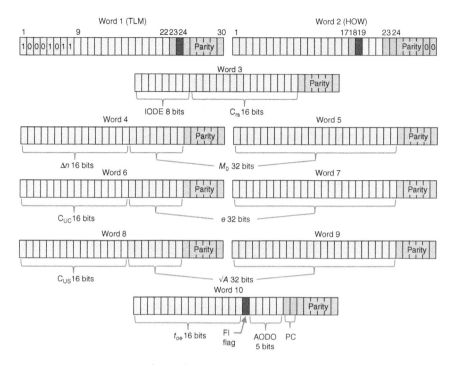

Figure 8.16 Word structure for subframe 2 of the navigation message.

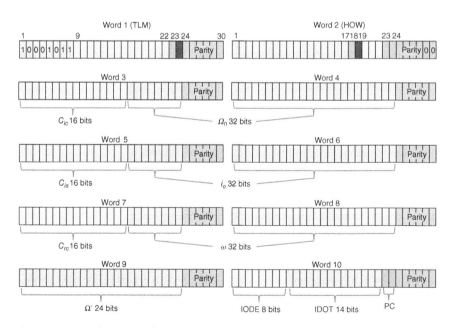

Figure 8.17 Word structure for subframe 3 of the navigation message.

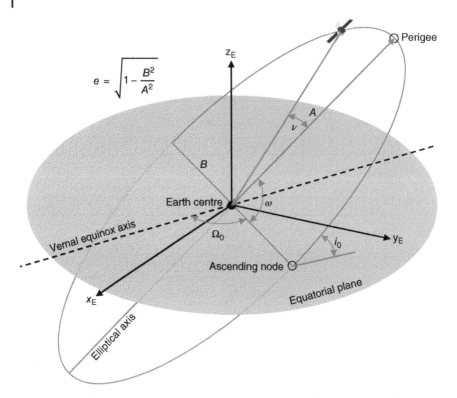

Figure 8.18 Ephemeris parameters, A, e, Ω_0, i_0, ω, and ν that specify the position of a satellite in an Earth-centered Earth-fixed (ECEF) reference frame.

The main ephemeris parameters can be described with reference to Figure 8.18, which shows a specific SV elliptical orbit. The satellite orbits are very close to circular but must be described as ellipses to achieve the desired accuracy of position fixing. The parameters are specified relative to a Cartesian Earth-centered Earth-fixed (ECEF) reference, which is the one drawn in Figure 8.18. The ellipse itself is specified by two parameters, that is, the semi-major axis, A, contained in words 8 and 9, subframe 2, and the ellipticity e, contained in words 6 and 7, subframe 2. Note that for computational convenience further down the line it is the value of $\sqrt{A}$ that is actually transmitted. Having specified the shape, three angles are required to orient the ellipse with respect to the Earth, which is assumed to be at one of the foci. The first is the right ascension, Ω_0, contained in words 3 and 4 of subframe 3, which is the angle between the Vernal equinox and the vector that runs between the Earth and the ascending node, that is, the point at which the orbit crosses the Earth's equatorial plane. The Vernal equinox is the point in the Earth's orbit at which the Sun is directly overhead the equator and the axis that runs from this point to the Sun provides an absolute reference within the Solar System. The second angle is the inclination, i_0, contained in words 5 and 6, subframe 3, which is the angle at which the orbit crosses the equatorial plane. This is nominally 55° as pointed out in the description of the constellation in Section 8.2.2. Finally, the direction of the semimajor axis relative to the vector between the Earth and the ascending node, denoted ω, needs to be specified and this is contained in words 7 and 8 of subframe 3. Words 9 and 10 of subframe 3 also contain the rate of change of Ω and i_0, labeled Ω' and IDOT, respectively, in Figure 8.17.

The parameters A, e, Ω_0, i_0, and ω completely specify the shape and orientation of the orbit, but the position of the spacecraft within the orbit needs to be defined and this can be done by a single angle, ν, relative to the vector from the Earth to perigee, which is called the true anomaly. Note that the word anomaly here is taken from Astronomy to mean the angle relative to perigee as opposed to the normal usage to mean something strange. This parameter is not directly contained within the navigation message but needs to be calculated using the transmitted parameters: the mean anomaly, M_0 (words 4 and 5, subframe 2), the ephemeris reference time, t_{oe} (word 10, subframe 2), e (words 6 and 7, subframe 2), and A (words 8 and 9, subframe 2) using the algorithm described below.

Figure 8.19 shows a plan view of a satellite elliptical orbit with the Earth at a focus and the true anomaly, ν, indicated. If we draw a circle of radius A (the semimajor axis) that is centered on the ellipse we can also define an eccentric

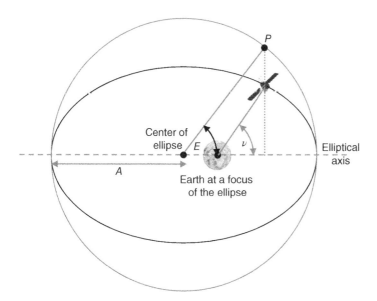

Figure 8.19 Definition of the true anomaly, ν, and the eccentric anomaly, E.

anomaly, E, indicated in Figure 8.19. This is the angle between the elliptical axis and a line from the center of the ellipse to a point, P, on the circle defined by a line that runs perpendicular from the elliptical axis through the satellite. The satellite in the elliptical orbit has a varying angular speed whereas a satellite in a circular orbit would have a constant angular speed. The angle that is transmitted by the satellite is the mean anomaly, M_0, which is the angle between the elliptical axis and a line from the Earth to an imaginary satellite that is in a circular orbit whose angular speed is the same as the *average* angular speed of the real satellite in its elliptical orbit. To obtain the relationship between M_0 and ν, we start with Kepler's third law, that is:

$$T = \frac{2\pi}{\sqrt{\mu}} A^{3/2} \tag{8.10}$$

where T is the orbital period, and $\mu = GM_\oplus$, where G is the gravitational constant. Thus, the mean angular speed (described as the mean motion, n_0, in GPS) is given by:

$$n_0 = \frac{2\pi}{T} = \sqrt{\frac{\mu}{A^3}} \tag{8.11}$$

The value of M_0 is the mean anomaly at the ephemeris reference time, t_{oe}, and the mean anomaly, M, a time, t, later is:

$$M = M_0 + n_0(t - t_{oe}) \tag{8.12}$$

where t is the running GPS time. Although n_0 can be computed from the transmitted value of A using Equation (8.11), satellite tracking by the ground segment provides a correction, Δn, that is contained in word 4, subframe 2 so rather than using n_0 in Equation (8.12), the quantity n is used given by:

$$n = n_0 + \Delta n \tag{8.13}$$

From M, the eccentric anomaly, E, can be found by a recursive solution of Kepler's equation:

$$M = e - e\sin E \tag{8.14}$$

and from E the true anomaly ν, follows from:

$$\nu = \tan^{-1}\left(\frac{\sqrt{1 - e^2}\sin E}{\cos E - e}\right) \tag{8.15}$$

The true anomaly, ν, and the orbital parameters, A, e, Ω_0, i_0, and ω define the precise point in space of the satellite and t_{oe} provides the reference time for the transmission. This information combined with the timing of the received signal provided by the PRN code enables the receiver to determine a pseudo-range from a specific SV.

8.3.6   Precision P(Y) Code

The precision code is also a Gold code, which is streamed at a chip rate of 10.23 MHz on both the L1 and L2 channels and is modulated by the navigation message as illustrated in Figure 8.7. The sequence, however, is much longer, consisting of 6.1871×10^{12} chips, which at 10.23 MHz would take just over a week to transmit but it is re-initialized after exactly one GPS week. The length of the code gives it a huge correlation gain and the higher chipping rate provides intrinsically more accurate timing. The length of the code is immaterial in terms of getting a match with a receiver-generated template as the excellent autocorrelation properties of Gold codes mean that any short section of a long code can be used to lock on. The navigation message is exactly as described above but since it is transmitted on both L1 and L2 channels, a pseudo-range can be obtained at two different frequencies, which allows an accurate correction of ionospheric propagation error as described in Section 8.5.2. Recently, the navigation message has also been made available on the L2 channel for general use providing the availability of real-time ionospheric error correction for everyone as described in Section 8.3.7.

Commercially available civilian GPS receivers do not have the hardware to generate the precision code, but the algorithm for producing it is known so there is nothing to stop someone building a receiver to use it without authorization. Alternatively, it is possible for a hostile agent to transmit fake GPS signals (spoofing) locally that overwhelm the satellite signals and provide false position information. To guard against this possibility, it is possible to send an encrypted version of the precision code known as the Y-code. This is notified by setting bit 19 in the HOW of each subframe to "1," which then alerts the receiver to use the decryption key provided to authorized users.

8.3.7   Additional GPS Signals

The description above is for the legacy system that has operated from the start of the GPS system in which the C/A code on the L1 channel only was available for general users while the P code was available on channels L1 and L2 for military or other authorized users. Up till 1 May 2000, the civilian code accuracy was deliberately degraded to around 100 m by randomly changing the clock corrections, a process called selective availability. Following an announcement by President Bill Clinton this was switched off on the above date and overnight the accuracy of the system for general users improved by an order of magnitude. For a while the threat remained that selective availability could be reinstated at any time but in September 2007, it was announced by the US Government that the selective availability feature would be absent in all satellites from block III onwards. For civilian users, the system has steadily improved with the addition of extra signals that have become available since block II RM starting in 2005 and are described below.

8.3.7.1 L2C Signal

Transmitting a user code available to all users modulated by a navigation message on the L2 channel enables real-time corrections for ionospheric delay (see Section 8.5.2) and also provides redundancy in case of interference. The capability required new hardware on the satellites, which was installed from block II RM starting in 2005. The L2C code is not a single PRN stream like the C/A code but consists of two separate sequences, each transmitted at 511 500 chips s^{-1} but multiplexed together into a single stream at a chip rate of 1023.0 kHz like the C/A code. The two PRN ranging codes are referred to as the civil-moderate (CM) code, which is 10 230 chips long repeating every 20 ms and the civil-long (CL) code, which is 767 250 chips long and repeats every 1500 ms. The CL code is not modulated by a navigation message but is transmitted to provide strong correlation with the internally generated template by the receiver and this strong correlation is then imposed on the CM code, which is multiplexed with the CL code.

The CM code is modulated by an upgraded navigation message referred to as CNAV, which has a much more flexible design than the legacy navigation message, now referred to as LNAV. It is organized into packets 300 bits long, each taking 12 seconds to transmit and each packet contains a message ID number that informs the system what data is contained in the message. The packet structure is illustrated in Figure 8.20 and the table in the figure shows what data is included according to the message type ID.

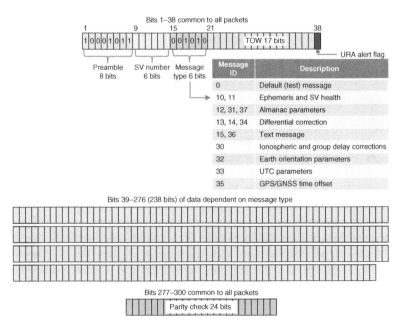

Figure 8.20 Structure of a message packet in the CNAV message sent on the L2C signal.

The first 38-bit block is common to all packets and starts with an 8-bit preamble that is the same as in the initial TLM word of the legacy navigation message. This is followed by 6 bits containing the SV number, 6 bits containing the message type ID, and the 17 bits containing the TOW count, which is synchronized to the start of the next 12-second message packet. The final bit 38 is a flag, which if set to "1" indicates that the URA components reported in the corresponding message packets are suspect. The initial 38-bit block is followed by a 238-bit block of data whose contents depend on the message type ID as shown in the table and finally the packet finishes with a 24-bit parity check block.

The packets can be sent in any order so that parameters can be updated at intervals that are appropriate for their expected periods of change. There are also two new types of message, that is a test data packet for message type ID = "000000," which fills the data block with alternating "1"s and "0"s and a text message packet for message type ID = "001111" or "100100" so that the system can send text messages to be displayed on user's receivers. There are only 15 types of message currently defined, whereas the 6 bits of the message type ID can specify up to 64 different types so there is plenty of flexibility to introduce new types of message. The basic method for finding a pseudo-range with the L2C signal is exactly the same as in the legacy system. That is, the PRN code is used to provide a timing to a fraction of a chip of the transit time uncorrected for the receiver clock offset and the navigation message contains the information to determine the exact position of the satellite when the signal started. Full details of all the possible CNAV messages are given in Ref. [1].

8.3.7.2 L5 Safety of Life Signal

From block II F satellites (from 2010) onwards, a signal transmitted at a new frequency of 1176.45 MHz ($=115f_0$) has been available with a ranging code and a navigation message. The new frequency is in the protected aviation UHF band in the range 960–1215 MHz and the signal has been developed with the aviation community in mind with the aim of improving accuracy to the level where GPS can be used for precision approaches without augmentation (see Section 8.7). For ranging, two PRN codes are transmitted in quadrature (see Section 4.3.2) and are labeled the I5 code (in-phase) and Q5 code (quadrature). The chip streams are transmitted at a rate of 10.23 MHz and are each 10 230 bits long giving a repetition period of 1 ms. The I5 branch is modulated by a similar CNAV message to the L2C signal and a 10-bit Neuman–Hoffman (NH) code [3] with a data rate of 1 kHz. NH codes are short codes designed to be imposed on data streams to achieve correlation. This is a similar system to that used in transmissions by the European Galileo GNSS. The Q5 chip stream is not modulated by the CNAV message but just a second 20-bit NH code clocked at 1 kHz. The details of the PRN codes are described in Ref. [4]. The advantages of the L5 signal include a factor of 2 increase in transmitted power, better autocorrelation of the PRN codes providing more accurate timing, the flexible CNAV message structure, and a transmission frequency in a protected aeronautical band.

8.3.7.3 L1C Signal

L1C is a new signal broadcast on the L1 1575.42 MHz frequency and is downward compatible with the legacy C/A signal. It will be implemented on the block III GPS satellites with the first one due to launch in May 2018. In a similar manner to the L5 signal, L1C will use overlay codes to improve correlation and sharper timing. It will also be broadcast at greater power and will have improved interoperability with the Galileo GNSS.

8.3.7.4 L3 and L4 Signals

The GPS satellites also broadcast on two additional frequencies not used for navigation known as the L3 signal and the L4 signal at frequencies 1381.05 MHz ($135f_0$) and 1379.913 MHz ($1214/9f_0$), respectively. The satellites are platforms for optical, X-ray, and electromagnetic pulse (EMP) sensors designed to detect nuclear explosions to check for noncompliance with test-ban treaties and any observations are reported on the L3 channel. The L4 channel is used to study the benefit of ionospheric corrections using an additional frequency.

Interesting Diversion 8.1: X-Ray Pulsar Navigation

Recently an innovative navigation system for deep space probes, based on accurate timing signals emitted naturally by rotating neutron stars, known as pulsars, has been tested by NASA. This shares some features in common with GNSS and could provide SVs exploring the Solar System and beyond with an autonomous accurate navigation system. For deep space probes that can have missions lasting for several years billions of miles from the Earth, the only navigation available is tracking from the Earth by, for example, NASAs deep space network, which is a worldwide system of communication facilities with large parabolic dish antennas. Because of the huge distances involved, however, tracking can be several hours out of date due to the time it takes a return signal to travel from the Earth to the space probe. Thus, there has long been a requirement to develop an onboard autonomous navigation system so that space probes can adjust their own track at critical phases of a mission, for example, during a fly-by of a moon or planet.

An idea that emerged in the 1990s was to use precise timing signals produced by pulsars, which are rapidly spinning neutron stars formed by a supernova explosion at the end of a star's life (Figure ID8.1.1). They have an intense magnetic field that accelerates jets of particles out along the two magnetic poles to produce intense beams of radiation covering wavelengths from radio waves to X-rays. If the magnetic field is not aligned with the spin axis, the radiation beams are swept around as the star rotates and when the beams cross the line of sight from the Earth, a pulse is detected.

Figure ID8.1.1 (a) Artist's impression of a pulsar emitting intense directional beams along the magnetic poles. The neutron star is rapidly spinning and if the magnetic poles do not coincide with the spin axis, the beams are swept as the star rotates and will be detected as a pulse when they cross an observer's line of sight. *Source*: NASA.

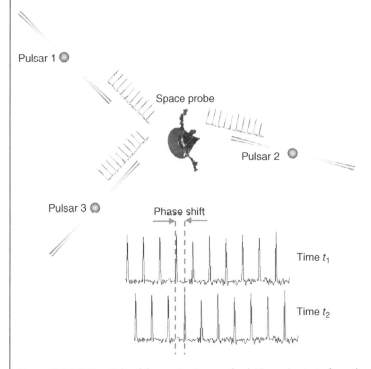

Figure ID8.1.2 Principle of the navigation method. The pulse train from three pulsars is measured at time t_1 and then again at time t_2. The phase shift determines the distance moved between t_1 and t_2 in three dimensions.

(Continued)

Interesting Diversion 8.1: (Continued)

Figure ID8.1.3 (a) NICER X-ray detector deployed on the International Space Station (ISS). *Source*: NASA. (b) A single grazing incidence X-ray lens of the NICER array. *Source*: NASA.

Many pulsars have pulse periods in the millisecond regime with an extremely stable time structure so that the arrival of time of pulses can be predicted to microsecond accuracy months into the future, which rivals atomic clock precision. In recent years a number of millisecond pulsars (MSPs) have been identified that emit X-rays, which produce a more stable time structure since X-rays do not interact significantly with the interstellar medium whereas the radio pulses can accumulate delays over sufficiently long distances. This is analogous to the GNSS ionospheric propagation errors discussed in Section 8.5.2.

The principle of the navigation method is illustrated in Figure ID8.1.2. The space probe measures the X-ray pulse train from three pulsars at time t_1 and then again at a later time t_2 to determine the phase shifts resulting from the change in arrival times of the pulses due to the spacecraft having moved. The phase shift in the three sets of pulses is entered into a computer algorithm to calculate the distance traveled between t_1 and t_2 in three dimensions. Thus, the method does need a reliable starting position and also accurate astrophysical data on the natural shifting phase of the pulses from effects such as the proper motion of the pulsar.

The method was tested recently by a team at NASA using an X-ray detector built specifically to measure X-ray pulses from MSPs known as the neutron-star interior composition explorer (NICER) installed on the ISS and shown deployed in Figure ID8.1.3a. It is an array of grazing incidence lenses, one of which is shown in Figure ID8.1.3b, that concentrate X-rays from pulsars onto detectors. Since the ISS is in low-Earth orbit, it is able to use GNSS for accurate position fixing. Starting from a known position obtained by GNSS, the experiment was able to switch over to X-ray pulsar navigation and track the ISS orbit to less than 10 miles as it orbited at 17 500 miles h^{-1}. Eventually it is estimated that the method will be able to provide autonomous onboard navigation for SVs to within 100 m.

8.4 Control Segment

The control segment for the GPS system is shown in Figure 8.21 and consists of a master control station (MCS) in Colorado, an alternate control station in California, ground antennas, monitoring stations, and tracking stations. The MCS provides command and control of the GPS constellation, uploads navigation messages, and ensures the health and accuracy of the SV constellation. It receives navigation information from the monitor stations, utilizes this information to compute the precise locations of the SVs in space, and then uploads this data to the SVs, which they pass on to users in the transmitted navigational message. In the event of a satellite failure, the MCS can reposition satellites to maintain an optimal GPS constellation. To change the orbit of a satellite, the satellite must be marked unhealthy, which, in the legacy system, is done by setting the appropriate value ("111100") in word 3, subframe 1 of the navigation message (see Figure 8.15). This is passed on in the transmitted message to users so receivers will not use the SV for their position calculation. Then the maneuver can be carried out, and the resulting orbit tracked from the ground. Finally, the new ephemeris is uploaded into subframes 3 and 4 of the navigation message and the satellite marked healthy again. In addition, the updates from the MCS synchronize the atomic clocks on board the SVs to each other, and adjust the ephemeris of each SVs' internal orbital model.

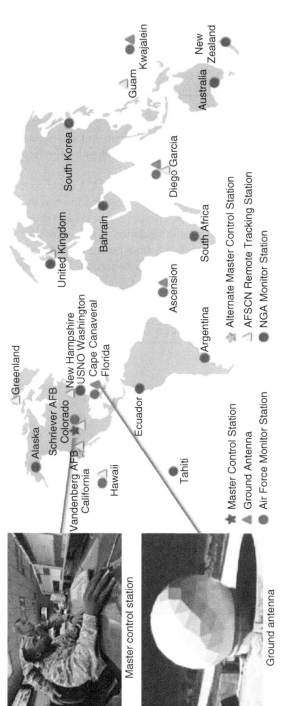

Figure 8.21 Control segment of the GPS system.

8.5 Sources of GPS Errors

8.5.1 Geometric Dilution of Position

Since a GPS position fix is calculated from the intersection of spheres (see Figure 8.1) and the precise radius of each one has an associated error, the corresponding positional error depends on the geometry of the overlapping region as illustrated in Figure 8.22. The ideal arrangement of satellites is a tetrahedron but that is impossible to be achieved as it requires at least one out of sight behind the Earth. The best arrangement for a grouping on a single hemisphere is to have one directly overhead and three spaced equally (i.e. at 120° to each other) on the horizon. In practice, an SV is not used if it is within 5° of horizon due to refraction of the signal and the receiver chooses the group that is closest to the ideal hemispherical geometry subject to the above rule. This is one of the main reasons why increasing the number of satellites in view improves accuracy as it becomes possible to find an arrangement closer to the ideal geometry.

8.5.2 Ionospheric Propagation Error

A significant source of error is getting a position fix is due to the ionosphere slowing down the passage of the UHF waves. The speed of light in a vacuum (c) slows down to c/n in a medium with a refractive index, n. The refractive index is given by:

$$n = \sqrt{\epsilon} \tag{8.16}$$

(a)　　　　　　　　　　　　　　　　　(b)

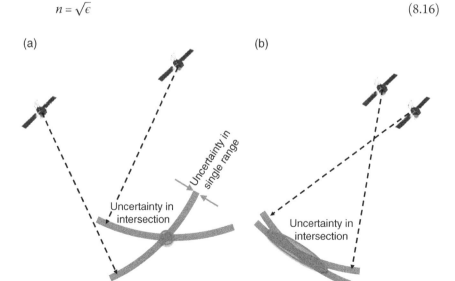

Figure 8.22 (a) Good satellite geometry for obtaining a position fix. (b) Poor satellite geometry for obtaining a position fix.

where ε is the dielectric constant of the medium, which for a wave of angular frequency ω $(= 2\pi f)$ given by:

$$\varepsilon = 1 - \frac{\omega_P^2}{\omega^2} \tag{8.17}$$

where ω_P is the plasma frequency, which in a charged plasma like the ionosphere is given by:

$$\omega_P = \sqrt{\frac{Ne^2}{\varepsilon_0 m}} \tag{8.18}$$

In Equation (8.18), N is the electron density, e is the electron charge, ε_0 is the permittivity of free space, and m is the electron mass. Beermat Calculation 8.3 provides an estimate of the error in the uncorrected range produced by the change of light speed under typical ionospheric conditions.

Beermat Calculation 8.3

On a specific day, the electron density in the ionosphere is $N = 6 \times 10^{11}$ m^{-3} and its effective thickness is 200 km. Estimate the error in range that would occur if it was not corrected for the change in light speed.

Putting in the values for e (1.6×10^{-19} C), ε_0 (8.85×10^{12} Fm^{-1}), m (9.1×10^{-31} kg), and N into Equation (8.18) gives a plasma frequency, ω_P, of 44 MHz. Using this value in Equation (8.17), the dielectric constant of the ionosphere at $f = 1575.42$ MHz ($\omega = 9899.9$ MHz) is 0.99998054, thus from Equation (8.16) the refractive index is $n = 0.99999027$. The speed of light is slowed down by a factor of c/n in the atmosphere, that is, by about 3000 m s^{-1}. If it traverses 200 km of ionosphere, the resulting time delay is 6 ns corresponding to a range error of 2 m.

For vehicles using the C/A code and the single L1 channel, a model is used to introduce a correction due to the ionosphere and the algorithm used to calculate the ionospheric delay is given in Ref. [1]. The ionosphere is, however, a chaotic system and there can be significant changes in ion density and thickness day to day and if the model is being used, ionospheric propagation delay usually produces the largest error in a GPS position fix.

If two frequencies can be used, then a comparison of ranges at the two frequencies provides a real-time correction for the actual conditions in the ionosphere prevailing on the day. This capability has been available to the military via the P(Y) code on both the L1 and L2 channels since the origin of GPS and has become available to general users since 2005 on the L1 and L2C channels. If a pseudo-range R_{L1} is measured on L1 and a pseudo-range R_{L2} is measured on L2, then the pseudo-range, R, corrected for ionospheric delay, is given by [1]:

$$R = \frac{R_{L2} - \gamma R_{L1}}{1 - \gamma} \tag{8.19}$$

where

$$\gamma = \left(\frac{f_{L1}}{f_{L2}}\right)^2 \qquad (8.20)$$

Putting in the values of the frequencies, $f_{L1} = 154f_0$ and $f_{L2} = 120f_0$, gives $\gamma = (77/60)^2 = 1.64694$.

8.5.3 Other Sources of Error

Table 8.2 gives a summary of possible sources of error in a single GPS range measurement not including geometric dilution of position. It is quite difficult to predict the accuracy of a position measurement on a given day as it depends on the state of chaotic systems such as the ionosphere and the weather, the number of satellites in view from the users' position, and local terrain features. It also depends on whether a user is using entirely the legacy C/A positioning service on the L1 channel only or one of the newer services on the L2C or L5 channel allowing measurements at multiple frequencies. The sure way to get a reliable position fix with a known, fixed, error is to use the augmentation systems described in the following section.

Table 8.2 Summary of sources of GPS errors.

Type of error	Explanation	Maximum range error (m)
Ionospheric delay	Covered in Section 8.5.2	5
Ephemeris error	SV position is checked every 12 hours and ephemeris parameters are updated if necessary. On a given pass the ephemeris parameters may have drifted.	2.5
SV clock error	SV clock is checked every 12 hours and clock correction parameters are updated if necessary. On a given pass the clock correction parameters may have drifted	1.5
Multipath reception	Reflection of radio waves off features around the receiver providing more than one timing signal	0.6
Troposheric propagation error	Weather-dependent changes in light speed in the atmosphere due to changes in the refractive index of the atmosphere from changes in density, humidity, etc. Can be reduced by measuring the range at two different frequencies	0.5
Receiver noise error	Jitters in the correlation between the received and internally generated PRN	0.3

8.6 Relativity Corrections Required for GPS

The time dilation effect predicted by special relativity and the effect of curved space time on clock rate predicted by general relativity are significant effects on the nanosecond timing required for GPS position fixes and have to be taken into account. In fact, putting atomic clocks on satellites was originally proposed as a test of relativity. Special relativity predicts that for an object moving at velocity v relative to a stationary observer, a given time, t_0, for the observer corresponds to a time, t, on the moving object given by:

$$t = \frac{t_0}{\sqrt{1 - \frac{v^2}{c^2}}} \tag{8.21}$$

where c is the speed of light (3×10^8 m s^{-1}). The GPS SVs are moving at an orbital speed of 3900 m s^{-1} so the clocks on board are running slow, compared to the same clock on the ground by a factor of ($1 - 8.4 \times 10^{-11}$). Over an entire day the clock loses about 7 μs compared to a stationary clock.

According to general relativity, the rate of passage of time depends on the local curvature of space time and on the Earth's surface deep in the gravity well, the curvature is greater than at the height of the satellite. The prediction is that a GPS SV clock will gain 45 μs day^{-1} due to this phenomenon. So, the combined relativistic effects predict that the SV clock will gain 38 μs (380 000 ns) per day. If this was not corrected the position error would accumulate at about 10 km day^{-1}.

Since the corrections due to special and general relativity can be calculated precisely, they do not amount to an error. As discussed in Section 8.3, the overall clock correction is given by (Equation 8.9):

$$\Delta t_{SV} = a_{f0} + a_{f1}(t - t_{OC}) + a_{f2}(t - t_{OC})^2 + \Delta t_r$$

where a_{f0}, a_{f1}, a_{f2}, and t_{OC} are all parameters transmitted in the navigation message and Δt_r is the relativistic correction. As shown in Ref. [1], the relativistic correction in seconds is given by:

$$\Delta t_r = Fe\sqrt{A}\sin E \tag{8.22}$$

where e is the ellipticity of the orbit, A is the semimajor axis, and E is the eccentric anomaly. The parameters e and A are transmitted directly in the navigation message and E can be found from other transmitted parameters as described in Section 8.3.5. The constant F is given by [1]:

$$F = \frac{2\sqrt{\mu}}{c^2} \tag{8.23}$$

where the constant $\mu = GM_\oplus$, G is the gravitational constant, and c is the speed of light. Putting in all the values gives $F = -4.442807633 \times 10^{-10}$ s m$^{-1/2}$.

8.7 Augmentation Systems

The ICAO specification for the overall system accuracy of the Standard Positioning Service (SPS), that is, GNSS without further corrections is:

- Horizontal   ±13 m
- Vertical   ±22 m
- Time   ±40 ns

This is well within the requirements for enroute navigation but it is not good enough for precision approaches, which are still mostly based on ILS. In order to improve the accuracy further with a view to implementing GNSS-guided precision approaches, augmentation systems have been developed to correct errors. These are mostly based on ground stations with precisely surveyed positions continuously monitoring their position using all GNSS constellations available and measuring any difference between the GNSS position fix and the known location. The corrections provided to aircraft can either be implemented in a wide area augmentation system (WAAS) where a map of deviations is broadcast by satellite or as a local area augmentation system (LAAS) where a single station provides local information to aircraft in the area. For example, this could be a station at an airport runway threshold providing guidance to landing aircraft. Sometimes augmentation is referred to as Differential GPS (DGPS) and the terms WAAS and LAAS are replaced by satellite-based augmentation systems (SBAS) and ground-based augmentation systems (GBAS).

8.7.1 Wide Area Augmentation Systems (WAAS)

There are several WAAS in operation, the original being established in the United States, which uses a series of 38 receiver sites. Each site receives signals from all GPS satellites in view and transmits this information to a WAAS master site, where the major sources of GPS errors are analyzed. The master site then develops a correction message, which is transmitted to two geosynchronous satellites. These retransmit the correction message to WAAS-enabled aircraft receivers that apply the corrections. These are in the form of a map showing the required adjustment in each area and it typically improves GPS accuracy to around the 5 m level, which is sufficiently good for Category 1 approaches, that is, a decision height of 200 ft. and a runway visibility of ≥550 m (see Table 7.2). Other WAAS systems include the European geostationary navigation overlay service (EGNOS), the multifunctional satellite augmentation system (MSAS) in Japan, and the GPS aided geo-augmented navigation (GAGAN) system in India.

8.7.2 Local Area Augmentation Systems (LAAS)

For Cat II/Cat III approaches, higher precision is required and this is achieved with ground-based LAAS installations. On the site is a precisely surveyed GNSS receiver, which determines its position from all satellites in view position and compares it with the known position of the site. The error is formatted and a correction is transmitted to an aircraft via a VHF link in the VOR band (108–118 MHz). The aircraft picks this up as an SV (*pseudolite*) on the site. Ionospheric and tropospheric errors are eliminated and the system allows precision CAT III approaches.

8.7.3 Aircraft-Based Augmentation Systems (ABAS) and Receiver Autonomous Integrity Monitoring (RAIM)

In all aircraft, there are systems on board that provide additional positional information, for example, the radio beacons described in Chapter 7 or inertial guidance systems (see Chapter 9). Even in the simplest aircraft there is an altimeter that provides a more accurate measurement of altitude (better than ±10 m at low level) than the ICAO specification for raw GNSS. These systems can check the data from satellites and thus corrections can be applied without external references. This is an aircraft-based augmentation system (ABAS).

Four satellites are required for a position fix and if more satellites are in view, an intelligent receiver can compare positions derived from different combinations to determine whether any are providing unreliable data. With only four SVs being received there is no way to check whether one has failed and the positional information has become degraded. A receiver needs at least five SVs in view to provide a monitor for four that provide position and decide if one is rogue and deselect it. If this happened the monitoring would then become unavailable so the CAA recommends that six SVs are in view so that if one fails the monitoring function will continue. This is termed receiver autonomous integrity monitoring (RAIM). A pseudolite in a LAAS can be used as one of the six SVs to provide full RAIM during an approach.

Many airports now publish GNSS-guided precision approach plates that are normally called RNAV (area navigation) approaches to acknowledge that multiple navigation systems may be in use to provide the required navigational performance (RNP – see Section 1.8.4). An example of an RNAV (GNSS) approach on to runway 09L at Heathrow airport is shown in Figure 8.23. Note the lack of angular lines coming from the runway and the final approach path is a line defining a lateral and vertical glide slope. In the case of a missed approach the guidance is provided by the conventional navigation systems, that is, the ILS localizer, DME, and the CHT NDB. This plate is a perfect demonstration of the reluctance to remove older working systems and combines twenty-first century satellite technology with 1920's LF radio technology.

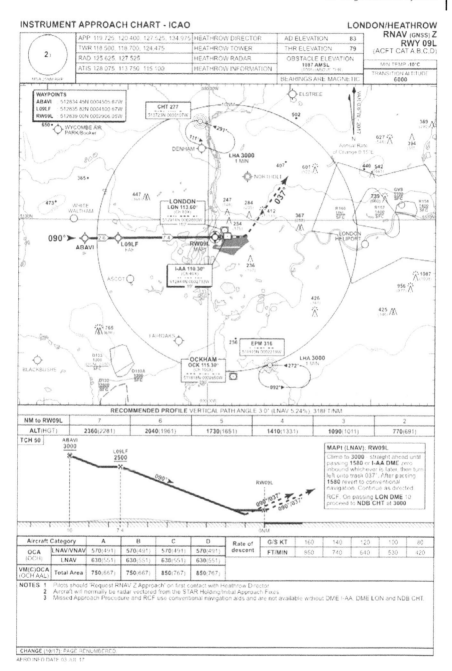

Figure 8.23 RNAV (GNSS) precision approach plate for runway 09L, Heathrow airport.

8.8 GPS Cockpit Instrumentation

The GPS cockpit instrumentation generally needs no setting up simply to report position, which is normally on a moving map display. As soon as the receiver is switched on, it automatically locates and locks on to satellite signals and starts the process of shifting the PRN code templates to match the movement of the satellites and the internal software streams navigational data in response to the incoming navigation message. If the stored almanac parameters are still valid, position fixing begins immediately while in the worst-case scenario of all parameters lost, there will be a 12.5-minute wait while the system downloads the latest almanac parameters from the entire navigation message. Normally the user does not need to consider the positions of the satellites in view but that information is available. For example, Figure 8.24 shows the satellite page on the multifunction display in a light aircraft glass cockpit indicating there are 10 satellites in view shown by the signal-level bars and positions displayed on a constellation map. The bars filled in green indicate the satellites with locked-on PRN codes and that are providing position data, which in this case is GPS satellites 2, 5, 7, 8, 9, 13, 20, 27, 28, and 30. The bars filled in light blue indicate satellites with locked-on PRN codes that are not being used to provide position data, in this case because they are too close to the horizon. The hollow

Figure 8.24 Satellite page on a glass cockpit display.

signal bars show satellites from which the receiver is currently collecting data prior to them being used. In Figure 8.24, these are satellites with PRN number 120 and 123, which are both EGNOS SBAS satellites. Also indicated with no signal bar is a satellite with PRN number 126, which is an EGNOS SBAS satellite. The zero signal indicates that the receiver is looking for the satellite but has so far not identified the PRN code. The "D" in the signal bars identifies which satellites are being used to improve positional accuracy via SBAS.

The screen also shows the estimated position uncertainty (EPU), which is the radius of a circle centered on the current estimated GPS horizontal position within which the true position of the aircraft has a 95% probability of lying, in this case, 0.01 nautical miles or about 18.5 m. The EPU has the same definition as the actual navigation performance (ANP) introduced in Section 1.8.4. The parameter is important in the evolution of air traffic control systems in which the calculated ANP will have to be less than a "required navigational performance" (RNP) for each phase of flight. In an airliner, the GNSS position is combined with all other navigation sources, including inertial guidance and radio beacons in the flight management system that calculates actual navigational performance (ANP) after Kalman filtering of all the inputs as described in Chapter 9.

Related to the EPU are the horizontal and vertical figures of merit (HFOM and VFOM) that are the distances within which there is 95% confidence that the reported position is the true one, in this case 10 and 20 ft, respectively. The highlighted box on the right-hand side indicates that RAIM is activated so the receiver is continuously checking the integrity of the incoming data by cross-correlating the satellite position fixes (ABAS).

The GNSS system can store a flight plan in the form of a sequence of waypoints whose positions are stored on a database and will then display the tracks for the pilot to follow or it can drive the autopilot to follow the flight plan. Figure 8.25a shows a flight plan entered into the system and Figure 8.25b is

(a)

(b)

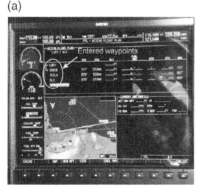
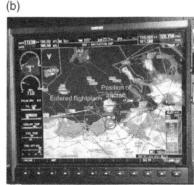

Figure 8.25 (a) Entering a flight plan as a series of waypoints into the flight management system. (b) The flight plan indicated on the GNSS moving map display showing all tracks and waypoints. The position of the aircraft is highlighted and the mauve track is the currently active leg.

the plan displayed on the moving map showing the tracks and waypoints. The position of the aircraft on the route is highlighted and the mauve line shows the active leg of the route.

8.9 Spoofing, Meaconing, and Positioning, Navigation, and Timing (PNT) Resilience

An issue that has started to be of some concern is the possibility of transmitting fake GNSS signals from the ground to produce erroneous position fixing in a local area. Given the small signal received from an SV, it is easy to transmit a stronger local signal, which could initially replicate the actual PRN code and navigation message from an SV. The vehicle could then be guided off course either by gradually phase-shifting the PRN code in the right way or by modifying parameters in the navigation message, for example, the clock corrections. This type of intervention is referred to as spoofing or meaconing (a derivative of "masking" and "beacon"). It was claimed by Iran that they did this to capture a US drone in 2011. Currently, in the case of the P code, it is possible to encrypt the PRN (whence it is called the "Y" code) to prevent this but no such measure is available for general users.

In addition to this concern is the vulnerability of the space segment. If a satellite fails, there is no fast way to repair it and while the system could easily cope with the removal of one satellite, there is the possibility of a number being removed by a particularly intense solar flare or by hostile action. Despite the fundamental vulnerability of GNSS there is an increasing reliance on it and it has been described as a single point of failure for critical infrastructure. This concern has led to the search for alternative systems to achieve so-called positioning, navigation, and timing (PNT) resilience. One possible alternative is ADS-B secondary surveillance radar, which is based on ground transmitters and was discussed in Section 5.4.12. Another is a new version of the LORAN system described in Section 1.8.1 known as enhanced LORAN (e-LORAN). This has been developed with an accuracy of about 8 m (similar to unenhanced GPS) and new sets of e-LORAN transmitters have been installed in the United States and the United Kingdom since 2014. The e-LORAN networks are low frequency (90–100 kHz) systems and so rely on surface wave propagation whose range depends on the power of the transmitter (see Equations 4.4 and 4.5). The transmitters used have powers up to 4 MW and are very difficult to spoof.

Problems

1   In the GPS constellation, the semimajor axis, A, and the ellipticity, e, of the orbit of an SV reported in its navigation message are 26 559 km and 0.02,

respectively. Estimate the maximum error that could occur in position fixing by this SV if it was assumed the orbit was circular.

2 Calculate the proportion of the Earth's hemisphere in the line of sight of a satellite in a geostationary orbit.

3 Calculate the distance of a satellite 10° above the horizon from a vehicle with a GNSS receiver. Hence determine the received power density at the vehicle assuming the SV transmitting power is 25 W and the antenna gain is 13 dB.

4 Estimate the precision of the orbital position of an SV that can be specified by a 32-bit binary number.

5 During a period of high solar activity, the electron density in the ionosphere reaches 10^{12} m^{-3} and the effective thickness is 200 km. Calculate the ionospheric delay of the GNSS L1 carrier at 1575.42 MHz.

References

1 Global Positioning Systems Directorate (2013). Global positioning system directorate systems engineering and integration. Interface Specification IS-GPS-200, Doc. IS-GPS-200H, 24 September 2013.
2 Gold, R. (1967). Optimal binary sequences for spread spectrum multiplexing. *IEEE Transactions Informatics and Theory* **13**: 619–621.
3 Shanmugam, S.K., Mongrédien, C., Nielsen, J., and Lachapelle, G. (2008). Design of short synchronisation codes for use in future GNSS systems. *International Journal of Navigation and Observation* **2008**: 14.
4 Global Positioning Systems Directorate (2013). Global positioning system directorate systems engineering and integration. Interface Specification IS-GPS-705, Doc. IS-GPS-705D, 24 September 2013.

9

Inertial Navigation and Kalman Filtering

9.1 Basic Principle of Inertial Navigation

Inertial navigation is unique among navigation systems in that it requires no external references and can be carried out entirely by the measurement of acceleration within the vehicle. The initial position and state of motion of the vehicle has to be known but then all subsequent tracks and locations can be determined. The principle is illustrated in Figure 9.1 for a vehicle starting at rest and then moving on a flat surface for 600 s with two accelerometers fitted orthogonal to each other aligned along x and y directions. As shown in Figure 9.1a, an initial acceleration is applied in the x-direction that builds quickly to 1 m s^{-2} and then slowly decreases as the wind resistance builds up reducing after about three minutes to zero. Integrating the acceleration with respect to time yields the velocity as a function of time illustrated in Figure 9.1b and during the acceleration phase the calculation shows that vehicle's x-velocity increases to $v_x = 52.6$ m s^{-1}. Integrating the velocity with respect to time gives the distance traveled and Figure 9.1c shows how the distance increases in the x-direction with time.

After 300 s, a short steering acceleration of 2 m s^{-2} is applied in the y direction (green curve, Figure 9.1a), which produces a y-velocity of 57.5 m s^{-1} and an increasing movement in the y-direction as shown by the green curves in Figures 9.1b and c. This is followed at 450 s by an opposite steering acceleration of -2 m s^{-2}, which removes the y-motion and steers the vehicle back to the x-direction. Plotting x vs. y for the 600 s of motion yields the track displayed in Figure 9.1d, which shows that after 10 minutes the vehicle is at the new coordinates (29.1, 9.1) km. The track and position of the end points are derived entirely from knowing the initial position and measuring any subsequent acceleration. Adding a third accelerometer in the z-direction would provide the track in three-dimensional space.

The first inertial navigation system (INS) was developed for the German V2 rockets during World War 2 and post war the focus of development in INS was for use in missiles with the cold war driving the design of smaller, lighter, and

Aircraft Systems: Instruments, Communications, Navigation, and Control,
First Edition. Chris Binns.
© 2019 John Wiley & Sons, Inc. Published 2019 by John Wiley & Sons, Inc.
Companion website: www.wiley.com/go/binns/aircraft_systems_instru_communi_Navi_control

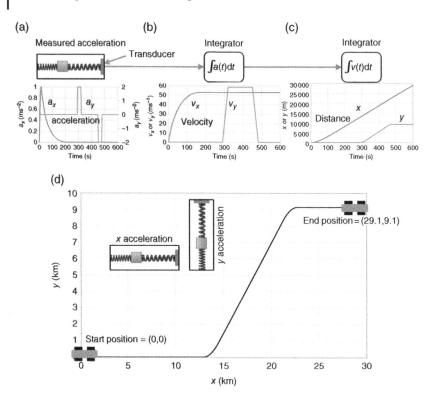

Figure 9.1 (a) Accelerations applied to a vehicle starting at rest at (0, 0). (b) *x* and *y* velocities of the vehicle. (c) *x* and *y* positions of the vehicle. (d) Track of the vehicle.

more accurate systems. The first use in aviation of INS was in military aircraft in the late 1950s, but it was not until the late 1960s that systems sufficiently inexpensive to be used in commercial aircraft were developed. The first was the Delco Carousel, which was described in Section 1.8.3 and was first installed in commercial aircraft in the late 1960s. Early systems, described in Section 9.2, operated the accelerometers on a gimbaled stabilized platform that stayed accurately parallel to the Earth's surface while the aircraft pitched and rolled in order to ensure that gravity was not measured as an acceleration. Modern "strapdown" systems that first appeared in 1978 have the unit mounted to the aircraft frame and measure the rotation of the aircraft so that gravity can be removed in software. These are described in Section 9.3. Subsequent sections describe sources of error in inertial navigation and cockpit instrumentation. Section 9.7 describes how all navigation methods, including INS, GNSS, and radio beacons are combined to yield an integrated resilient navigation system that is able to report the confidence in position determination in all phases of flight.

9.2 Gimbaled Systems

9.2.1 Stabilized Platforms

If only linear accelerometers are used, the platform on which they are mounted must be exactly perpendicular to the Earth's gravity vector as the gravitational force is indistinguishable from an acceleration. Indeed, this was the starting point of the train of thought that led Albert Einstein to develop the general theory of relativity. Even a small component of gravity present along the orthogonal directions would produce errors in position that would rapidly accumulate due to the double integral required to convert acceleration to distance. The method used in the original systems was to mount the platform on a three-level gimbal system illustrated in Figure 9.2. The platform would be initialized while the aircraft was at rest by rotating about the longitudinal and lateral axes using torque motors till the accelerometers indicated zero. This orientation would then be maintained perpendicular to the gravity vector as the aircraft pitched and rolled using gyroscopes mounted on the platform. The entire system would be attached to the aircraft frame on isolastic mountings to minimize the effect of vibrations.

The gyros used in this type of system are very high rigidity space gyros (see Section 3.1.4) with negligible true wander ($<0.01°$ h^{-1}), which are not tied to

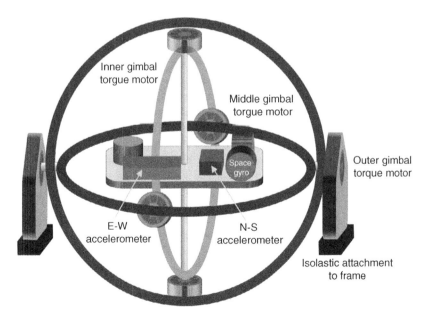

Figure 9.2 Stabilized platform INS in which the accelerometers are mounted on a common platform that is kept perpendicular to the gravity vector using the space gyros attached to the platform.

the Earth vertical and their axis is fixed only relative to the distant stars. Thus, they will naturally change their spin axis alignment due to Earth rate as the Earth spins on its axis and transport wander as the aircraft moves over the spherical surface of the Earth (see Sections 3.1.7 and 3.1.8). The platform has to be maintained perpendicular to the gravity vector by using the torque motors to compensate for both these effects and the method used is described in Section 9.2.3. Detectors on the gyros measure the movement of the axes with respect to the gimbaled mountings as the aircraft maneuvers and as a result of Earth rate and transport wander. The accelerometers in modern units are micro-electromechanical systems (MEMS) described in Section 3.2.6.

9.2.2 Obtaining Latitude and Longitude

Assuming that the platform is kept corrected for Earth rate and transport wander and is aligned before takeoff along East–West and North–South direction, the measured acceleration outputs from the North–South and East–West channels can be integrated to obtain the North–South velocity (denoted v) and the East–West velocity (denoted u). Integrating v and u will then give the change in North–South and East–West positions. If the initial position is known as a latitude and longitude then the distance traveled can be used to continuously update the current position in terms of latitude and longitude. For latitude, this is straightforward as the number of nautical miles traveled can simply be added as minutes to the original latitude (by definition, 1 nautical mile = 1 minute change in latitude along a great circle).

For longitude, because of meridian convergence, the change of longitude equals the distance traveled along East–West in nm divided by cosine(latitude) or equivalently, the distance traveled in nautical miles along East–West multiplied by secant(latitude). Thus, to calculate longitude, the distance traveled along East–West (departure – see Section 6.1.2) is obtained from the East–West channel and fed into a secant multiplier whose other input comes from the latitude obtained from the North–South channel as shown in Figure 9.3. Combining the outputs at the velocity stage gives the current

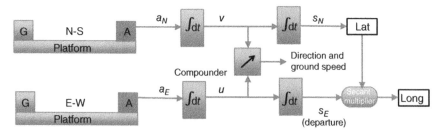

Figure 9.3 Obtaining direction, ground speed, latitude, and longitude from a 2-channel N–S, E–W accelerometer on a platform kept parallel to the Earth's surface. "G" represents the space gyroscopes and "A" represents the accelerometers.

direction and ground speed. All references are with respect to the ground since wind will initially produce an acceleration that is detected by the two channels and its effect on the ground position is automatically calculated.

9.2.3 Correcting the Platform Orientation for Earth Rate and Transport Wander

The alignment of the platform has to be maintained along North–South and East–West and also kept accurately perpendicular to the gravity vector. The gimbaled mounting ensures that the roll pitch and yaw of the aircraft do not affect the alignment, but the aircraft is traveling around a spinning sphere while the gyros maintain alignment relative to inertial space. Thus, corrections must be applied to the platform orientation by torque motors to compensate for Earth rate and transport wander.

The effect of Earth rate can be separated into a component that will tilt the platform with respect to the gravity vector and a component that will rotate the platform as the Earth spins. Figure 9.4a shows a platform at the equator aligned along North–South and as the Earth rotates the alignment will remain but the platform must be tilted about the North–South axis at a rate of $360°/$ sidereal day $= 15.04° \ h^{-1}$ to maintain a perpendicular gravity vector. Note that it is the sidereal day that needs to be used as it is the rotation relative to the distant stars that is relevant (see Section 8.2.1). A platform near the pole, however (Figure 9.4b), will remain perpendicular to gravity but must be rotated at $15.04° \ h^{-1}$ about the vertical to maintain alignment along North. At a general latitude, ϕ (Figure 9.4c), the platform will have to be tilted about the North–South axis at a rate, ω_{ER}:

$$\omega_{ER} = 15.04 \cos \phi \ (\text{degrees per hour}) \tag{9.1}$$

and rotated at a rate, $\dot{\theta}_{ER}$,

$$\dot{\theta}_{ER} = 15.04 \sin \phi \ (\text{degrees per hour}) \tag{9.2}$$

The sense of rotation in 9.2 is anticlockwise for the Northern hemisphere and clockwise for the Southern hemisphere. Earth rate is present whether the aircraft is parked or moving and the relevant torque motors must provide the rotation and tilt from start-up. If the aircraft is moving through different latitudes, the rate of tilt and rotation need to be continuously adjusted for the changing latitude and this can be controlled directly by the latitude output from the INS.

In addition, if the aircraft is moving tilt and rotations have to be added to the Earth rate due to transport wander (Figure 9.4d). The tilt component arises from the INS platform moving away from gravity normal as the aircraft travels over the surface of a sphere and can be resolved into components along u and v. For each component, to maintain the gravity vertical requires the platform to be tilted at a rate ω_u and ω_v given by (Figure 9.4e):

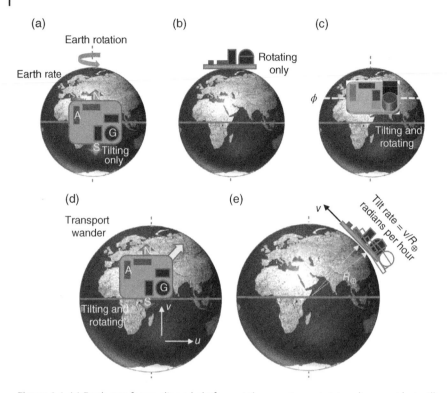

Figure 9.4 (a) Earth rate for an aligned platform at the equator maintains alignment but will tilt the platform relative to the gravity normal. (b) Earth rate for an aligned platform at the pole will rotate the platform but alignment relative to gravity will remain. (c) At a general latitude, the platform must be rotated and tilted to remove Earth rate. (d) Transport wander will produce a rotation (due to u) and a tilt (due to u and v) of the platform. (e) Illustration of the rate of tilt given by the v component, which will be $\underline{v}/R_\oplus$ radians per hour if v is in knots and $R_\oplus$ is in nautical miles.

$$\omega_u = \frac{u}{R_\oplus} \times \frac{180}{\pi} \, (\text{degrees per hour}) \tag{9.3}$$

$$\omega_v = \frac{v}{R_\oplus} \times \frac{180}{\pi} \, (\text{degrees per hour}) \tag{9.4}$$

where $R_\oplus$ is the radius of the Earth in nautical miles (= 3440) and u, v are in knots. The tilt compensations due to transport wander are referred to as Schüler tuning after the German engineer, Maximilian Schüler.

In order to maintain the same North reference, the platform has to be rotated due to meridian convergence (see Section 6.3.2.1), at a rate, $\dot{\theta}_{TW}$:

$$\dot{\theta}_{TW} = u \frac{sec\,\phi}{60} \, (\text{degrees per hour}) \tag{9.5}$$

where u is in knots and if u is along East, the rotation is anticlockwise while if it is along West the rotation is clockwise. The divergence of the rotation rate as ϕ approaches 90° at the pole predicted by Equation (9.5) is reasonable since precisely at the pole, an infinitesimal movement will rotate the direction of North by 180°.

Beermat Calculation 9.1

An aircraft fitted with a gimbaled INS is at a latitude of 30° flying North–West at 400 knots ground speed. Calculate the rate of rotation required to maintain the INS platform aligned along North–South in degrees per hour as it passes over the latitude.

If the ground speed is 400 knots and the aircraft is flying North–West, then $u = v = 400/\sqrt{2} = 282.8$ knots.

Since the aircraft is in the Northern hemisphere, the rotation to cancel Earth rate is anticlockwise and from Equation (9.2) has magnitude: $15.04 \times \sin(30°) = 7.52° \, h^{-1}$.

Since u is along West, the rotation rate to cancel transport wander is clockwise and from 9.3 has magnitude:

$$u \times \sec(30°)/60 = 282.8/30 = 9.42° \, h^{-1}.$$

Thus, the total rotation rate required is $1.9° \, h^{-1}$ clockwise. Note that this will continuously change as the aircraft moves through different latitudes.

The set of tilts and rotations to compensate for Earth rate and transport wander described by Equations (9.1)–(9.5) depend entirely, apart from constants, on the three variables, ϕ, u and v, which are all outputs of the INS platform. These can thus be fed back via suitable controllers to the torque motors that tilt and rotate the platform so it becomes self-correcting to maintain a North reference and gravity vertical as the Earth spins and the aircraft travels over the sphere as illustrated in Figure 9.5.

9.2.4 Initializing the Platform

Before being switched into navigation mode, the platform needs to be aligned while the aircraft is not moving. Leveling the platform is straightforward as it simply has to be tilted till both North–South and East–West accelerometers read zero, that is, no component of gravity is measured. The gyroscopes can then be caged and spun up to operating speed and uncaged. The alignment to true North is achieved by a process known as gyrocompassing. At any latitude, ϕ, the platform in an aircraft at rest, when aligned to true North, must be tilted at a rate $15.04 \times \cos \phi$ (Equation 9.1) and rotated at a rate $15.04 \times \sin \phi$ (Equation 9.2) in order to prevent a component of gravity being measured by the accelerometers. Thus, rotating the platform till these conditions are met will achieve the alignment to North reference. The process does, however, require

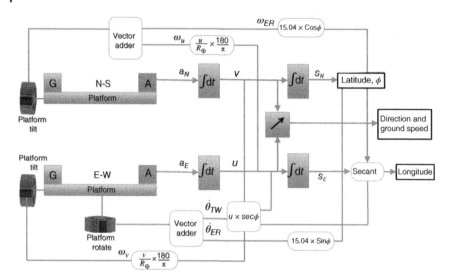

Figure 9.5 Feedback from ϕ, u, and v outputs from the INS used to rotate and tilt platform to maintain the North reference and gravity vertical in the presence of Earth rate and transport wander.

that ϕ is known and the current latitude and longitude are both entered by the pilot when the unit is powered up. These values should agree with the position data that were stored at the previous power off and if they do not match, a warning is issued. The process takes an amount of time that depends on the latitude. For example, at the equator, initialization is relatively fast, taking about five minutes, since the rate of tilt is maximum but due to the cos ϕ term, the tilt becomes increasingly slow as the latitude increases resulting in more time required for alignment. Above the Arctic circle this reaches 17 minutes and for latitudes higher than about 72°, it becomes impractical to gyrocompass the INS without additional inputs, for example, from GPS. The initialization time of the INS is one of its disadvantages since the aircraft cannot be moved or loaded while it is in process. On the other hand, it is a feature of the system that it is able to reference itself by detecting the rotation of the Earth.

9.3 Strapdown Systems

Maintaining a stabilized platform that shields the horizontal accelerometers from gravity requires high-precision mechanical structures that are expensive to build and maintain and are difficult to miniaturize in order to reduce weight. In a strapdown system the platform containing the accelerometers is attached directly to the aircraft structure (hence the name) and the attitude of the platform is accurately measured so that the component of gravity present in each accelerometer can be calculated. This can then be subtracted from the measured acceleration to produce an output as if the accelerometer was on a stabilized platform. All the velocity and position calculations can then be carried out as described above. Although straightforward in conception, this implementation had to wait for two enabling technologies. One was sufficiently accurate measurements of the attitude of the platform and the preferred option is to measure the rate of rotation of the platform by rate gyros and from this compute the final angle relative to a given aircraft axis. The other was sufficiently fast processing power in a computer that was practical in an aircraft given that measurements from all accelerometers and gyroscopes have to be taken at 1000s of times a second and processed at that rate. The first requirement was met by ring laser gyros described in Section 3.2.4 and the second by the microelectronics revolution that led to rapid computing power on a single chip. The corrections for Earth rate and transport wander are the same in strapdown systems as in stabilized platforms but the computed values are deducted from the measured accelerations in software. Initializing the platform by gyrocompassing also involves the same process as described above. In a strapdown system a third accelerometer is required orthogonal to the North–South and East–West accelerometers found in a stabilized platform system.

The earlier stabilized platform INSs (Figure 9.6) were stand-alone units that provided navigation data only but the newer strapdown systems are referred to as inertial reference systems IRSs. They are normally incorporated in an air data inertial reference unit (ADIRU) that combines data from the air data computer

Figure 9.6 Stabilized platform INS (Delco Carousel) with the case open to reveal the gyro-stabilized Earth-centered platform containing the accelerometers. *Source:* Reproduced with permission of National Air and Space Museum.

with aircraft attitude and navigation data from the IRS in a single unit (see Section 2.13). An ADIRU is a solid-state system that is more compact than a stabilized platform INS. In a commercial airliner, there are normally two independent ADIRUs providing redundant data so that they monitor each other for integrity and the positional information is passed to the flight management computer (FMC) that combines it with other navigation sources.

9.4 Accelerations Not due to Changes in Aircraft Motion

Assuming the platform is kept perfectly level with respect to gravity, there will still be accelerations measured by the accelerometers that are not directly due to changes in the state of motion. The most significant one arises from the Coriolis force described in Section 3.2.5, which is due to the North–South component of velocity, v. Viewed from inertial space, an aircraft flying directly North follows a curved path due to it traveling on a spinning sphere. The path followed is the same as if the aircraft was experiencing an apparent force that imparted an acceleration, a_C, given by:

$$a_C = 2\omega_E v \sin\phi \tag{9.6}$$

where ω_E is the rotation rate of the Earth in radians per second. This acceleration is sensed by the East–West accelerometer and must be subtracted digitally to not be interpreted as an aircraft acceleration. As shown by Beermat Calculation 9.2, in the most unfavorable circumstances at typical airliner speeds, the accumulated error per hour due to the Coriolis force can be up to 50 nautical miles.

Beermat Calculation 9.2

An aircraft fitted with a gimbaled INS is at a latitude of 30° flying North at 400 knots ground speed. Calculate the error in departure after one hour if the Coriolis force is not accounted for.

The rotation rate of the Earth is 2π radians per sidereal day = 7.29×10^{-5} rad s^{-1} and $v = 400$ knots = 205.8 m s^{-1}. The accumulated distance error in time, t, would be $0.5 \times a_C \times t^2$. Thus, from Equation (9.6) and setting $t = 3600$ s gives an error of 97.26 km = 52.57 nautical miles.

There are other accelerations with respect to inertial space that are below the noise limit of the accelerometers including the acceleration/deceleration of the Earth in its orbit around the Sun as it follows an elliptical path. In addition, since the Earth is not a perfect sphere, there will be small changes in the direction of the gravity vector as an aircraft travels over the surface. These can be modeled and subtracted using the WGS84 geoid described in Section 6.4, which brings errors below the noise threshold.

9.5   Schüler Oscillations

A consequence of the Schüler tuning feedback to compensate for transport wander as an aircraft moves is that any tilt error induces oscillations in the measured acceleration, velocity, and position. This is best illustrated in the case of a stabilized platform INS in an aircraft at rest and considering just one channel. Figure 9.7a shows the East–West accelerometer and Schüler tuning feedback of a platform that is at rest and perfectly perpendicular to the gravity vector. In this case, there is no measured acceleration and no derived velocity so the $u/R_{\oplus}$ feedback provides no rotation and the platform remains still. If the platform is now tilted by an angle, θ, as shown in Figure 9.7b, there will be a measured horizontal acceleration in the direction indicated given by $g\sin\theta$, which will produce an increasing value of East–West velocity, u. This will be interpreted as Westwards motion by the Schüler tuning, which will tilt the platform in the sense that reduces θ and after a given period the tilt angle and thus the acceleration reduces to zero as shown by the blue line in Figure 9.7c. During this period the velocity, represented by the red line in Figure 9.7c, will increase and reach a maximum value as the acceleration and tilt reach zero. The finite value of the velocity, however, will induce the Schüler tuning to keep tilting the platform so that an acceleration in the opposite sense appears, which reduces the velocity and eventually reverses the tilt. Thus, the initial tilt will initiate an oscillation of the platform that, as shown in Figure 9.7c, has a period of about 84 minutes. The calculation in Figure 9.7c was performed assuming an initial platform tilt of 0.1° and Figure 9.7d is a plot of the corresponding variation in distance, which has an amplitude of about 6 m.

The period of the oscillations can be derived analytically as follows. Consider a tilt bias, $\Delta\theta$, that produces an acceleration error, Δa, which can be written for small bias angles:

$$\Delta a = g\Delta\theta \tag{9.7}$$

This will produce a velocity error Δv given by:

$$\Delta v = \int \Delta a\, \mathrm{d}t = g\int \Delta\theta\, \mathrm{d}t \tag{9.8}$$

The velocity error will produce an error in the platform tilt rate, $\Delta\omega$, given by, using Equation (9.8):

$$\Delta\omega = \frac{\Delta v}{R_{\oplus}} = -\frac{g}{R_{\oplus}}\int \Delta\theta\, \mathrm{d}t \tag{9.9}$$

The minus sign has been inserted since, as shown above, an increasing velocity will result in a decreasing tilt error. Since the angular rate, $\Delta\omega$, can also be written:

$$\Delta\omega = \frac{\mathrm{d}}{\mathrm{d}t}\Delta\theta \tag{9.10}$$

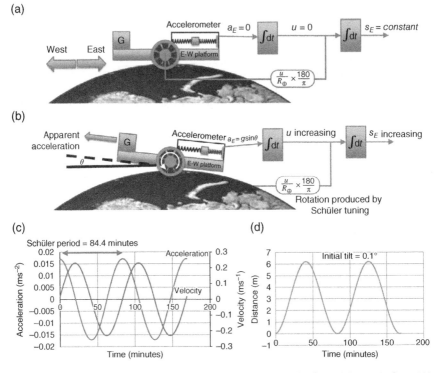

Figure 9.7 (a) East–West channel with Schüler tuning feedback of a stabilized platform INS perpendicular to gravity. There is no measured acceleration, no derived velocity so the platform remains still. (b) If the platform is tilted, gravity will give an apparent horizontal acceleration in the direction shown, which will produce a velocity signal and feedback that reduces the tilt. At zero tilt, there is a finite velocity so the feedback will take the tilt past zero producing an opposite acceleration so an oscillation is set up. (c) and (d) show the acceleration velocity and position with a 0.1° initial tilt.

We can write, using Equations (9.9) and (9.10):

$$\frac{d}{dt}\Delta\theta = -\frac{g}{R_\oplus}\int \Delta\theta dt \tag{9.11}$$

Differentiating this gives:

$$\frac{d^2}{dt^2}\Delta\theta = -\frac{g}{R_\oplus}\Delta\theta \tag{9.12}$$

which has the solution:

$$\Delta\theta = \Delta\theta_0 \cos\sqrt{\frac{g}{R_\oplus}}t \tag{9.13}$$

This shows that the tilt error will oscillate with an angular frequency, ω_s, given by:

$$\omega_s = \sqrt{\frac{g}{R_\oplus}} = 1.24088 \times 10^{-3}\,\text{rad s}^{-1} \qquad (9.14)$$

The corresponding linear frequency, f_s, is given by $\omega_s/2\pi = 1.9749 \times 10^{-4}$ Hz. Thus, the oscillation period is 5063.4 s = 84.4 minutes. The expression for the angular frequency written in Equation (9.14) is identical to that for a pendulum of length $R_\oplus$.

The Schüler oscillation has been described in terms of tilt bias in a stationary stabilized platform but Schüler tuning is also used in strapdown systems though the transport wander is compensated digitally. An accelerometer bias will still induce the same oscillations since the system will generate an incorrect compensation for transport wander that will act to reverse the accelerometer bias, however, the oscillations will be superimposed on a growing positional error (see Section 9.7).

9.6 Earth-Loop Oscillations

Another type of oscillation with a 24-hour period known as an Earth loop occurs if a platform has an initial heading error as illustrated in Figure 9.8a for a platform at the equator with an error in the North reference. As the Earth rotates the platform senses that it travels the inclined path represented by the red dashed line rather than the equator. After six hours the North reference is pointing at true North so the heading error has been reduced to zero but a latitude error has appeared. After 12 hours (Figure 9.8b) the heading error has been reversed and the latitude error has returned to zero, thus a 24-hour cycle of heading and latitude error has been established with the two curves phase shifted by 90° as shown in Figure 9.8c. This type of error can occur if there is a bias in one of the horizontal axis gyroscopes or gyrocompassing has produced an inaccurate North reference. The Earth-loop oscillations will appear at all latitudes but the amplitude will be different for a given initial error.

9.7 Summary of Inertial Guidance Errors

The 84.4 minute Schüler and 24-hour Earth-loop oscillations are referred to as bounded errors as they do not grow with time and oscillate about a mean value. There are also errors that produce a miscalculation of position that grows with time and these are referred to as unbounded. The predominant sources of inaccuracies are sensor bias from accelerometers or optical rate gyros, random walk

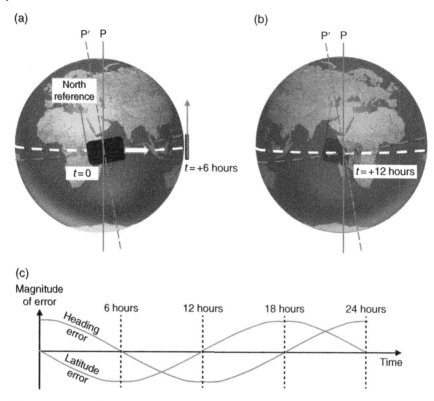

Figure 9.8 (a) Platform with a heading error at the equator. As the Earth rotates, the platform will sense that it is traveling along the red dashed line rather than the equator. After 6 hours, the heading error will be reduced to zero but a latitude error will be displayed. (b) After 12 hours, the heading error will be reversed and the latitude error will be back at zero. (c) The heading and latitude error follow a 24-hour cycle.

errors, environmental factors, and true wander in the mechanical gyros in the case of stabilized platform systems. The presence of unbounded errors in inertial guidance systems makes them a more precise source of information near the beginning of a flight. Thus, when combining multiple navigational sources by Kalman filtering (see Section 9.9), the weighting given to inertial guidance information needs to decrease with time.

9.7.1 Sensor Bias

Sensor bias is when the sensor (accelerometer or rate gyro) produces a nonzero output when still and since this signal is integrated twice to determine the distance, a positional error that grows as the square of the elapsed time occurs. The one-dimensional distance error, s_e, given by an accelerometer bias, a_b, after a time t is given by:

$$s_e = \frac{1}{2}a_b t^2 \tag{9.15}$$

For example, over a 2-hour flight an accelerometer bias of 1×10^{-5} m s^{-2} (1 μg) will have accumulated an error of 259.2 m. The numbers demonstrate the level of precision required in the sensors in an INS. Bias in rate gyros will have a similar effect since they will produce an effective acceleration bias. In either case there will also be Schüler oscillations superimposed on the increasing error signal since the incorrect Schüler tuning to compensate transport wander will act to reverse the acceleration bias and initiate the oscillations.

9.7.2 Random Walk Position Error Produced by Sensor Noise

Any measurement system has inherent random noise and if the random noise signal is averaged with respect to time then the result is zero. In an inertial guidance system, however, the noisy signal is integrated twice to obtain a position, which is updated regularly and in which each position is offset from the previous one. Thus, for a perfectly still platform, each position is a random offset along North–South or East–West from the previous position and the result is a random walk in which the total area of position covered is much bigger than a single random positional step due to noise. This is illustrated in Figure 9.9, which shows a random walk of 25 000 steps in which each point is a random unit displacement along North–South or East–West from the previous one. This is not a true representation of the random walk error in an inertial guidance system since the step size will also vary randomly but it illustrates the principle that the area covered is vastly greater than a single step. For this example, the linear dimension across the whole area covered is around 150 steps. The total area covered, however, will increase with the number of steps and it can be shown that the total random walk positional error in an inertial guidance system grows as $t^{3/2}$ [1].

9.7.3 Environmental Factors

Environmental factors such as pressure, temperature, and mechanical stress can cause changes in sensor bias of both gyroscopes and accelerometers. This can be due to, for example, a change in the optical path length in the case of laser-based rate gyros or changes to the stiffness of components in MEMS accelerometers. There is often a highly nonlinear relationship between sensor bias and temperature making it difficult to obtain a generic calibration. Most systems include

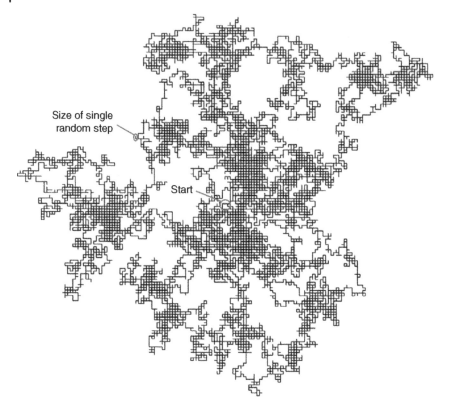

Figure 9.9 Random walk of 25 000 steps in which each step is a random unit displacement along North–South or East–West. The area covered is vastly greater than a single step and will grow with the number of steps.

temperature sensors so that real-time compensation of changes in temperature can be achieved.

9.7.4 True Wander

True wander is a feature of mechanical gyroscopes discussed in Section 3.1.2 that is not observed in optical gyroscopes, so it is an issue only in the older stabilized platform systems. Any mechanical imperfections or bearing friction in a rotating mass gyroscope will generate torques perpendicular to the spin axis, which will then produce precession according to Equation 3.3. This will produce a movement of the spin axis relative to inertial space that is described as true wander to distinguish it from the "apparent" wander produced by Earth rate and transport wander in which the spin axis stays aligned with inertial space but the frame of reference changes.

Any true wander in the gyroscopes of a stabilized platform system will tilt the platform and generate accelerometer bias due to a component of gravity in both axes, which will generate the positional errors described in Section 9.7.1. As will be found by solving problem [1], a wander of 0.005° will produce an accumulated position error after one hour of about 3 nautical miles.

9.8 Cockpit Instrumentation

The cockpit instrumentation will be described using the Honeywell ADIRU and control panels installed in the Boeing 737 as it is typical of the basic functions found in other systems and aircraft. Figure 9.10 shows the two instrument units, which are normally mounted in the panel above the pilot's head. The upper one is the IRS display unit (ISDU), which can be used to enter the current aircraft coordinates at start-up and the other is the mode selector unit (MSU), which

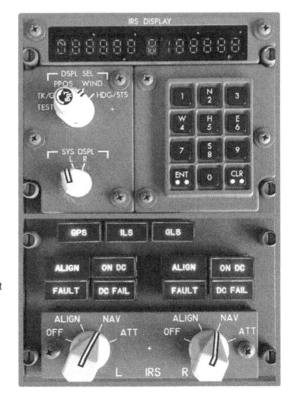

Figure 9.10 Cockpit instrumentation in a B737 NG aircraft used to control the Honeywell ADIRU. *Source:* Reproduced with permission of Chris Brady from http://www.b737.org.uk.

can be used to select the various operating modes of the IRS by the bottom switches. The panels displayed in Figure 9.10 control two independent IRSs.

At power-up the two switches on the MSU are set to "NAV", the top switch on the ISDU is set to "'PPOS' and the current aircraft coordinates can be input using the keypad in the format N or S XX° XX.X" E or W XXX° XX.X' (see Section 6.1.1) followed by "ENT." It is also possible to send the current coordinates obtained by GPS directly to the ISDU using the control display unit (CDU) that is used to enter information to the flight management system (FMS). Having entered the current position, the top switch on the ISDU is set to the "HDG/STS" (heading/status) position to check for any fault codes that will appear on the IRS display.

One of the MSU switches is then turned to "ALIGN" to start the gyrocompassing alignment process described in Section 9.2.4, which determines the direction to true North by sensing the platform tilt along North–South and East–West caused by the rotation of the Earth. Gyrocompassing takes between 5 minutes at the equator and 17 minutes for extreme latitudes and during this process the aircraft cannot be moved or loaded. In addition, any systems that required ADIRU data such as the fight controls, weather radar (WXR), the ground proximity warning system (GPWS), and the traffic collision and alert system (TCAS) cannot be tested without generating fault codes. Thus, the alignment process needs to be inserted carefully in the start-up checklist for the aircraft to allow tasks that do not interfere with gyrocompassing to continue.

The switch on the MSU that was set to "ALIGN" is then set to "NAV" and the system starts streaming positional and other data to the FMS. The other two positions on the MSU are "OFF", which is self-explanatory and "ATT," which disconnects computing and loses alignment. This is used if the positional data from the IRS becomes unreliable and in this mode the system continues to give attitude information and limited heading information.

The ISDU selects what information appears on the display and the bottom switch selects one of two displays in case one becomes faulty. The settings on the top switch are:

"TEST":	Tests the LEDs on all the seven-segment display windows.
"TK/GS":	The left display shows the ground track and the right display shows the ground speed.
"PPOS":	Described above and used to enter current position coordinates at start-up.
"WIND":	The display shows the direction and speed of the wind, computed from the difference between the heading and the ground track.
"HDG/STS":	Display shows heading or status, that is, the display shows any error codes.

9.9 Kalman Filter

9.9.1 Basic Principle of the Kalman Filter

The various navigation sources that have been described, including radio navigation beacons (Chapter 7), GNSS (Chapter 8), and inertial guidance systems (this chapter) all produce navigational data that predict the position of the aircraft with an associated error. This error depends on various factors, for example, the distance from a radio beacon, the number of satellites in view in the case of GNSS, or the time that the aircraft has been flying in the case of Inertial navigation. Whenever a quantity with an associated uncertainty is being measured, the normal method to find the "true" value of the quantity is to make a large number of measurements and then find the mean, which can be assumed to be the true value. Rather than just averaging the data, a more accurate method is to plot a histogram of the data and analyze the distribution.

For example, suppose two thermometers were measuring the temperature of a room, whose true temperature is 20 °C and one had noise fluctuations with an amplitude of around 0.1 °C and the other had noise fluctuations with an amplitude of around 0.5 °C. We could separate the measured temperature range into a number of "bins" of chosen width and plot a histogram of the number of results in each bin as illustrated in Figure 9.11. The envelope shape describing the histogram is known as a normal distribution and is given by:

$$P \propto \exp - \frac{(x - \mu)^2}{2\sigma^2} \tag{9.16}$$

where P is the probability of measuring a specific value, x, μ is the mean, which can be taken to be the best estimate of the true value, and σ is the standard deviation. This is a measure of the spread of the data and its value is the deviation from the mean within which we will find 68.3% of the measurements. The quantity σ^2 is known as the variance and indicates the total spread of all the measurements. We can fit a normal curve using a least squares method to each of the measured distributions and the noteworthy point here is that given enough data points both the accurate and the inaccurate thermometers will yield the same mean, or true, value to any specified level of accuracy.

The difference with noisy navigational data is that we cannot rest at one point, take a large number of measurements, and then move. The data is continuously streaming in as the aircraft moves and we need a way of combining the data from different sources with different associated errors in real time as each measurement arrives. Going back to the above thermometer example, if we were to display a continuously updating temperature as the data came in, we must be able to analyze the incoming data in such a way that the confidence

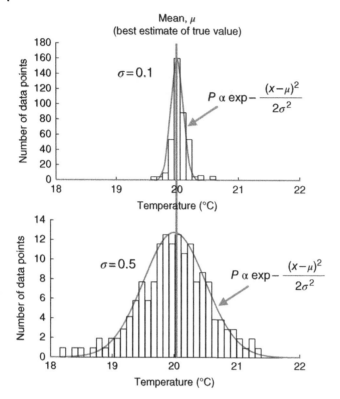

Figure 9.11 Histogram data from two thermometers with different accuracies measuring the same temperature of 20 °C. Given enough data points, the histogram of their distribution will both provide the same value of μ, which is the best estimate of the true value.

in the displayed value increases as the number of measurements increase. Kalman filtering provides a way of finding this optimized confidence value.

In practice, there are multiple values that need to be filtered for navigation data including, x, y, z position and velocities, but to illustrate the principle we will consider a measurement of a single value such as a temperature as in the above example. The algorithm for continuously updating a measured value from an incoming data stream with an associated uncertainty is shown in Figure 9.12. It is assumed that the data arrives at a regular interval and each new measured value triggers an iteration of the filter. The process starts with an original estimate of the value (which can simply be the first measurement) and an original estimate of the error in the value. The displayed value is insensitive to these original estimates as the output will converge to close to the true value after a few iterations.

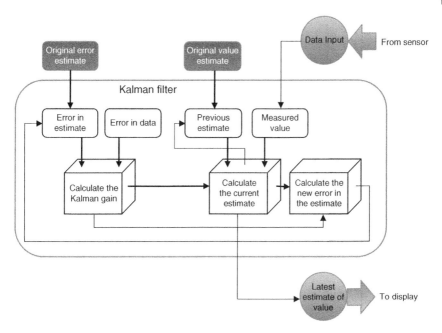

Figure 9.12 Flow diagram of the operations in a Kalman filter for the data stream from a single sensor.

Within the Kalman filter are three main calculations, that is, the Kalman Gain, K_G, the current estimate of the value, Q_E, and the error in that estimate ΔQ_E. The Kalman gain calculation requires two inputs, that is, the error in the estimate (starting with the original value) and the error in the data, ΔQ_M, which can either be a set value from the known characteristics of the sensor or it can be obtained by comparing consecutive measurements. The second calculation determines the current estimate of the value from three inputs, that is, the estimate from the previous iteration, the measured value, and the Kalman gain. The value of the gain is used to determine how much weight to put on the previous estimate relative to the measured value in order to calculate the new estimate. The current estimate is passed to update the displayed value and is also fed back to the previous estimate box for the next iteration. The third calculation is the new error in the estimate, which has inputs from the current estimate and from the Kalman gain and is fed back to the error in the estimate box to calculate the Kalman gain in the next iteration. The process produces an output that rapidly converges to the true value of the measured quantity Q_T with a known uncertainty and the following sections describe the calculations within the filter.

9.9.2 Kalman Filter for One-Dimensional (Single Value) Data

To begin with, the operation of the Kalman filter for a single noisy data value will be considered and later will be generalized to the simultaneous processing of multiple data values. The first of the calculations within the filter is the Kalman gain, K_G, which is given by:

$$K_G = \frac{\Delta Q_E}{\Delta Q_E + \Delta Q_M} \tag{9.17}$$

where ΔQ_M is the error in the measured value of the quantity and ΔQ_E is the error in the estimated value of the quantity. The Kalman gain has a value between 0 and 1 and low values indicate that the error in the measurement is high and will put more weight on the estimated quantity, while high values indicate that the error in the measurement is small and the gain will put more weight on the measured value. Having obtained the Kalman gain, the estimate of the quantity, Q_E^n, in the current iteration, n, is calculated using:

$$Q_E^n = Q_E^{n-1} + K_G\left(Q_M - Q_E^{n-1}\right) \tag{9.18}$$

where Q_E^{n-1} is the estimate of the quantity from the previous iteration. Finally, the error in the current estimate, ΔQ_E^n, is obtained from the equation:

$$\Delta Q_E^n = \frac{\Delta Q_M \Delta Q_E^{n-1}}{\Delta Q_M + \Delta Q_E^{n-1}} \tag{9.19}$$

where ΔQ_E^{n-1} is the estimate in the error from the previous iteration. Using Equation (9.17), this can also be written as:

$$\Delta Q_E^n = \left(1 - K_G\right)\Delta Q_E^{n-1} \tag{9.20}$$

The calculated values of the estimate and the error in the estimate of the quantity in the current iteration are then fed back as shown in Figure 9.12 to be combined with the measured data in the next iteration to obtain new values of the estimate and error.

As an example, returning to the thermometer measuring the temperature of a room (Figure 9.11), the data from the noisy thermometer will be passed through a Kalman filter to obtain an "averaged" output as the data streams in real time. In this case, the value of ΔQ_M is a known constant quantity (0.5 °C) and we just need to choose an initial estimated value for the temperature, Q_E^0, and an initial error in the estimate, ΔQ_E^0. As explained above, the values chosen for these are not important as the filter will provide an output that quickly settles to close to the true value. Figure 9.13a compares the raw data with the data after filtering for an initial choice of $Q_E^0 = 25$ and $\Delta Q_E^0 = 1$. It is seen that the filtered data quickly converge to close to the true value with an error in the estimate that decreases with iteration number shown in Figure 9.13b. This is achieved with a large error in the original estimate and demonstrates that the value assigned to the initial estimates is not critical.

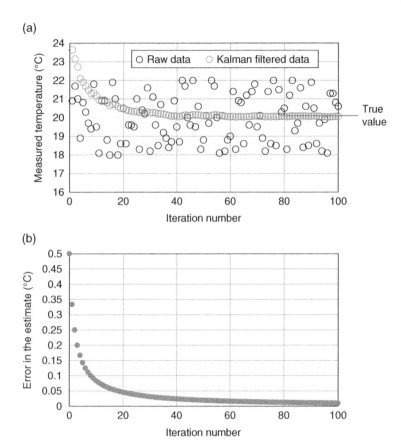

Figure 9.13 (a) Comparison of raw data (black hollow circles) from a noisy thermometer with the Kalman filtered estimate (red hollow circles) of temperature as a function of iteration number. (b) Error in the estimated temperature as a function of iteration number.

9.9.3 Kalman Filtering of Multiple values

The simple example above demonstrated the power of Kalman filtering to rapidly converge towards a true value of a measured quantity by averaging data in real time as it comes in from a noisy sensor. In navigation, it is necessary to incorporate data from several sources measuring several quantities, each with its own associated error, which may change with time and determine an estimate and associated error in the estimate for all the quantities. To do this it is necessary to switch to a matrix format to describe the quantities being measured and the associated manipulations as it becomes straightforward to generalize to any number of variables. Matrix algebra is also particularly suited to rapid computation electronically [2].

In a matrix representation, instead of a single value, Q_E, the set of values to be estimated is described by the column vector X, for example,

$$X = \begin{bmatrix} x \\ y \\ z \end{bmatrix} \tag{9.21}$$

which in this case represents the three position coordinates, x, y, z. The vector X is usually described as the state vector, which can be the measured state vector in the current iteration, X_M^n, and the estimated or predicted state vector in the current iteration, X_P^n. The predicted state vector is updated at each iteration using the matrix equation:

$$X_P^n = AX_P^{n-1} + Bu^n + w^n \tag{9.22}$$

where X_P^{n-1} is the predicted state vector from the previous iteration, u^n is the control variable matrix, which includes information on how the variables are controlled by, for example, gravity and enables us to predict how the state vector will evolve. The term w^n is the predicted state noise matrix and A and B are known as adaptation matrices, which convert X and u matrices into equivalent formats as illustrated by the example below.

As an example of how to evaluate Equation (9.22), consider a mass falling under the influence of gravity. In this case a suitable state vector is:

$$X = \begin{bmatrix} y \\ \dot{y} \end{bmatrix} \tag{9.23}$$

where y and $\dot{y}$ are the vertical position and velocity of the object. A suitable adaptation matrix, A is:

$$A = \begin{bmatrix} 1 & \Delta t \\ 0 & 1 \end{bmatrix} \tag{9.24}$$

where Δt is the time interval corresponding to an iteration. Thus, the first term in Equation (9.22) would be given by:

$$X_P^n = AX_P^{n-1} = \begin{bmatrix} 1 & \Delta t \\ 0 & 1 \end{bmatrix} \begin{bmatrix} y^{n-1} \\ \dot{y}^{n-1} \end{bmatrix} = \begin{bmatrix} y^{n-1} + \Delta t \dot{y}^{n-1} \\ 0 + \dot{y}^{n-1} \end{bmatrix} \tag{9.25}$$

In other words, ignoring the acceleration due to gravity, the new position of the object would be $y + \Delta t \dot{y}$ and the velocity would remain at $\dot{y}$. The effect of gravity on the body is included in the second term on the right-hand side of Equation (9.22), that is Bu^n and a suitable adaptation matrix, B is:

$$B = \begin{bmatrix} \frac{1}{2}\Delta t^2 \\ \Delta t \end{bmatrix} \tag{9.26}$$

to represent the acceleration, which produces a change in position $\frac{1}{2}at^2$. The u matrix is simply the single value $[g]$, that is, the acceleration due to gravity, which is the same at every iteration, so the second term in Equation (9.22) is:

$$Bu^n = \begin{bmatrix} \frac{1}{2}\Delta t^2 \\ \Delta t \end{bmatrix} [g] = \begin{bmatrix} g\frac{1}{2}\Delta t^2 \\ g\Delta t \end{bmatrix} \tag{9.27}$$

Thus, it is clear that when the state vector X is updated by Equation (9.22), the variable y goes to $y + \Delta t \dot{y} + g\frac{1}{2}\Delta t^2$ and the variable $\dot{y}$ goes to $\dot{y} + g\Delta t$, that is the new state vector is:

$$X_p^n = AX_p^{n-1} + Bu^n = \begin{bmatrix} y^{n-1} + \Delta t \dot{y}^{n-1} + g\frac{1}{2}\Delta t^2 \\ \dot{y}^{n-1} + g\Delta t \end{bmatrix} \tag{9.28}$$

The last term, w^n is a vector that allows an estimate of uncertainly to be included for the estimate of every variable in the state vector. As illustrated by the above example, the matrices A and B do not introduce specific information but are simply used to line up the variable with the correct multiplier to generate suitable equations to predict the new state.

Having seen how to update a state vector, the full implementation of a multidimensional Kalman filter is shown in Figure 9.14 with the various matrices and vectors described below.

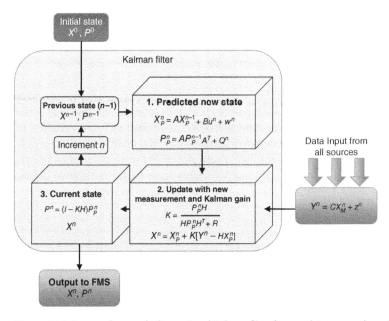

Figure 9.14 Process for a multidimensional Kalman filter for an arbitrary number of variables in the state vector, X. The rest of the matrices are described in the text.

The state vector X, which has already been introduced contains all the variables of interest and evolves with time with the current iteration labeled by the superscript n. In a simple aviation navigation context, for example, it may be:

$$X^n = \begin{bmatrix} \text{Latitude} \\ \text{Longitude} \\ \text{Altitude} \end{bmatrix} \tag{9.29}$$

or in a fully integrated GPS/IRS and attitude reference system may be, for example:

$$X^n = \begin{bmatrix} \text{Latitude} \\ \text{Longitude} \\ \text{Altitude} \\ \text{IAS} \\ \text{Roll angle} \\ \text{Pitch angle} \\ \text{Roll rate} \\ \text{Pitch rate} \\ \text{Yaw rate} \\ \text{E} - \text{W acceleration} \\ \text{N} - \text{S acceleration} \end{bmatrix} \tag{9.30}$$

The other matrix describing the current state is the state covariance matrix P^n, which specifies values of the error in the estimation of the variables in the current state vector. It is the matrix equivalent of the quantity ΔQ^n_E for a one-dimensional Kalman filter described in Section 9.9.2 and is a square matrix with the same dimension as the state vector. For example, in the case of an 11-variable state vector written in Equation (9.30) it will be an 11×11 matrix. The contents of the matrix are the variances (see Section 9.9.1), σ_i^2 in all the predicted variables of the state vector and covariances $\sigma_i \sigma_j$ describing how the uncertainty in one predicted variable affects that of another. Examples of the state covariance matrix in a given iteration, n for a 1-, 2-, and 3-variable state vector are shown below:

$$1 - \text{variable}: P_1^n = \begin{bmatrix} \sigma_1^2 \end{bmatrix}$$

$$2 - \text{variable}: P_2^n = \begin{bmatrix} \sigma_1^2 & \sigma_1 \sigma_2 \\ \sigma_2 \sigma_1 & \sigma_2^2 \end{bmatrix}$$

$$3 - \text{variable}: P_3^n = \begin{bmatrix} \sigma_1^2 & \sigma_1 \sigma_2 & \sigma_1 \sigma_3 \\ \sigma_2 \sigma_1 & \sigma_2^2 & \sigma_2 \sigma_3 \\ \sigma_3 \sigma_1 & \sigma_3 \sigma_2 & \sigma_3^2 \end{bmatrix} \tag{9.31}$$

If the uncertainty in the calculation of one variable has no effect on the uncertainty in another, for example, the calculation of altitude from a pressure measurement and the calculation of position from an acceleration measurement, then the relevant covariance term $\sigma_i\sigma_j = 0$. The state covariance matrix is updated at each iteration by the equation:

$$P_p^n = AP_p^{n-1}A^\mathrm{T} + Q^n \tag{9.32}$$

where the A matrix was described above and A^T is the transpose of A. The manipulation by A and A^T is simply format adaptation as illustrated for the state vector above. The change in P_p^n at each iteration is produced by adding Q^n, the process noise covariance matrix, which describes uncertainties in the calculated variables. The calculation itself contains no uncertainties but Q represents uncertainties that cannot be accounted for precisely when predicting the new state, for example, changes in wind velocity.

How the predicted state is updated, that is, box 1 in Figure 9.14 has been described above and now we will look at how the predicted state is combined with the measured data, that is, box 2 in Figure 9.14.

The Kalman gain is given by:

$$K = \frac{P_p^n H}{HP_p^n H^\mathrm{T} + R} \tag{9.33}$$

which is the matrix equivalent of Equation (9.17) that was appropriate for a one-variable system. The H matrix and its transpose H^T are used for adaptation in a similar manner to the A and B matrices for the state vector X and the state covariance matrix P. In many applications, H and H^T can simply be taken to be unit matrices with the same dimension as P. As with the one-dimensional example, the Kalman gain determines how much weight is given to the measured value of a variable in comparison to its predicted value when updating the state vector. In the multidimensional formulation, K is a square matrix that has a list of the Kalman gains for every variable in the state vector along the diagonal with off-diagonal terms equal to zero. The term R in Equation (9.33) is the measurement covariance matrix, which has the same dimensions as P and K and lists the errors in the measurement of all variables along the diagonal. It is here, for example, that one would introduce an error that increased with time in a position variable measured by an IRS system.

If the value in R for a specific measured variable was very small, the Kalman gain for that variable would be close to 1 and updates of the variables in X would predominantly come from the measured values. On the other hand, if the value in R for a specific measured variable was large, the Kalman gain would tend to zero and the updates would rely more heavily on the predicted values determined in box 1 in Figure 9.14. It is important therefore to introduce uncertainty into the state covariance matrix, P, as described in Equation (9.32) as it is relatively easy to let the errors in the predicted values go to zero in which case the Kalman gain would also go to zero and the state vector would be updated by prediction alone and ignore incoming data.

The measured variables that are inputs into box 2 are contained in the measured state vector X_M^n and are multiplied by another adaptation matrix, C, which in the simplest case is just a unit matrix. There is also an additional process error matrix, z, that contains additional errors generated by the process in a similar manner to the w matrix of box 1. These are not measurement errors, which are contained in R but additional errors that may be introduced by the process of filtering and enable some extra flexibility in the description of the process. Thus, the input from the sensors into box 2 of the filter in a specific iteration is the Y^n matrix given by:

$$Y^n = CX_M^n + z^n \tag{9.34}$$

and the new state vector is generated by combining the predicted and measured values taking into account the respective Kalman gains by:

$$X^n = X_P^n + K\left[Y^n - HX_P^n\right] \tag{9.35}$$

The final box sets the latest version of the state covariance matrix, P^n, into the correct format and outputs the current state vector, X^n and P^n to the FMS. The value of n is incremented and the current state vector and state covariance matrix are passed to box 1 to begin the process of finding the new state.

Within the FMS, the variables in the state vector X are used to generate the PFD and MFD while the state covariance matrix is used to predict actual navigation performance (ANP). In a modern PFD this is displayed along with the required navigational performance (RNP) specified for a specific phase of flight (see Figure 1.33).

Problems

1   A space gyroscope in a gimbaled INS has wandered off the correct axis of rotation by 0.005° and tilted the platform by this amount. Estimate the accumulated distance error in nautical miles after one hour.

2   A gimbaled INS has to be able to measure position to an accuracy of <200 m over an eight-hour flight. Assuming the error is predominantly from true wander, calculate the limits required of the true wander in degrees per hour.

3   The orbital period of a satellite decreases with the orbital radius. Show that a satellite in orbit just above the surface of a hypothetical perfectly spherical and airless Earth has the Schüler period.

4 An IRS is able to perform gyrocompassing at the equator in five minutes. Estimate the time required to gyrocompass the system at a latitude of 40° N.

References

1 Woodman, O.J. (2007). An introduction to inertial navigation. Technical report 696 of the University of Cambridge Computer Laboratory (UCAM-CL-TR-696), August. http://www.cl.cam.ac.uk/techreports (accessed 12 June 2018).
2 For an excellent detailed description of Kalman filters, see the lecture online series by Michel van Biezen at https://www.youtube.com/watch?v=CaCcOwJPytQ&spfreload=1 (accessed 12 June 2018).

Appendix A

Radiation from Wire Antennas

The analysis of radiation from wire antennas starts with Maxwell's equations, which describe how electric (E) and magnetic (B) fields behave in space and time. These are:

$$\nabla \times E = -\frac{\partial B}{\partial t} \text{ (Faraday's law)} \tag{A1.1}$$

$$\nabla \times B = \mu_0 J + \mu_0 \varepsilon_0 \frac{\partial E}{\partial t} \text{ (Ampere's law)} \tag{A1.2}$$

$$\nabla \cdot E = \frac{\rho}{\varepsilon_0} \text{ (Gauss's law)} \tag{A1.3}$$

$$\nabla \cdot B = 0 \text{ (Gauss's law for magnetism)} \tag{A1.4}$$

where E is the electric field strength, B is the magnetic field strength, ρ is the charge density, and J is the current density. The constants ε_0 (= 8.8542×10^{-12} C^2 N^{-1} m^{-2}) and μ_0 (= $4\pi \times 10^{-7}$ N A^{-2}) are the electric permittivity and magnetic permeability of free space, respectively. Charges that are static or moving with a uniform speed (steady current) do not produce electromagnetic radiation but an accelerating charge or a time-varying current generate radiation, which the above equations predict will propagate as a wave as shown below.

A. Static Electric Fields Produced by Static Charges

First, we will consider the time-independent electric and magnetic fields produced by static charges and steady currents.

In this case Faraday's law reduces to $\nabla \times E = 0$ and since $\nabla \times (\nabla \Phi) = 0$ for any function Φ, the electric field can be written as the gradient of a scalar potential, Φ, that is:

$$E = -\nabla \Phi \tag{A1.5}$$

Aircraft Systems: Instruments, Communications, Navigation, and Control,
First Edition. Chris Binns.
© 2019 John Wiley & Sons, Inc. Published 2019 by John Wiley & Sons, Inc.
Companion website: www.wiley.com/go/binns/aircraft_systems_instru_communi_Navi_control

and the negative sign is included to produce the convention that the electric field points from high to low potential. For a point charge, q, located at the origin, both Φ and $\mathbf{E}$ are spherically symmetric, so the electric field can be written as

$$\mathbf{E}(\mathbf{r}) = \hat{r}E(r) \tag{A1.6}$$

The relationship between $\mathbf{E}$ and q can be found by taking the volume integral throughout an enclosed volume, V, of both sides of Gauss's law (Equation A1.3), which gives:

$$\oiiint_V \nabla \cdot \mathbf{E} = \frac{1}{\varepsilon_0} \oiiint_V \rho = \frac{q}{\varepsilon_0} \tag{A1.7}$$

and using Gauss's divergence theorem illustrated in Figure A1.1 gives, using Equation (A1.6):

$$\oiiint_V \nabla \cdot \mathbf{E} = \oiint_S \mathbf{E} \cdot \hat{n} \, dS = \oiint_S \hat{n} \cdot \hat{r}E(r) dS = 4\pi r^2 E(r) \tag{A1.8}$$

Hence, from Equations (A1.7) and (A1.8):

$$E(r) = \frac{q}{4\pi\varepsilon_0 r^2} \tag{A1.9}$$

Thus, from Equations (A1.5) and (A1.6) the electric field due to a point charge, q, at the origin is:

$$\mathbf{E}(\mathbf{r}) = \frac{\hat{r}q}{4\pi\varepsilon_0 r^2} = -\nabla\Phi \left(\mathrm{Vm}^{-1}\right) \tag{A1.10}$$

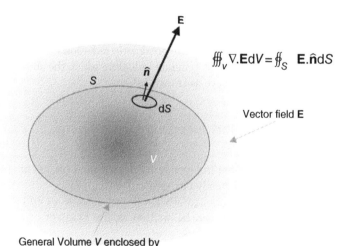

$$\oiiint_V \nabla \cdot \mathbf{E} dV = \oiint_S \mathbf{E} \cdot \hat{n} dS$$

Vector field **E**

General Volume *V* enclosed by surface *S* within the field

Figure A1.1 Illustration of Gauss's divergence theorem.

Since $\mathbf{E} = -\nabla\Phi$, the scalar potential due to the point charge can be obtained by integrating Equation (A1.10). The radial symmetry of the field suggests that the simplest expressions will be derived using spherical polar coordinates in which:

$$\nabla\Phi = \left(\hat{\mathbf{r}}\frac{\partial}{\partial r} + \hat{\boldsymbol{\theta}}\frac{1}{r}\frac{\partial}{\partial\theta} + \hat{\boldsymbol{\varphi}}\frac{1}{r\sin\theta}\frac{\partial}{\partial\varphi}\right)\Phi \tag{A1.11}$$

and since Φ is spherically symmetric, $\partial/\partial\theta = 0$ and $\partial/\partial\varphi = 0$, so that Equation (A1.11) reduces to:

$$\nabla\Phi = \hat{\mathbf{r}}\frac{\partial\Phi}{\partial r} \tag{A1.12}$$

demonstrating that using spherical polar coordinates is a good choice. Substituting Equation (A1.12) into Equation (A1.10) gives:

$$\hat{\mathbf{r}}\frac{\partial\Phi}{\partial r} = -\frac{\hat{\mathbf{r}}q}{4\pi\varepsilon_0 r^2} \tag{A1.13}$$

Therefore,

$$\Phi = -\frac{q}{4\pi\varepsilon_0}\int\frac{\mathrm{d}r}{r^2} = \frac{q}{4\pi\varepsilon_0|\mathbf{r}|} + \Phi_0 \ (\mathrm{V}) \tag{A1.14}$$

where Φ_0 is any background charge present, which we normally set to zero. This is for a charge at the origin but Equation (A1.14) can be generalized to describe a charge at position $\mathbf{r_q}$ by replacing $|\mathbf{r}|$ with $|\mathbf{r} - \mathbf{r_q}|$, which will be denoted just by r_q below. Thus, r_q is the distance from the source to a general point $\mathbf{r}$ (the observer). Equation (A1.14) describes the potential due to a point charge while the potential due to a distributed charge distribution $\rho(\mathbf{r})$ occupying a volume V_q is given by the integral:

$$\Phi = \oiiint_{V_q}\frac{\rho(r_q)}{4\pi\varepsilon_0 r_q}\mathrm{d}v \ (\mathrm{V}) \tag{A1.15}$$

The static electric field generated by $\rho(\mathbf{r})$ is obtained by computing the electric potential using Equation (A1.15) and then differentiating the potential as described by Equation (A1.10).

It follows from Gauss's law (A1.3) and (A1.5) that:

$$\nabla^2\Phi = -\frac{\rho}{\varepsilon_0} \tag{A1.16}$$

which is the *Poisson equation* so that Equation (A1.15) is also a solution to equations of the form (A1.16), which is a result we will use below.

B. Static Magnetic Fields Produced by Steady Currents

Static magnetic fields generated by a steady current density $\mathbf{J}$ are evaluated using Ampere's law (A1.2), which for no time variation is:

$$\nabla \times \mathbf{B} = \mu_0 \mathbf{J} \tag{A1.17}$$

Gauss's law for magnetism is (A1.4):

$$\nabla . \mathbf{B} = 0 \tag{A1.18}$$

and since the operator ∇. $\nabla \times \mathbf{A}$ will give zero for any vector $\mathbf{A}$, Equation (A1.18) will be satisfied by writing:

$$\mathbf{B} = \nabla \times \mathbf{A} \tag{A1.19}$$

where $\mathbf{A}(\mathbf{r}, t)$ is known as the vector potential and plays the same role for magnetic fields as the electric potential, Φ, does for electric fields. Substituting Equation (A1.19) into Ampere's law, (A1.17), gives:

$$\nabla \times \nabla \times \mathbf{A} = \mu_0 \mathbf{J} \tag{A1.20}$$

We can simplify this using the vector identity:

$$\nabla \times (\nabla \times \mathbf{A}) = \nabla(\nabla . \mathbf{A}) - \nabla^2 \mathbf{A} \tag{A1.21}$$

Since specifying the curl of a vector field, as in Equation (A1.19) has no effect on its divergence, the quantity ∇. $\mathbf{A}$ can be chosen arbitrarily without changing the value of $\mathbf{B}$, which is what we are interested in. This is a property known as *gauge invariance* and it allows us to choose ∇. $\mathbf{A}$ to simplify equations. This is one of the advantages specifying magnetic and electric fields in terms of potentials. In this case we specify ∇. $\mathbf{A} = 0$, so that using Equation (A1.20) in Equation (A1.19) gives:

$$\nabla^2 \mathbf{A} = -\mu_0 \mathbf{J} \tag{A1.22}$$

This is known as the *vector Poisson equation* in analogy to the scalar Poisson equation (A1.16) above. Each component (r, θ, ϕ in spherical polar coordinates) is like a scalar Poisson equation and so will have a similar solution to Equation (A1.15), thus, after rearranging the constants, we write the solution to Equation (A1.22) as:

$$\mathbf{A} = \oiiint_{V_q} \frac{\mu_0 \mathbf{J}(r_q)}{4\pi r_q} \mathrm{d}v \; \left(\mathrm{Vsm}^{-1}\right) \tag{A1.23}$$

For a given static current density distribution we can evaluate $\mathbf{A}$ and thus the magnetic field $\mathbf{B}$.

C. Dynamic Electric and Magnetic Fields Produced by Time-Dependent Charges and Currents

To evaluate the dynamic electric and magnetic fields produced by time-dependent charges and currents, we need to include the time dependences in Faraday's and Ampere's laws ((A1.1) and (A1.2)).

Substituting Equation (A1.19) into Faraday's law (A1.1) gives:

$$\nabla \times \mathbf{E} = -\frac{\partial(\nabla \times \mathbf{A})}{\partial t} \tag{A1.24}$$

or

$$\nabla \times \left(\mathbf{E} + \frac{\partial \mathbf{A}}{\partial t} \right) = 0 \tag{A1.25}$$

Since $\nabla \times \nabla\Phi = 0$ for any scalar function Φ, we can write:

$$\mathbf{E} + \frac{\partial \mathbf{A}}{\partial t} = -\nabla\Phi \tag{A1.26}$$

or

$$\mathbf{E} = -\frac{\partial \mathbf{A}}{\partial t} - \nabla\Phi \tag{A1.27}$$

where earlier Φ is the scalar potential but in a dynamic system, it is a function of time. Thus, in an environment of changing current, the dynamic electric field is derived from two contributions, that is, the instantaneous value of $\Phi(t)$ and the time derivative of $\mathbf{A}(t)$.

Substituting Equation (A1.19) into Ampere's law (A1.2) and applying the vector identity (A1.21) gives:

$$\nabla \times (\nabla \times \mathbf{A}) = \mu_0 \mathbf{J} + \mu_0 \varepsilon_0 \frac{\partial \mathbf{E}}{\partial t} = \nabla(\nabla.\mathbf{A}) - \nabla^2 \mathbf{A} \tag{A1.28}$$

Substituting for $\mathbf{E}$ from Equation (A1.27) gives:

$$\mu_0 \mathbf{J} - \mu_0 \varepsilon_0 \frac{\partial^2 \mathbf{A}}{\partial t^2} - \mu_0 \varepsilon_0 \nabla \left(\frac{\partial \varphi}{\partial t} \right) = \nabla(\nabla.\mathbf{A}) - \nabla^2 \mathbf{A} \tag{A1.29}$$

that is,

$$\nabla^2 \mathbf{A} - \nabla \left(\nabla.\mathbf{A} + \mu_0 \varepsilon_0 \frac{\partial \varphi}{\partial t} \right) - \mu_0 \varepsilon_0 \frac{\partial^2 \mathbf{A}}{\partial t^2} = -\mu_0 \mathbf{J} \tag{A1.30}$$

Specifying the curl of a vector field, as in Equation (A1.19) has no effect on its divergence (gauge invariance), so as before $\nabla.\mathbf{A}$ can be chosen arbitrarily without changing the value of **B**, in the case of static fields we chose $\nabla \cdot \mathbf{A} = 0$, but here we specify:

$$\nabla \cdot \mathbf{A} = -\mu_0 \varepsilon_0 \frac{\partial \varphi}{\partial t}, \tag{A1.31}$$

which is known as the *Lorentz gauge*. Using Equation (A1.31), Equation (A1.30) simplifies to:

$$\nabla^2 \mathbf{A} - \mu_0 \varepsilon_0 \frac{\partial^2 \mathbf{A}}{\partial t^2} = -\mu_0 \mathbf{J} \tag{A1.32}$$

Substituting Equation (A1.27) into Gauss's law (A1.3) and using the Lorentz gauge (A1.31) gives a similar equation for the scalar potential:

$$\nabla^2 \Phi - \mu_0 \varepsilon_0 \frac{\partial^2 \varphi}{\partial t^2} = -\frac{\rho}{\varepsilon_0} \tag{A1.33}$$

So an additional benefit for choosing the Lorentz gauge is that it produces complete symmetry between the space and time dependence of the scalar and vector potential. Equations (A1.32) and (A1.33) are known as the vector and scalar inhomogenous Helmholtz equations, respectively. Setting the right-hand side to zero gives the homogenous versions. They are quite revealing in that they are both easily recognized as wave equations that yield a wave traveling at a speed $c = 1/\sqrt{\mu_0 \varepsilon_0}$, that is, the speed of light. The **A** wave is generated by a time-dependent current density, **J**(t), while the Φ wave results from a time-dependent charge density, $\rho(t)$. After the charges and currents are removed the wave continues to propagate at the speed of light.

The boundary conditions for solutions of the Helmholtz equations are that they must be traveling waves when the sources **J**(t) and $\Phi(t)$ are zero and they must reduce to the static expressions for **A** and Φ if ρ and **J** are both static, that is, $\partial/\partial t = 0$. Wave equations such as (A1.32) and (A1.33) must have the same type of function describing the space and time dependence (e.g. $e^{i\mathbf{k} \cdot \mathbf{r}}$ and $e^{i\omega t}$) since for both variables the solution must involve a function that remains the same, apart from the constants, after differentiating twice. Thus, in general, the solution for space and time can be expressed as a single function of an argument containing both space and time converted to the same units such as $(t - \mathbf{r}/c)$, where c is the speed of propagation of the wave. So suitable solutions to the Helmholtz equations ((A1.32) and (A1.33)) are the static scalar and vector potentials (A1.15) and (A1.23) expressed in terms of the argument $(t - r_q/c)$, which also ensures that they reduce to the correct form in the case of static sources, ρ and **J**. Thus, the solutions for Φ and **A** to the wave equations can be written as

$$\Phi = \oiiint_{V_q} \frac{\rho\left(t - r_q/c\right)}{4\pi\varepsilon_0 r_q} dv \ (V) \tag{A1.34}$$

and

$$A = \oiiint_{V_q} \frac{\mu_0 J\left(t - r_q/c\right)}{4\pi r_q} dv \ \left(Vsm^{-1}\right)$$ (A1.35)

Note that r_q/c is the time taken for a wave to travel from the source to the observer so that a change at the source is not detected by the observer till a time r_q/c later, so these are described as *retarded* solutions.

D. Fields Produced by a Dynamic Infinitesimal Current Element (Hertzian Dipole)

We want to determine the radiation generated by a wire antenna carrying an oscillating current. Since Maxwell's equations are linear, the fields generated can be taken to be the sum of fields produced by a number of infinitesimal current elements joined in a configuration that reproduces the wire antenna. A single current element is known as a Hertzian dipole and we will now evaluate the electric and magnetic fields it produces by calculating the scalar and vector potentials using Equations (A1.34) and (A1.35). Figure A1.2 shows a Hertzian dipole as an element of a wire antenna aligned along the z-axis in a Cartesian coordinate frame, but we will determine the field in spherical polar coordinates and the relationship between the two coordinate systems is shown in the figure.

The current density required in Equation (A1.35) in this case is time-dependent, aligned along the z-axis and constant throughout the element with length δl and cross-sectional area A_c. It is thus given by:

$$J(t) = \frac{I(t)}{A_c} \hat{\mathbf{z}}$$ (A1.36)

Substituting this into Equation (A1.35) gives:

$$A = \oiiint_{V_q} \frac{\mu_0 I\left(t - r_q/c\right)}{4\pi r_q A_c} dv$$ (A1.37)

Since the current is spatially uniform over the volume in question, the integral simply yields the volume, $\delta l \times A_c$, as a multiplier, so Equation (A1.37) becomes:

$$A = \frac{\mu_0 I\left(t - \frac{r_q}{c}\right)\delta l}{4\pi r_q} \hat{\mathbf{z}}$$ (A1.38)

Equation (A1.38) is for a general time- and space-dependent current, but for antenna radiation the time variation of interest is a harmonic sinusoidal

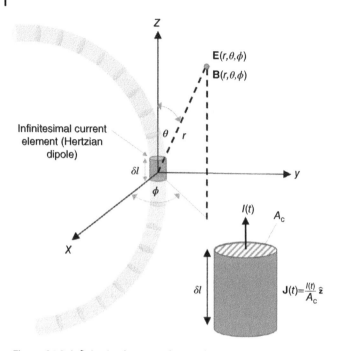

Figure A1.2 Infinitesimal current element (Hertzian dipole) within a current carrying antenna aligned along the z-axis in a Cartesian frame.

variation. The current can then be written in its *time harmonic* form (see below), which is:

$$I\left(t - \frac{r_q}{c}\right) = I_0 e^{i\omega t} = \tilde{I} \tag{A1.39}$$

The time dependence has become a frequency dependence represented by the wave vector, k. In spherical polar coordinates the unit vector $\hat{z}$ is transformed to:

$$\hat{z} = \cos\theta\hat{r} - \sin\theta\hat{\theta} \tag{A1.40}$$

Substituting Equations (A1.40) and (A1.39) into Equation (A1.38) gives:

$$\mathbf{A} = \left(\cos\theta\hat{r} - \sin\theta\hat{\theta}\right)\frac{\mu_0\delta l\tilde{I}e^{-ikr}}{4\pi r} \tag{A1.41}$$

Note that the subscript "q" has been removed from r and it is understood that r is the distance from the current element. From Equation (A1.41) it is straightforward to obtain the magnetic field from $\mathbf{B} = \nabla \times \mathbf{A}$ (that is, the definition of **A** – Equation (A1.19)). For a vector function **A**, written in spherical polar coordinates as:

$$\mathbf{A} = A_r \hat{\mathbf{r}} + A_\theta \hat{\boldsymbol{\theta}} + A_\phi \hat{\boldsymbol{\phi}} \tag{A1.42}$$

The curl in spherical polar coordinates is given by:

$$\nabla \times \mathbf{A} = \hat{\mathbf{r}} \frac{1}{r \sin\theta} \left[\frac{\partial}{\partial\theta} (A_\phi \sin\theta) - \frac{\partial A_\theta}{\partial\phi} \right] + \hat{\boldsymbol{\theta}} \frac{1}{r} \left[\frac{1}{\sin\theta} \frac{\partial A_r}{\partial\phi} - \frac{\partial}{\partial r} (r A_\phi) \right]$$
$$+ \hat{\boldsymbol{\phi}} \frac{1}{r} \left[\frac{\partial}{\partial r} (r A_\theta) - \frac{\partial A_r}{\partial\theta} \right] \tag{A1.43}$$

Comparing Equations (A1.42) and (A1.41), the components of **A** are:

$$A_r = \cos\theta \frac{\mu_0 \delta l \tilde{I} e^{-ikr}}{4\pi r}$$
$$A_\theta = -\sin\theta \frac{\mu_0 \delta l \tilde{I} e^{-ikr}}{4\pi r} \tag{A1.44}$$
$$A_\phi = 0$$

since A does not vary with ϕ, any term containing $\partial/\partial\phi = 0$, so Equation (A1.43) reduces to:

$$\nabla \times \mathbf{A} = \hat{\boldsymbol{\phi}} \frac{1}{r} \left[\frac{\partial}{\partial r} (r A_\theta) - \frac{\partial A_r}{\partial\theta} \right] = \mathbf{B} \tag{A1.45}$$

Substituting the components A_r and A_θ from Equation (A1.44) into Equation (A1.45) gives:

$$\mathbf{B} = \frac{ik\mu_0 \delta l \tilde{I}}{4\pi r} \sin\theta e^{-ikr} \left[1 + \frac{1}{ikr} \right] \hat{\boldsymbol{\phi}} \tag{A1.46}$$

The electric field can be obtained from Equation (A1.46) and Ampere's law (Equation (A1.2)), which relates the **E** and **B** fields. However, the general form shown in Equation (A1.2) is not convenient to use and we can obtain a version that is simpler to apply in this case in which we have already specified that the time dependence of sources is harmonic and sinusoidal (Equation (A1.39)). For harmonic sources the space- and time-dependent electric and magnetic fields, $\mathbf{E}(\mathbf{r}, t)$ and $\mathbf{B}(\mathbf{r}, t)$, can be written as

$$\mathbf{B}(\mathbf{r}, t) = \mathrm{Re}\left\{ B_s(r) e^{i\omega t} \right\} \tag{A1.47}$$

and

$$\mathbf{E}(\mathbf{r}, t) = \mathrm{Re}\left\{ E_s(r) e^{i\omega t} \right\} \tag{A1.48}$$

where E_s and B_s are phasors, that is, complex numbers that depend on **r** only, and Re denotes the real part of the expression. Using this form in Ampere's equation (A1.2) gives:

$$\text{Re}\left\{\nabla \times \mathbf{B_s}(r)e^{i\omega t}\right\} = \mu_0 \mathbf{J} + \mu_0\varepsilon_0\,\text{Re}\left\{\mathbf{E_s}(r)\frac{\partial}{\partial t}e^{i\omega t}\right\}$$

$$= \mu_0 \mathbf{J} + \mu_0\varepsilon_0\,\text{Re}\left\{E_s(r)i\omega e^{i\omega t}\right\} \tag{A1.49}$$

The $\mu_0\mathbf{J}$ term is zero since we are only interested in the electric field generated by the changing magnetic field while the **J** refers to any *additional* current density in the region. Since we are taking the real part of both sides we can drop the Re notation and the subscripts and the term $e^{i\omega t}$ cancels out, so Equation (A1.49) reduces to:

$$\mathbf{E}(r) = \frac{\nabla \times \mathbf{B}(r)}{i\omega\mu_0\varepsilon_0} \tag{A1.50}$$

This is the harmonic form of Ampere's law. Since $\omega = ck$ and $c = 1/\sqrt{\mu_0\varepsilon_0}$, Equation (A1.50) can be written as

$$\mathbf{E}(r) = \frac{1}{i\omega\sqrt{\mu_0\varepsilon_0}}\nabla \times \mathbf{B}(r) \tag{A1.51}$$

The components of **B** in spherical polar coordinates are:

$$B_r = 0$$

$$B_\theta = 0 \tag{A1.52}$$

$$B_\phi = \frac{ik\mu_0\delta l\tilde{I}}{4\pi r}\sin\theta e^{-ikr}\left[1 + \frac{1}{ikr}\right]$$

and from Equation (A1.43),

$$\nabla \times \mathbf{B} = \hat{\mathbf{r}}\frac{1}{r\sin\theta}\left[\frac{\partial}{\partial\theta}\left(B_\phi\sin\theta\right)\right] + \hat{\boldsymbol{\theta}}\frac{1}{r}\left[-\frac{\partial}{\partial r}\left(rA_\phi\right)\right] \tag{A1.53}$$

So after evaluating the differentials and rearranging, Equation (A1.51) gives:

$$\mathbf{E} = \frac{ikZ_0\delta l\tilde{I}}{4\pi r}e^{-ikr}\left\{2\cos\theta\left[\frac{1}{ikr} + \frac{1}{(ikr)^2}\right]\hat{\mathbf{r}} + \sin\theta\left[1 + \frac{1}{ikr} + \frac{1}{(ikr)^2}\right]\hat{\boldsymbol{\theta}}\right\} \tag{A1.54}$$

where

$$Z_0 = \sqrt{\frac{\mu_0}{\varepsilon_0}} = 377\Omega \tag{A1.55}$$

is the impedance of free space. Equations (A1.46) and (A1.54) describe the electric and magnetic fields radiated by the Hertzian dipole at all length scales but for distances $r \gg \lambda$, where λ is the wavelength of the electromagnetic waves, any term containing r^{-2} or r^{-3} can be ignored. This is known as the *far field* of the dipole and in the context of aviation VHF communication and radio navigation frequencies, where $\lambda \sim 2$ m, an aircraft parked on the apron communicating with the tower is already in the far-field regime.

The far-field electric and magnetic fields radiated by the dipole are, from Equations (A1.46), (A1.54), and (A1.55):

$$\mathbf{B} = \frac{ik\mu_0\delta l\tilde{I}}{4\pi r}\sin\theta e^{-ikr}\hat{\boldsymbol{\phi}} \tag{A1.56}$$

$$\mathbf{E} = \frac{ikZ_0\delta l\tilde{I}}{4\pi r}\sin\theta e^{-ikr}\hat{\boldsymbol{\theta}} \tag{A1.57}$$

Thus, the magnetic field points entirely in the $\hat{\boldsymbol{\phi}}$ direction while the electric field points entirely in the orthogonal $\hat{\boldsymbol{\theta}}$ direction as illustrated in Figure 4.25.

Appendix B

Theory of Transmission Lines and Waveguides

Wires used to carry oscillating signals from one place to another are generally referred to as transmission lines and can simply be a pair of wires as shown in Figure A2.1a or more commonly, a central conductor surrounded by hollow conducting cylinder (a coaxial cable) as shown in Figure A2.1b. When carrying a signal, there is an oscillating voltage between the wires and generally one of the conductors (the central one in the case of the coaxial cable) carries the current to the load. Either arrangement exhibits a capacitance per unit length, C, and an inductance per unit length, L.

Consider initially the line characterized in unit lengths each with a discrete capacitance C and a discrete inductance L forming the infinite network as illustrated in Figure A2.2a. The entire transmission line can be represented by a single impedance load at the input terminals, Z_c (Figure A2.2b) and since it is an infinite network adding a single unit in parallel to Z_c at the input will not change the load (Figure A2.2c). For a signal input with an angular frequency ω the impedance of L is $Z_1 = i\omega L$ and the impedance of C is $Z_2 = 1/i\omega C$, thus with reference to Figure A2.2 we can write:

$$Z_c = Z_1 + \cfrac{1}{\cfrac{1}{Z_2} + \cfrac{1}{Z_c}} = Z_1 + \frac{Z_2 Z_c}{Z_2 + Z_c} \tag{A2.1}$$

giving

$$Z_c^2 - Z_1 Z_c - Z_1 Z_2 = 0 \tag{A2.2}$$

Equation (A2.2) can be solved using the quadratic formula to give:

$$Z_c = \frac{Z_1}{2} \pm \sqrt{\frac{Z_1^2}{4} + Z_1 Z_2} \tag{A2.3}$$

Aircraft Systems: Instruments, Communications, Navigation, and Control,
First Edition. Chris Binns.
© 2019 John Wiley & Sons, Inc. Published 2019 by John Wiley & Sons, Inc.
Companion website: www.wiley.com/go/binns/aircraft_systems_instru_communi_Navi_control

(a)

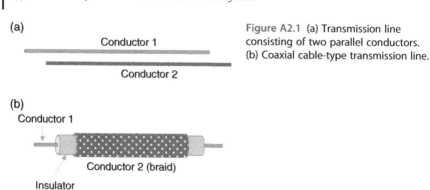

Figure A2.1 (a) Transmission line consisting of two parallel conductors. (b) Coaxial cable-type transmission line.

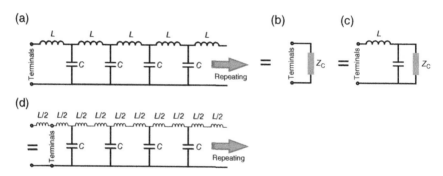

Figure A2.2 (a) Transmission line with an inductance and a capacitance per unit length represented as an infinite network of discrete inductors and capacitors. (b) The transmission line can be represented as a single impedance load, Z_c, at the input. (c) Since it is an infinite network, adding a single unit of the network in parallel to Z_c will not change the impedance. (d) Redrawing the network as shown and moving the terminals to the new position indicated will not change the load impedance.

The $Z_1/2$ term at the start is half the impedance of the first inductor and it is clear that redrawing the network as shown in Figure A2.2d and repositioning the terminals as indicated will eliminate the first half inductance but will not change the load. Thus, the $Z_1/2$ term can be eliminated in Equation (A2.3) and since having a negative impedance makes no sense, Equation (A2.3) can be written as

$$Z_c = \sqrt{\frac{Z_1^2}{4} + Z_1 Z_2} \qquad (A2.4)$$

and putting in the values for the impedances Z_1 and Z_2 gives:

$$Z_c = \sqrt{\frac{L}{C} - \frac{\omega^2 L^2}{4}} \qquad (A2.5)$$

For frequencies, $\omega < \sqrt{4/LC}$, Z_c is real and the signal propagates normally down the transmission line but if $\omega > \sqrt{4/LC}$ the signal is attenuated and does not propagate. The network thus acts as a low-pass filter with a cut-off frequency:

$$\omega_0 = \sqrt{\frac{4}{LC}} \tag{A2.6}$$

If we now evolve the network toward a transmission line with continuously distributed inductance and capacitance, taking shorter and shorter lengths to quantify L and C, these individually tend to zero while their ratio L/C stays constant. Thus, in the limit of a continuous line, Equation (A2.5) becomes:

$$Z_c = \sqrt{\frac{L}{C}} \tag{A2.7}$$

where L and C are now the inductance and capacitance per unit length. The quantity Z_c is known as the characteristic impedance of the transmission line and is purely real, that is, a pure resistance, with typical values in the range 50–100 Ω. There is now no cut-off frequency and the transmission line will, in principle, work at any frequency, though as shown below at sufficiently high frequencies it is better to dispense with the central conductor in transmission lines of the type shown in Figure A2.1b and pass electromagnetic waves through a hollow conducting tube.

Consider a rectangular metallic pipe with the z-axis along the length of the pipe and the x- and y-axes parallel to the sides as shown in Figure A2.3a. Let us assume that the electric field, $\mathbf{E}$, of the propagating wave is polarized along the y-direction (transverse electric or TE mode), that is $E_x = E_z = 0$. Since the electric field must be zero at the walls, the variation of E_y with x is as shown in Figure A2.3a and can be written as

$$E_y = \sin k_x x \tag{A2.8}$$

where

$$k_x = \frac{n\pi}{a} \tag{A2.9}$$

with n being an integer. Figure A2.3b shows the variation of E_y across the waveguide for $n = 1$ and other possible modes are shown in Figure A2.3c, but for the moment we will consider only the $n = 1$ (fundamental) mode. For a plane wave of frequency ω, the variation of E_y with z and time, t, can be written as

$$E_y = e^{i(\omega t - k_z z)} \tag{A2.10}$$

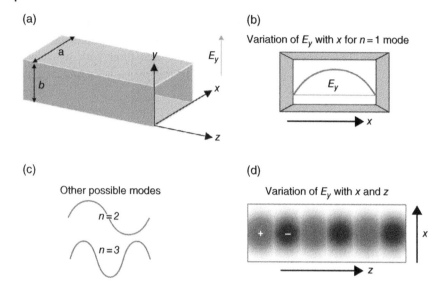

(a)

(b) Variation of E_y with x for $n = 1$ mode

(c) Other possible modes

(d) Variation of E_y with x and z

Figure A2.3 (a) Geometry of rectangular waveguide and a transverse electric field. (b) Variation of E_y with x across the waveguide for the $n = 1$ (fundamental) mode. (c) Variation of E_y with x for higher-order modes. (d) Variation of E_y with x and z for the fundamental mode.

which represents a wave traveling down the waveguide at a speed $c = \omega / k_z$. For the transverse wave, E_y does not vary with y, so the three-dimensional variation of E_y in the waveguide is given by:

$$E_y = e^{i(\omega t - k_z z)} \sin k_x x \qquad (A2.11)$$

The variation of E_y with x and z is illustrated in Figure A2.3d.

Maxwell's equations (see Appendix 1) specify that E_y must satisfy the wave equation:

$$\frac{\partial^2 E_y}{\partial x^2} + \frac{\partial^2 E_y}{\partial y^2} + \frac{\partial^2 E_y}{\partial z^2} = \frac{1}{c^2} \frac{\partial^2 E_y}{\partial t^2} \qquad (A2.12)$$

Substituting Equation (A2.11) into Equation (A2.12) gives:

$$k_x^2 + k_z^2 = \frac{\omega^2}{c^2} \qquad (A2.13)$$

and since $k_x = \pi/a$ for the fundamental mode, Equation (A2.13) becomes:

$$k_z = \pm \sqrt{\frac{\omega^2}{c^2} - \frac{\pi^2}{a^2}} \qquad (A2.14)$$

with plus and minus values representing the wave propagating in either direction along the waveguide. For frequencies below the critical value ω_c, given by:

$$\omega_c = \frac{\pi c}{a} \tag{A2.15}$$

corresponding to a wavelength $\lambda = 2a$, the value of k_z becomes imaginary. In this case the amplitude of the wave decays exponentially with z and does not propagate, so ω_c is known as the cut-off frequency. Rearranging Equation (A2.15) and allowing for higher orders from Equation (A2.9) shows that to propagate a given order, n, the waveguide width, a, must satisfy:

$$a > \frac{n\lambda}{2} \tag{A2.16}$$

Thus, as far as the fundamental mode is concerned, in a waveguide of width a, waves with wavelengths below $2a$ will propagate normally while those at longer wavelengths will be attenuated. Generally, a waveguide used in, for example, a radar installation is designed to work at a single frequency and will have a dimension that propagates the fundamental ($n = 1$) mode but cuts off all higher modes, which can be achieved by making the waveguide width, $\lambda/2 < a < \lambda$. In addition, the rectangular cross-section ensures that the polarization is preserved since waves polarized in the orthogonal direction with an E_x component will be cut off. It is also possible to have waves polarized in the z direction, whose magnetic field is polarized along the x- or y-direction. These are known as transverse magnetic or TM modes.

Appendix C

Effective Aperture of a Receiving Antenna

Consider a perfect parabolic dish antenna with a geometrical area, A, that is receiving a radio signal with a power density, S (W m^{-2}), and it converts this without loss to an output power P (W). Then, we can write:

$$P = A \times S \tag{A3.1}$$

The incoming signal will have a spectrum, S_f, so considering a narrow frequency range between f and $f + \Delta f$:

$$P_f \Delta f = A S_f \Delta f \tag{A3.2}$$

where P_f is the output power at f.

Conversely, if an antenna is receiving a signal with power density, S_f, and converts it into an output power, P_f, it can be said to have an *effective area*, A_e, given by:

$$A_e = \frac{P_f}{S_f} \tag{A3.3}$$

A directional antenna produces a redistribution of emitted power to concentrate the emission in one direction but in order to conserve energy the total power radiated must remain the same as a hypothetical isotropic antenna with the same efficiency. The gain, $G(\theta, \phi)$, is defined by the power emitted along direction (θ, ϕ) relative to an isotropic antenna and for an antenna with an efficiency equal to 1:

$$\langle G \rangle = 1 \tag{A3.4}$$

and this is true for all antennas with an efficiency of 1. In the same way, for a given signal, all lossless receiving antennas will produce the same power output, which from Equation (A3.3) means that all antennas have the same average effective area when averaged over all directions. This effective area can be determined using the following thought experiment illustrated in Figure A3.1.

Aircraft Systems: Instruments, Communications, Navigation, and Control,
First Edition. Chris Binns.
© 2019 John Wiley & Sons, Inc. Published 2019 by John Wiley & Sons, Inc.
Companion website: www.wiley.com/go/binns/aircraft_systems_instru_communi_Navi_control

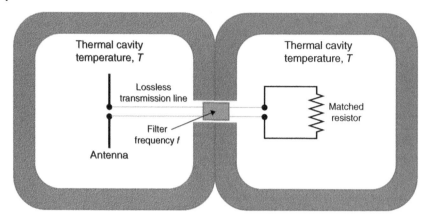

Figure A3.1 Hypothetical setup constructed to calculate the effective aperture of an antenna.

Consider an antenna immersed in a thermal cavity at a temperature, T, and in thermal equilibrium with it. The spectrum of radiation received is converted to an electrical signal and passed via a lossless transmission line to a matched resistor (so there is no power loss) in a second cavity at the same temperature. The connection contains a filter that only passes the frequency range, f to $f + \Delta f$. If $A_e(\theta, \phi)$ is the effective aperture in the direction (θ, ϕ) the total spectral power collected by the antenna is, from Equation (A3.2):

$$P_f \Delta f = \frac{1}{2} \int A_e(\theta, \phi) S_f \Delta f \, d\theta d\phi \qquad (A3.5)$$

where the factor ½ has been included because in a black body the radiation is unpolarized while normally antennas emit polarized radiation and in reception are only sensitive to one polarization. In a black body, S_f is isotropic and according to the Rayleigh–Jeans approximations for long wavelengths (valid in the microwave region) it can be written as

$$S_f \Delta f = \frac{2k_B T}{\lambda^2} \Delta f \qquad (A3.6)$$

where k_B is Boltzmann's constant and λ is the wavelength corresponding to the frequency, f. So, Equation (A3.5) becomes:

$$P_f \Delta f = \frac{k_B T}{\lambda^2} \int A_e(\theta, \phi) \Delta f \, d\theta d\phi \qquad (A3.7)$$

There can be no net power flow between the cavities, otherwise thermal equilibrium would not be maintained, so the power produced by the antenna must equal the Nyquist–Johnson spectral power produced by the resistor, that is:

$$P_f \Delta f = k_B T \Delta f$$

Thus, from Equation (A3.7),

$$\frac{k_B T}{\lambda^2} \int A_e(\theta, \phi) \Delta f \, d\theta d\phi = k_B T \Delta f \tag{A3.8}$$

or

$$\lambda^2 = \int A_e(\theta, \phi) d\theta d\phi \tag{A3.9}$$

What we want is the isotopic average of A_e, that is, $\langle A_e \rangle$ and using this in Equation (A3.9) gives:

$$\lambda^2 = \langle A_e \rangle \int d\theta d\phi = 4\pi \langle A_e \rangle \tag{A3.10}$$

or

$$\langle A_e \rangle = \frac{\lambda^2}{4\pi} \tag{A3.11}$$

This is the angular average of the effective aperture of an antenna irrespective of its directionality and is the same for all antennas. In the case of a directional antenna we are interested in the effective aperture in the direction of the maximum emission, which is $G_{max} \langle A_e \rangle$, so from Equation (A3.11):

$$A_e = \frac{\lambda^2 G_{max}}{4\pi} \tag{A3.12}$$

For a parabolic receiving antenna with a geometric area, A, one can write, from Equation (A3.12):

$$G = \frac{4\pi A}{\lambda^2} \eta \tag{A3.13}$$

where the "max" subscript has been dropped as an antenna's gain is normally given in the direction of maximum emission and η is the efficiency (see Equation 4.44).

Appendix D

Acronyms

ABAS	Aircraft-Based Augmentation System
ACARS	Aircraft Communications Addressing and Reporting System
ADC	Air Data Computer
ADF	Automatic Direction Finder
ADIRS	Air Data Inertial Reference System
ADIRU	Air Data Inertial Reference Unit
ADS-B	Automatic Dependent Surveillance – Broadcast
ADS-R	Automatic Dependent Surveillance – Rebroadcast
AFM	Aircraft Flight Manual
AI	Attitude Indicator
ANP	Actual Navigation Performance
ASI	Airspeed Indicator
BCPFSK	Binary Continuous Phase Frequency Shift Keying
BPSK	Binary Phase Shift Keying
BTDF	Bellini–Tosi Direction Finder
CAS	Calibrated Airspeed
CDI	Course Deviation Indicator
CDU	Control Display Unit
CPDLC	Controller–Pilot Data Link Communications
CRC	Cyclic Redundancy Check
DDM	Difference in the Depth of Modulation
DGPS	Differential GPS (GPS with augmentation)
DI	Direction Indicator
DME	Distance Measuring Equipment
EAS	Equivalent Airspeed
ELT	Emergency Locator Transmitter
ES	Extended Squitter
FAA	Federal Aviation Administration
FADEC	Full Authority Digital Engine Control
FBW	Fly by Wire

Aircraft Systems: Instruments, Communications, Navigation, and Control,
First Edition. Chris Binns.
© 2019 John Wiley & Sons, Inc. Published 2019 by John Wiley & Sons, Inc.
Companion website: www.wiley.com/go/binns/aircraft_systems_instru_communi_Navi_control

FEC	Forward Error Correction
FIS-B	Flight Information Service – Broadcast
FL	Flight Level
FMC	Flight Management Computer
FMS	Flight Management System
FRUIT	False Replies from Unsynchronized Interrogator Transmissions
GBAS	Ground-Based Augmentation System
GLONASS	GLObal NAvigation Satellite System (Russia)
GNSS	Global Navigation Satellite System
GPS	Global Positioning System (US)
GPWS	Ground Proximity Warning System
HF	High Frequency
HSI	Horizontal Situation Indicator
IAS	Indicated Airspeed
ICAO	International Civil Aviation Organisation
IGRF	International Geomagnetic Reference Field
ILS	Instrument Landing System
IMC	Instrument Meteorological Conditions
INS	Inertial Navigation System
IRS	Inertial Reference System
ISA	International Standard Atmosphere
ISDU	IRS Display Unit
LAAS	Local Area Augmentation System
LF	Low Frequency
LFR	Low Frequency Range
LNAV	Lateral Navigation
LPV	Localizer Performance with Vertical Guidance
LSS	Local Speed of Sound
MEMS	Micro Electrical Mechanical Systems
MF	Medium Frequency
MLS	Microwave Landing System
MSL	Mean Sea Level
MSU	Mode Selector Unit (for IRS)
NDB	Nondirectional Beacon
NUC	Navigational Uncertainty Category
OAT	Outside Air Temperature
OBI	Omnibearing Indicator
OBS	Omnibearing Selector
PBN	Performance-Based Navigation
PFD	Primary Flight Display
PPI	Plan Position Indicator (Radar)
PRT	Pulse Repetition Time (Radar)

PRF	Pulse Repetition Frequency (Radar)
QFE	Q-code for altimeter base level set to the measured pressure at an aerodrome
QNH	Q-code for altimeter base level set to the calculated pressure at local sea level
RA	Radio Altimeter
RAIM	Receiver Autonomous Integrity Monitoring
RAT	Ram Air Temperature
RCC	Rescue Coordination Centre
RDF	Radio Direction Finder
RDH	Reference Datum Height
RLG	Ring Laser Gyro
RMI	Radio Magnetic Indicator
RNAV	Area Navigation
RNP	Required Navigational Performance
ROC	Rate of Climb
ROD	Rate of Descent
SAR	Search and Rescue
SARPS	Standards and Recommended Practices
SAT	Static Air Temperature
SBAS	Satellite-Based Augmentation System
SHF	Super High Frequency
SPS	Standard Pressure Setting (base level on altimeter set to 1013 hPa)
SSR	Secondary Surveillance Radar
STOL	Short Takeoff and Landing
SV	Space Vehicle (satellite)
TAS	True Airspeed
TAT	Total Air Temperature
TCAS	Traffic Alert and Collision Avoidance System
TIS-B	Traffic Information Service – Broadcast
TRSB	Time-Referenced Scanning Beam
UAT	Universal Access Transceiver
UHF	Ultra High Frequency
VHF	Very High Frequency
VMC	Visual Meteorological Conditions
VNAV	Vertical Navigation
VOR	VHF Omnidirectional Range
VSI	Vertical Speed Indicator
WAAS	Wide Area Augmentation System
WXR	Weather Radar

Index

Aircraft Systems: Instruments, Communications, Navigation, and Control,
First Edition. Chris Binns.
© 2019 John Wiley & Sons, Inc. Published 2019 by John Wiley & Sons, Inc.
Companion website: www.wiley.com/go/binns/aircraft_systems_instru_communi_Navi_control

Printed and bound by CPI Group (UK) Ltd, Croydon, CR0 4YY

16/04/2025

14658602-0005